QUANTUM 22

THE HEBREW LIVING LETTERS

Quantum 22, The Hebrew Living Letters by Robin Main ©2025
All rights reserved.

LEGAL DISCLAIMER: No part of this book may be reproduced or transmitted in any form or by any means, electronic or mechanical, including photocopying, recording or by any information storage and retrieval system, without written permission from the author.
You may not reprint, resell or distribute the contents of this book without express written permission from the author.

Without limiting the author's exclusive rights, any unauthorized use of this publication to train generative artificial intelligence (AI) technologies is expressly prohibited.

First Edition.

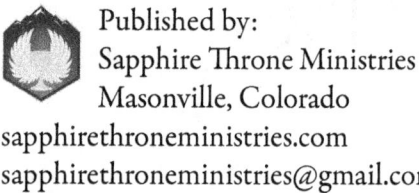

Published by:
Sapphire Throne Ministries
Masonville, Colorado
sapphirethroneministries.com
sapphirethroneministries@gmail.com

Managing Editor: Robin Shukle
Design: Liz Mrofka
whatifpublishing.com

Cover Image: "Quantum 22" by Robin Main
06-19-2023

ISBN: 978-0-9985982-8-4

QUANTUM 22™

THE HEBREW LIVING™ LETTERS

ROBIN MAIN

DEDICATION

Yeshua

ALEF-TAV (את)

The Word of God.

My Resting Place. My Beloved.

Also, to those made in God's image

to the genuine seekers of God's Kingdom,
His righteousness, and His peace

to the hungry souls longing for goodness

to those desperate for the truth spoken in love

to those discovering the ultimate treasure of
who they truly are

You are the light of the world.

CONTENTS

Acknowledgments 9

Preface 11

How to Read This Book 15

BUILDING BLOCKS OF CREATION

Three Spheres and the Hebrew Living Letters 19

Layers of Meaning 23

    The Hebrew Living Letters and Ascension in Christ 24

Water into Wine 27

    ALEF-TAV (את) 27

    The Infinity Cross 29

    Three Spheres Active During Creation 30

    Miraculous Waters of MEM (מ) 34

    Where the Water and Wine Meet 35

    Quantum Entanglement with the Quantum One 36

    The Seven Spirits of God 41

    Transformed Ones 49

    Quantum Levels 52

DNA—The Multidimensional Code of Life 55

    DNA is a Language 55

    DNA in the Darkest Part of a Cell 57

    Shadow DNA 58

Burnt Offerings and Ten Copper Chariot Lavers.................................... 60

Redemption of DNA.. 62

Antenna to God... 63

Jacob's Ladder and the Pillar of Righteousness..................................... 64

DNA is Light... 64

Creation's Quantum Elements 69

Hebrew Living Letter Principles... 81

Alphabetic Order of the Hebrew Alphabet 82

The Sacred Letters... 84

Numerical Value.. 85

Simple and Compound Letters 86

Hebrew Letters at the End of Words 87

Methods of Interpreting.. 88

Deciphering Pictographs... 91

SUBSTANCE OF CREATION

ALEF (א) – 1 – Mystery of Oneness.. 97

BET (ב) – 2 – House of Creation 105

GIMEL (ג) – 3 – Divine Completeness.. 113

DALET (ד) – 4 – Door to I AM that I AM 121

HEI (ה) – 5 – Divine Breath, Revelation & Light..................................... 129

VAV (ו) – 6 – Power to Connect Heaven & Earth 139

ZAYIN (ז) – 7 – Sword of Time 147

CHET (ח) – 8 – Gateway of Life.. 155

TET (ט) – 9 – Concealed Goodness... 163

YUD (י) – 10 – Divine Point Energy 173

KAF (כ) – 11 – Conformed to His Image..................................... 179

LAMED (ל) – 12 – Shepherd of Love.. 189

MEM (מ) – 13 – Fountain of Divine Wisdom................................ 201

NUN (נ) – 14 – Kingdom Life & Action.. 211

SAMECH (ס) – 15 – Endless Glory.. 223

AYIN (ע) – 16 – Spiritual Light & Sight....................................... 233

PEY (פ) – 17 – Kingdom Speech.. 243

TSADIK (צ) – 18 – Righteous Ones.. 253

KOOF (ק) – 19 – Emulation of Holiness...................................... 265

RESH (ר) – 20 – Awakened to the King of Kings......................... 275

SHIN (ש) – 21 – Complete Perfection of El Shaddai.................... 285

TAV (ת) – 22 – Seal of Truth... 293

ADDITIONAL RESOURCES

The "Hebrew Living™ Letters Speak" Class YouTube Video Recordings 307

Glossary 309

Recommended Reading 321

Bibliography 323

Endnotes 327

About the Author 345

Other Books by Robin Main 346

ACKNOWLEDGMENTS

Thankful for the constant and generous support, encouragement, prayers, and various other assists:

Kristy Hall

Cody Main

Special gratitude to Holly Brown for her invaluable contributions and reinforcement that has been integral to bringing *Quantum 22*™ to the world.

PREFACE

May the divine language of Eden echo in your soul, calling you home.

It's the dawn of a new day. It's time for the Sons of Light to arise. Step into the deed of Eden. The land that time forgot. Step into the foundation of redemption and restoration. Step into the reset back to pristine Eden and beyond. Hear the song of creation. See the *Hebrew Living™ Letters* come out of the Lion of the Tribe of Judah's mouth. Receive the bursting forth of new life. Receive the reset and knitting of the One New Man in the Messiah's DNA. Receive His seed of immortality. Receive the divine exchange. Let the *Hebrew Living™ Letters* move in and out of your light body to repair your DNA. Hear the song of the Heavenly Father's love. He's playing "Your Song."

The letters of the Hebrew Alphabet are the notes of His song. You are one-of-a-kind. You are a distinct note combination that makes up an incredibly beautiful and unique song. You can take a piece of a person's DNA and in it, you have the whole, which is unlike any software program.[1] When you began, there was a union of a sperm and an egg physically while spiritually God spoke into the zygote in your mother's womb: "Let there be light."[2] The DNA in your first cell contained all the genetic information for a new individual organism. . . y-o-u. This first cell also forms your heart. There are molecular biologists who convert the protein sequences in a person's DNA into classical music. You are His favorite song. He thinks you are magnificent.

Many Hebrews believe that God actually sang the universe into existence—a song of desire (*shir taev*). Scripture tells us that your body is the temple of the Holy Spirit (1 Cor. 6:19), which corresponds to all the songs that accompanied everything done in God's Temple. "A knowledge of the sacred letters and what they represent was always a prerequisite for one who undertakes to build God's Temple. In commenting on Proverbs 30:1, the Midrash says of King Solomon: 'The words of the man [i.e., Solomon] upon whom the Holy Spirit rested; to Isiel [a name the Midrash interprets as] the one who understood the letters of God' (Midrash Mishlei)."[3] The Hebrew Sages illustrate that the builder of the Tabernacle in the Wilderness—Bezalel—knew the art of combining the Hebrew letters: "Bezalel knew how to combine the letters with which heaven and earth were created" (Berachos 55a). Many Hebrew commentators explain how each part of God's Temple and its service symbolized a feature of the physical world since the purpose of human existence is to make earthly life worthy of God's Presence and to lead mankind to acknowledge Him. [4]

The *Dead Sea Scrolls* reveal that the "Songs of the Sabbath Sacrifice" were sung by the congregation of Israel to Melchizedek. As the community on earth joined heaven, they were led through a 13-week progressive experience of highest praise. The Seven Chief Princes of Highest Praise took center stage in the midst of their 13-week ecstatic celebration,[5] which corresponds to the Seven Spirits of God spotlighted in the "Water into Wine" chapter.

Isaac E. Mozeson has an excellent book called *THE WORD: The Dictionary That Reveals The Hebrew Source of English,* in which he provides evidence to the linguistic community for the Biblical thesis that all human languages are derived from a single Mother Tongue—Hebrew. He reveals "the three-letter Hebrew root is a chemical element composed of and charged by the dance of a pair of two-letter roots exchanging electrons." [6]

Because the majesty of Hebrew is only faintly visible in its offspring as well as the possibility of a bit of prejudice, many people won't consider this likelihood. However, Noah Webster's etymologies were full of English words traced to "Shemitic" sources. Mozeson states that it is "the unique molecular dynamism of Hebrew's two-letter roots, which make 'Proto-Semitic' the most logical candidate for the original human tongue." [7] Although I am not a linguistics expert, my gut tells me that Mr. Mozeson is right.

According to the Midrash (Hebrew Commentary), God taught Adam the Hebrew alphabet along with their accompanying numerical values, mathematical relationships, etc. Adam transmitted this knowledge to his sons. Adam's sons passed this down to their sons, and so on and so forth, until it was taught to Jacob at the School of Shem in Salem, later renamed Jerusalem. Jacob taught the secrets of the alphabet to Joseph, who used it to decipher dreams. The Hebrew mystics say that the entire cosmos was created from the 22 letters of the Hebrew Alphabet, called *otiyot yesod* or "foundational letters." [8]

Through the *otiyot yesod*, God formed substance out of chaos and brought forth existence from non-existence. In other words, the entire universe is created and sustained by the divine language of the Word of God, which elementally is made up of His letters. The LORD's spoken word that called all creation into existence is the same words and letters still echoing throughout it for its sustenance. *"And He is the radiance of His glory and the exact representation of His nature, and upholds all things by the word of His power" (Hebrews 1:3).*

At the primeval instance when *"let there be"* began, our Creator equipped Creation with the full bandwidth of His glory, which includes letters, pictures, numbers, symbols, languages, frequencies, math, science, etc. As a former computer scientist and analyst, I was led to incorporate some brilliant math and science in our *Quantum 22*™ pursuits, which include a basic view of chemistry, physics, geology, biology, and geometry. Thus, *Quantum 22*™ is in some ways as much a technical and analytical approach to the Hebrew letters, as a sincere spiritual quest. Significantly, this book uniquely examines the foundational Hebrew letters' connections to various elements in the Periodic Table.

We will especially take many *Quantum 22*™ deep dives into Scripture because these sacred letters are the substance of God's breath... His voice... His Word. If we truly know ALEF-TAV (את) and His ALEF-BET (א ב ג ד ה ו ז ח ט י כ ל מ נ ס ע פ צ ק ר ש ת), we can grasp ancient mysteries that unlock both our astonishingly beautiful physical reality as well as our gloriously splendid spiritual reality. You will notice that each Hebrew letter in this book is associated with various pictures, symbols, numbers, mathematics, and science in our journey back to Eden and beyond, so we can understand our world and ourselves as a crossover place between the spiritual and physical realm.

Hebrew is alphanumeric. The number value of Hebrew words as well as mathematics and science have been largely locked down in a very thick black tuxedo jacket. These aspects have been all dressed up, but for the most part corrupted with darkness seeking to cover their true God-given light. This lockdown can be tracked back to ancient Babylon and the mystery religions that scattered throughout the world when the people were disbursed from the Tower of Babel. In this Kingdom Day, YHVH is releasing His numbers encoded in His Hebrew letters back to being righteously governed. This is the position of His original pristine creation.

There should be no special exaltation of numbers, mathematics, or science. They are simply aspects of the truth encoded in His Word. Studying Hebrew words through looking at the number values of each letter is commonly called Gematria. I avoid using this term due to darkness encroachment twisting gematria for its own benefit. I caution everyone to only follow the Holy Spirit's leading when you investigate any aspect of Scripture, especially a Hebrew word's numeric value.

Your safety net is to do what Yeshua did (John 5:19). Stay within the confines of the perfect will of the Father of Lights. It is so easy to get off with numbers because they can be so powerful. The Greatest Mathematician and Scientist—The Creator Himself—used mathematical and scientific properties of the Hebrew Living Letters to create and sustain the Universe. Those held captive in darkness want to harness their mysteries and power for their own selfish gain and purpose. They unfortunately use numbers in Scripture, just like they use the sacred letters, as a means to their own evil end. This is demonic. It is the basis of many witchcraft tactics, including divination, spelling curses, etc. God's people must stay away from all occult witchcraft practices. An example of what to avoid is using Bible Codes to divine the future.

Know that I come at the letters of the Hebrew Alphabet from a different angle. I've always wanted to "officially" study the Hebrew language, but the Lord has continually told me that He wanted me in His School of the Holy Spirit (רוח הקודש). Many times, to see new things we need a new perspective. This book assumes, but doesn't require, a more than basic view of Scripture. Anyone can make a quantum leap understanding in a moment what took others a lifetime to learn.

Our quantum leap is all about the redemption of the Quantum 22 and the freedom that it brings. Due to the redemption of the Quantum 22 foundational Hebrew letters being the primary focus of this book, I was prayerfully and purposely led on what to include. Know that this is not an all-inclusive volume. Not all the Scriptures of a letter or a number are included, nor is all the science and math. This book is meant to whet your revelatory appetite, so you are inspired to dive deeper.

In *Quantum 22*, we start with "Three Spheres and the Hebrew Living Letters," which is a profound heavenly Ascension in Christ experience where a group was given a unique point-of-view to see how Creation came to be. More details about "Ascension in Christ" are in the "Hebrew Living Letter Principles" chapter. Know that this group checks all our Ascension in Christ experiences against the plumbline of the Bible. Please try to "go there" by the Spirit of the Living God when you read any of the Ascension in Christ experiences in this book. Many are included and are highlighted with a gray background.

After experiencing the "Three Spheres and the Hebrew Living Letters," we discuss the "Layers of Meanings" that each letter, word, phrase, verse, chapter, and book can contain. In this chapter, we also paint a general Biblical understanding for Ascension in Christ.

Then, we circle back to before the beginning to mediate upon our Creator, the ALEF-TAV (את), the First and the Last, the Almighty. It is here we explore the infinity cross and the Three Spheres active during Creation and today. Additionally, in the "Water into Wine" chapter, we imbibe the "Miraculous Waters of MEM (מ)" and "Where the Water and Wine Meet." Then, we flow into the vital and constant need for "Quantum Entanglement with the Quantum One" and the Seven Spirits of God's key to the transformation of mankind.

After blowing past our minds to the spirit, we shift to explore "DNA—The Multidimensional Code of Life." Various topics in this chapter include: "Shadow DNA," "DNA is Light," "Antenna to God," "Jacob's Ladder and the Pillar of Righteousness," and the actual "Redemption of DNA."

Next, we shift to "Creation's Quantum Elements" where we explore several mysteries of quantum physics that link the spiritual to the natural. We will also go into some aspects of how the material Periodic Table of Elements relate to the ethereal Word of God as well as intersecting with His Hebrew Living Letters.

Then, before we delve into each sacred letter from ALEF (א) to TAV (ת) separately, we detail some fundamentals to understand the Hebrew letters and various methods for deciphering deeper truths in our "Hebrew Living Letter Principles" chapter.

Finally, when we immerse ourselves in the atmosphere of each Hebrew letter, we start out each section with The Father of Lights view on the redemption of each Hebrew Living Letter. Then, we unearth specific particulars for each Hebrew Letter and how it relates to element(s) in the Periodic Table as well as some other magnificent connections, including the mystery in the numbers.

Behold, the mysteries of the Kingdom of Heaven and and the Kingdom of God contained within the Hebrew Living Letters. We are being transformed into ancient mysteries. We are each a Word of God—a sound made visible. Let Wisdom and the Quantum 22 teach us.

∞

HOW TO READ THIS BOOK

This book is unprecedented. It is mind-blowing and earth-shattering. It will amaze you, inspire you, and challenge you to greater heights and depths. *Quantum 22* will take you into great riches that the glory of God has concealed that are now being revealed: *"It is the glory of God to conceal a matter, But the glory of kings is to search out a matter"* (Proverbs 25:2 NASB).

"Oh, the depth of the riches, both of the wisdom and knowledge of God! How unsearchable are His judgments and unfathomable His ways!" (Romans 11:33 NASB). Even though we will be mining the depths of His boundless riches, it is accessible to whosoever wills. All you have to do is desire it and press into the upward call of God in Christ (Phil. 3:14). I know this; because when I taught children's church, I taught them the depths of the Word of God in bite-sized pieces in a simplified manner. These kindergarteners through 6th graders were like ravenous lions devouring the meat of the Word. They always wanted more and more and more, because they were pointed to the Anchor of Hope and the Word implanted, which was able to save their souls (Jam. 1:21; Heb. 6:19).

To get the most out of this book, your job is to position your heart and mind rightly before applying any matter to yourself. King David puts it this way: *"Examine me, LORD, and put me to the test; Refine my mind and my heart. For Your goodness is before my eyes, And I have walked in Your truth"* (Psalms 26:2-3 NASB). One difficult truth is that all of us are refined in the fires of life. Our goal should be to come forth as gold: *"But He knows the way I take; When He has put me to the test, I will come out as gold"* (Job 23:10 NASB).

Quantum 22 is all about ALEF-TAV (את) and His Hebrew Living Letters. The glorious realities of ALEF-TAV (את) and His letters are locked up with His Word. A close and personal relationship with the Almighty *(El Shaddai)* is the foundation upon which the revelation of the Hebrew Living Letters and the Word of God rests. Like any close and personal relationship, it requires you to invest of yourself. You must invest your time, your attention, your affections, et cetera. What we behold, we become; and we become by walking it out:

> *"Beloved, now we are children of God, and it has not appeared as yet what we will be. We know that when He appears, we will be like Him, because we will see Him just as He is. And everyone who has this hope set on Him purifies himself, just as He is pure"* (1 John 3:2-3 NASB).

> *"Whoever follows His word, in him the love of God has truly been perfected. By this we know that we are in Him: the one who says that he remains in Him ought, himself also, walk just as He walked"* (1 John 2:5-6 NASB).

So first, each of us must set our hearts and minds to know our Creator and His truths about creation that He has put in place *"in the beginning" (Genesis 1:1)*. Then, we must search it out through discernment and application.

This means that we choose, by our own will, life and blessing first. We choose all the goodness created during the first six days of Creation and the seventh day of rest; and then, we set our hearts and minds to know everything that aligns with everlasting life. To a Hebrew, the concept "to know" is the watchword of Israel—The *Shema*—and it carries the meaning of hearing (listening), understanding (digesting), and doing (actions aligning).

When any particular aspect is highlighted in this book—a Hebrew Living Letter, a Scripture, a concept, etc.—each of us needs to utilize our God-given discernment mechanism. If we discern in our discerner that something is good, we need to seek to apply it in our own lives. Understand that every single human soul has been given the gift of discernment, which is dependent on God, not man: *"For since the creation of the world His invisible attributes, that is, His eternal power and divine nature, have been clearly perceived, being understood by what has been made" (Romans 1:20 $_{NASB}$)*.

Perhaps you are unsure about something, but you are being drawn to it and it doesn't "feel" bad or off. I teach people that a key to accelerated growth in Christ is to only reject the things that are bad or evil: *"Abhor what is evil. Cling to what is good." (Romans 12:9)*. If you are unsure of something and your discernment is telling you that it is not necessarily bad, I suggest that you put it on the pantry shelf inside yourself and ask the Lord to speak to you about it. When I say speak to you, He speaks in many different ways, including through situations. This pantry shelf concept helps a person to keep what is good inside themselves, as they seek the Lord of Love to confirm it. Remember that there is great grace in learning the Way and the Truth and the Life (John 14:6). He is truly a good and loving God. I pray that you get to know Him intimately like I do. I even pray that my ceiling be your floor.

If you don't know how to apply a truth that's being highlighted to your life, simply ask the Lord to teach you His truth. Once again, the man after God's own heart shows us the way: *"Show me Your ways, O LORD; Teach me Your paths. Lead me in Your truth and teach me, For You are the God of my salvation; On You I wait all the day" (Psalms 25:4-5)*.

One last suggestion, please pray before you read this book. Ask the Lord to speak to you personally. Ask Him to let you discern what is good. Ask Him to know the power of ALEF-TAV (את) and the strength of His love and life in His joyous Hebrew Living Letters. Ask Him and the Father of Lights above for every good and perfect gift that comes from above (Jam. 1:17).

As an extra special bonus, there are 22 unlisted video links to the *Hebrew Living Letters Speak* classes included in the back of the book. These are the online interactive classes from which *Quantum 22* was birthed. These videos are delightful but not as complete as this book.

Unless otherwise specified, the Biblical translation used is the New King James Version (NKJV).

Building Blocks of Creation

THREE SPHERES AND THE HEBREW LIVING LETTERS

"¹ God, who at various times and in various ways spoke in time past to the fathers by the prophets, ² has in these last days spoken to us by His Son, whom He has appointed heir of all things, through whom also He made the worlds; ³ who being the brightness of His glory and the express image of His person, and upholding all things by the word of His power, when He had by Himself purged our sins, sat down at the right hand of the Majesty on high" (Hebrews 1:1-3).

Like a comet, we ascend in Christ by the Blood of the Lamb to the intersection place where the Heavenly Father and the Son are one, where there is fullness of joy and love. Light expands from us and pulsates out. The vibrancy of the white light is a very high frequency. The brightest light appears to be coming down like some sort of pole. Then, it wraps around us, as a whirlwind.

We behold three spherical lights in front of us. We understand that this is the source of the light. It is the Father, the Son, and the Holy Spirit. There is a horizontal beam behind the three spheres of light that is a soft lemony color. The pale lemon-yellow band of color appears to go behind, above, and below the three spheres. We understand that the pastel lemon color encompasses all three, which speaks of a particular type of unity. It is their goodness uniting them.

We hear in the Spirit: "How about a lemony treat?" Lemon yogurt. Lemon bar. Lemon sorbet. Lemonade. Each treat has a different texture, but they are unified in their common ingredient. We are directed to taste the various lemony treats, which supernaturally brings up our energy levels as well as opens our spiritual eyes. It's all refreshing and energizing. Our physical eyes feel clearer and our mouths vibrate. God is working in more than one realm.

We understand that our mouth is a portal, and we get a connection to the Hebrew letter PEY (פ). Heh, PEY! The first thing that we notice is that all the Quantum 22 letters have a sense of humor before we behold them flowing in, like a dance, as they are welcomed to the goodness table. They are all on a tabletop in front of the Three Lights. The tabletop seems like a stage where the Quantum 22 are doing a unified Rockette Dance.

TAV (ת) tips his hat and walks like he is a ringmaster. We notice that TAV is a masculine letter. We understand that what we are beholding is a Redeemed Three Ring Circus.

There is a sense of multiplication with each Hebrew letter. We hear in the spirit the word "introducing," and we see something coming off of them. It is what the Hebrew Living Letters are birthing. These sacred letters have the ability to bring both expression and framework. These Hebrew letters frame our reality. Three Hebrew letters are a reflection of the Three Light Spheres in the unity of goodness.

We hear: "All the world is a stage and all of us are actors in it." We also hear: "Infinite dimension" and we are being shown the infinite reality of it all.

Each Hebrew Living Letter can come at the beginning, the middle, or the end of a base word—root word. When the combination of the letters changes, it creates a much deeper meaning due to the depth within each quantum letter. We then hear the phrases "infinite possibilities" and "infinite combinations." This is our created reality.

The Father, the Son, and the Holy Spirit are delighted to show us their Hebrew letters. Each have multiple dimensions, which is the framework for what they are expressing: language, pictures, symbols, mathematics, science, signs, revelation, communication, etc. Their expressions are individualized for whatever is needed. All dialects on earth have these Hebrew Living Letters.

We also sense that Wisdom is in the midst of the lemony goodness. We behold the letters stacked single-filed vertically. They are one on top of each other. Together, they form a vertical pillar that goes upward, which is encompassed by a circle. The Hebrew letters start to spin, As the encircled letters spin, it gets bigger and bigger. There's a Being in the midst of a whirlwind. It is "The Word." Hebrew is the patriarch. Messiah Yeshua (Jesus Christ) is the Word of God (Rev. 19:13). The Being in the midst of the whirlwind is an outline of a man with prisms of crystals. We see clear light and color shards; the image of the man is mainly translucent. We also hear the roaring of the whirlwind.

We see a void. The Hebrew phrase for "void" in Genesis 1:2 is *tohu bohu,* which means empty and void. *Tohu bohu* describes the condition of the earth immediately before the creation of light in *Genesis 1:3*— *"Let there be light!"* When we focus on the whirlwind spinning, we also see a void (darkness). We behold lights coming off Yeshua, which we understand are His Hebrew Living Letters. They go off into the void. This brings a creative light to the void, which manifests whatever He wants, whatever He wills. Hebrews 1:2 reveals that God has spoken to us in His Son through whom He made the world(s). Then, Hebrews 1:3 tells us Messiah Yeshua upholds all things by the word of His power.

The Lord tells us: "I created a blank canvas for all to create." We get the concept of co-creating. The Lord has made space for us.

We receive a vision of Yeshua being the Word; and then, hear the word "layers." We see the Hebrew Living Letters layered on top of one another again. It is a movement of dimensions. We understand each sacred letter can be on the top, the bottom, diagonal, etc. It can be any combination; and it is a fractal that is infinitely self-similar, which goes on forever.

The Hebrew letters spin and they all go in all different directions. One pattern reveals letters that are not spinning but are circled around 360 degrees. Another pattern shows the letters spinning slowly and it is very dimensional.

The Ascension in Christ group focusing on the perfect will and perfect heart of the Father feels led to step into the pattern rotating very slowly. We first hear an ancient primordial creaking sound. We know that this pattern is birthing from the deep. After we stepped through the wall of this letter grouping, there is a calmness in the midst. We behold the flashing of life past, present, and future—the eternal now. We see light going from one side of the letter grouping to the other and back again. There is energy inside. There is life inside, and we are in the middle of it. We feel a vibration in our feet moving up and down. It tightens around us. We feel its vibration, but we are no longer in our physical bodies in the ascended realms. We seem to be interacting with our own vibration.

We understand that the ancient creaking sound is connected to the "nothing," "emptiness," and "void" of Job 26:7.

- *"He stretches out the north from the empty place, and hangs the earth upon nothing"* (Job 26:7 $_{Lamsa's\,Aramaic}$).
- *"He spreads out the North over a void; He suspends the earth upon nothingness"* (Job 26:7 $_{Tanach}$).

The Lord is showing us moments of Creation. When we look at our feet, we cannot see an end. Our existence would not exist without this void in Messiah Yeshua. The Father, the Holy Spirit, and Messiah Yeshua created this void. Behold, the Lamb slain before the foundation of the world (Rev. 13:8; 1 Pet. 1:18-19). We see ourselves "in Christ" before the foundation of the world. We see ourselves on the Cross in Christ suspended above the void. We understand that the void, which made Creation possible, was Yeshua's crucifixion before the foundation of the world.

The Biblical Feast of Passover *(Pesach)* in the kingdoms of our God is about passing into the depths of the Messiah. It's about entering into the deep—the deep below and light above the Crucified One.

We are aware of the whirlwind, the elemental light, and the Seven Spirits of God (i.e., manifesting the Nature of Christ) that are fractals related to us. They are aspects of who we were originally and who I AM becoming. Back to the future. In unity and oneness, I was there, as we all were. There are different colors, lights, and sounds for each individual person.

Creation came from Christ's Cross. Inside Him. The Kingdom of God within Him (Luke 17:21). There is a spiral of dimensions. I in Him and Him in me (John 17:23). The dimensions outside are heavenly dimensions, like the Kingdom of Heaven. It is a beautiful creative space. It is constantly creative and fluid, as He wills.

The realms and dimensions of Creation are in Yeshua. He is showing us realms, dimensions, and the infinite. We see spirals going vertically up. We observe concentric circles going out horizontally. At the intersection place, we can see within and without at the same time. The intersection place is a crossover place of heaven and earth. Passover is the Promise Land for all of Messiah Yeshua.

From Mel Gel to Mellow Yellow. The quantum letters are cracking up. They are laughing. Yeshua tells us: "As I drank wine on the Cross, it came through Me and out to Creation." We see little circles of light energy come through Yeshua and out. The Bread was an elemental sparkly substance—manna. It is the Bread of Heaven in Christ. This firmament is in Him, which is the essence of creation.

When we step into the truth of the Passover Seder Plate, it is dimensional and sweet. Taste and see. Citron—lemony goodness—points to the Feast of Tabernacles *(Sukkot)*. We enter into all Biblical Feasts "in Christ". The feast of Passover *(Pesach)* is highlighted. We enter into the crossover place where dimensions are layered. God's Biblical Feasts portray the dimensional layering and dimensional gateways for all creation—life.

We see a deep green leaf of a lemon tree and are pulled into its cells. Then we are shown a snowflake. The letters of the Hebrew Alphabet follow the same pattern as the snowflake form. We understand that the lemon tree does the same thing as the snowflake. The sound is the seed of the Tree of Life. We are all fractal. The artists, the mathematicians, the scientists, and the linguists need to overlay ... infinite fractal ... infinity cross. There's a cube within the infinity cross. [9]

NOTE: "Mel Gel" is about gelling as one with Christ in the Righteous Order of Melchizedek. During a *Mystic Mentoring* Group Ascension, the Lord revealed:

"Be washed in the water of His Word, the oil of His Presence, and His Salt Covenant to become carriers of the frequency of His Light. Let the laser red light connect you like an umbilical cord to the Father's Heart, as an astronaut is connected to his life support by a hose. Let the blueish gel of His Presence be your bonding agent to love.

At the cross point of the red light (where the salt crystals rain down and ignite) and the blue light is a creative gel-like substance. YHVH breathes on us, and we cross over into a new dimension. We are in the gel as well as on fire—blazing flames of love.

Hear His voice: "Come be consumed with Me." The red laser beams penetrate our hearts. We connect to His lifeline—His heart to ours. As His priestly intercessors, YHVH leads His people to take something dark and rectify it by bringing it upward. As pillars of salt, we arc for His redemption and His protection. The "Mel Gel" that encompasses us forces us to slow down, move in Him and rest in Him. That's how the Righteous Order of Melchizedek gels." [10]

LAYERS OF MEANING

The Eternal Self-Existent One has placed an infinite amount of wisdom in each letter of the Hebrew Alphabet. The holy language of Hebrew has innumerable ways of communicating deep truths for each letter, each word, each phrase, each verse, each chapter, and each book. There are pictures (forms), sounds (frequencies), numbers (mathematics), compounds (science), etc.

Let's consider the many layers of meanings from the literal to mystical. A Hebrew knows that there are at least four levels that we can study the Word of God:[11]

[1] Peshat (P'shat, פשט) Level—*Peshat* is when a person studies the plain sense of a Biblical text. It is when a person looks at the simple, obvious, and literal meaning of what is presented. A person pulls out the who, what, when, where, and how of a story, as they read it in context. A *peshat* study of Scripture includes studying the basic facts, like how many animals were on Noah's Ark, two of every unclean animal, seven of every clean animal, etc.

Always remember that God is not redundant. When something is repeated in the Bible, it is a road sign to pay special attention. Do not ignore something repeated. Stop, look, and listen because this is emphasizing the Way and the Truth and the Life. This is a clue to dig deeper. Not one jot and tittle in God's Word is idle. Hebrews believe that there are at least seventy facets of wisdom that come with every word in Scripture. It takes a lifetime to study the Word of God.

[2] Derash (דָּרַשׁ) Level—*Derash* is the root verb for *Midrash*, which is an exposition or investigation of Scripture that goes into a Biblical text to grab principles, understanding, and concepts. *Derach* is all about seeking to understand the exact truth in Scripture, so we can align our behavior accordingly. *Derash* (דָּרַשׁ) points to enquiring through seeking with care.

To *midrash* is to speak with one another about Scripture in a way that engages the words of a Biblical text, the meaning behind the text, and the meaning beyond the text as well as focusing on each letter in each word. The term *Midrash* can also be used to refer to a separate body of commentaries on Scripture that have used this interpretative mode.

[3] Remez (רֶמֶז) Level—*Remez* means a "sign or hint" in Hebrew. *Remez* is the hidden or secretive meaning beyond the literal sense. It is the deeper esoteric, allegoric, or symbolic meaning of Scripture. *Remez* is when the Bible talks about something, but it never mentions it, like God in the Book of Esther.

[4] Sod (סוֹד) Level—*Sod* studies the Word of God on a mysterious level. Bible codes fit in the *sod* category. Sir Isaac Newton spent the majority of his life studying Bible codes in Torah. He did all his calculations by hand. Recall from the preface that everyone is warned against divining the future using Bible codes and every other witchcraft tactic. However, to righteously look deeper into something that already exists is always an admirable quality.

The divine council who meets with God in divine deliberations is connected to the Hebrew term "sod" due to *sod* being defined as "a confidential, a secret or plan, a circle of confidants, or council." *"A godlike being has taken his place in the council of God; in the midst of the divine beings He holds judgment" (Psalms 82:1)*. Scripture also says about him, *"Over [it] take your seat in the highest heaven; a divine being will judge the peoples" (Psalms 7:7-8)*. Dominion will pass from Belial and return to the Son of Light by the judgment of God, just as it is written concerning him, *"who says to Zion 'Your divine being reigns'" (Isaiah 52:7)*. Zion is the congregation of all the sons of righteousness, who uphold the covenant and turn from walking in the way of the people. "Your divine being" is Melchizedek, who will deliver them from the power of Belial. [12]

A person hits all four of these levels in a proper study of the Word of God. This is a way where we can get different dimensions of what's going on.

THE HEBREW LIVING LETTERS AND ASCENSION IN CHRIST

The "Three Spheres and the Hebrew Living Letters" is an Ascension in Christ experience that illustrates these four levels of *Peshat, Derash, Remiz,* and *Sod*.

Please allow me to share a basic Biblical explanation of Ascension in Christ. The pattern for Ascension in Christ comes from *Ephesians 4:9-10—"⁹ (Now this, 'He ascended'—what does it mean but that He also first descended into the lower parts of the earth? ¹⁰ He who descended is also the One who ascended far above all the heavens, that He might fill all things.).*"

The plain sense *(peshat)* understanding of Ephesians 4:9-10 is that it is speaking about Messiah Yeshua's death, burial, and resurrection. On the *derash* level, the Bible also speaks of God's people who say that they abide in Christ should walk in the same manner as our Pattern Son—Messiah Yeshua (1 John 2:6). We need to follow Him in His death, burial, and resurrection (Romans 6:1-6). Or in other words, follow Christ by first descending to the Kingdom of God within a believer (i.e., the current level a person operates in); and then, ascending "in Christ" by the Blood of the Lamb to focus on the perfect will of our Heavenly Father (Rev. 4:1-2; Heb. 10:19-20; John 5:19). This is on the *remez* (hidden) level and the *sod* (mysterious) level.

I've written an entire book explaining Ascension in Christ according to the Spirit of the Living God and Truth (i.e., the Word of God). It is called the *Ascension Manual*. [13]

The group that I "do" Ascension in Christ with is called *Mystic Mentoring*, which is solely connected to *Sapphire Throne Ministries*. It was established in 2015. Our first action when we begin to ascend in Christ is to step into Christ's Cross to die to our own Kingdom of Self. This is where we lay down all selfish ambition, interest, desire, etc. on God's Altar, which is basically anything that is not the perfect will of our Heavenly Father. We are then coated by the Blood of

Christ, as we step into Messiah Yeshua to be one with Him. The only legal and protected way to do Ascension in Christ is by the Blood of the Lamb: *"Therefore, brethren, having boldness to enter the Holiest by the Blood of Jesus"* (Hebrews 10:19). Then, as one, we arise by the Blood to the intersection place where the Father's and Son's love is one to corporately focus on the perfect will of the Father by the Holy Spirit, as we are "in Christ." *"And raised us up together, and made us sit together in the heavenly places in Christ Jesus"* (Ephesians 2:6).

"^{19} Then Jesus answered and said to them, 'Most assuredly, I say to you, the Son can do nothing of Himself, but what He sees the Father do; for whatever He does, the Son also does in like manner. . . . ^{30} I can of Myself do nothing. As I hear, I judge; and My judgment is righteous, because I do not seek My own will but the will of the Father who sent Me'" (John 5:19,30).

Once we have ascended on high "in Christ," we focus solely on reporting what we see, hear, or perceive by our spiritual senses, as everyone is solely focused on our Heavenly Father's perfect heart and perfect will.

WATER INTO WINE

The most fundamental part of our material realm is the Hebrew Living Letters connected to the Word of God. We could basically say that the Book of Creation is written in the language of the letters of the Hebrew Alphabet.

ALEF-TAV (את)

Even though ALEF-TAV (את) is a known element in Hebrew grammar, which serves to designate the direct object in a sentence, it can also carry a deeper, revelatory meaning that points to the Righteous Messiah directly. "*God conceals the revelation of His word in the hiding place of His glory. But the honor of kings is revealed by how they thoroughly search out the deeper meaning of all that God says*" (Proverbs 25:2 $_{TPT}$).

If we care to take advantage of our privilege to seek out the mysteries of the Kingdom of Heaven and the Kingdom of God, we will eventually press into the secret place of the Lord of All Creation (Matt. 13:11; Luke 8:10). Here you will find Messiah Yeshua (i.e., Jesus Christ) whose hidden name is ALEF-TAV (את)—the First and the Last. For those new to the ALEF-TAV concept, know that ALEF (א) is the first letter of the Hebrew Alphabet while TAV (ת) is the last; therefore, it can carry the general meaning of the first and the last. However, ultimately, ALEF-TAV (את) is the First and Last, the Almighty, who upholds all things by the power of the Word. Just as the Word of God is one of our righteous Messiah Yeshua's Names, so is ALEF-TAV (Rev. 1:8,11; Rev. 2:8; Rev. 22:13). As the Creator, He literally fulfills being the first and last of Creation; therefore, we can proclaim that Yeshua is Lord of All or Lord of All Creation.

ALEF-TAV (את) is also His hyperquantum substance that flowed out of the Lamb slain before the foundation of the world in the form of the Quantum 22 letters of the Hebrew Alphabet. In physics, hyperquantum pertains to values in which a quantum-mechanical equation approaches infinity. Not only do the Quantum 22 Hebrew Living Letters approach infinity, they are infinite because they are one with the Eternal, Self-Existent One.

When ALEF-TAV (את) built the House of Creation, it all began with the hyperquantum letters of the Word of God. An entire book and study guide has been written about ALEF-TAV and His Hebrew letters: *ALEF-TAV's Hebrew Living*™ *Letters: 24 Wisdoms Deeper Kingdom*

Bible Study. It is highly recommended to start with the ALEF-TAV Bible Study for a proper foundation before cracking open *Quantum 22*.

Please allow me to re-establish some fundamentals about ALEF-TAV (את) that have already been laid out in the *ALEF-TAV's Hebrew Living™ Letters* book. There is a hidden meaning behind the first verse in the Bible. In English, we read: *"In the beginning God created the heavens and the earth."* However, in Hebrew, it says: *"Bereshit barà Elohim **et** hashamaim veet haaretz."* After the words *"In the beginning God created"* (*Bereshit barà Elohim*) and before the word *"heavens"* (*hashamaim*), there is the word *et*. Notice that is seven Hebrew words, but only six have been translated. The untranslated word *et* is composed of the letters ALEF (א) and TAV (ת), which are the first and last letters of the Hebrew Alphabet. This is one of the reasons why the Hebrew sages say that God created the Hebrew Alphabet first; and then, He used them to create heaven and earth.

Throughout Scripture, there are random references to ALEF-TAV (את) that have been largely disregarded when the Jewish people read their Bible (except to designate the direct object in a sentence), or when Scripture has been translated into other languages, including English. So far, it seems that mainly Hebrew Mystics have tried to delve into the reason why those random ALEF-TAVs exist in the written word. Perhaps these scholars have overlooked the Messianic reference in *Isaiah 44:6 "Thus says the LORD, the King of Israel and His Redeemer, the LORD of hosts: I am the first* [ALEF] *and I am the last* [TAV], *and there is no God besides Me."* Fundamentally, we need to remember that Messiah Yeshua declares in *Revelation 1:8, "I am the Alpha and the Omega . . . who is and who was and who is to come, the Almighty."* Yeshua is the ALEF-TAV or as the Word of God says, I AM the ALEF-TAV, and there is no God besides Me.

Not only is Messiah Yeshua the Beginning and End of all creation, He is also the Creator and Sustainer of all things: *"For by Him were all things created that are in heaven, and that are in earth, visible and invisible, whether they be thrones, or dominions, or principalities, or powers: all things were created by Him and for Him: And He is before all things and by Him all things consist"* (Colossians 1:16-17).

It is an incredible mystery how Yeshua *"existed before all things"* and is *"the first born of all creation"* at the same time. When one understands that the array of individual spiritual forces ordered via the Hebrew Alphabet were resident in Yeshua before those letters or that language was created, one can get a glimpse into the vast mystery of Yeshua being the firstborn of all creation. *"In the beginning was the Word, and the Word was with God, and the Word was God"* (John 1:1). The Word is made up of letters. *"In the beginning God created ALEF-TAV"* (Genesis 1:1). The individual spiritual forces inherent in Christ were birthed from His essential, intrinsic essence, which was originally included wholly in the Word. The firstborn of all creation is the very thing that constitutes every living substance. It is the protoplasm of the universe. It is the individual spiritual forces that originally belonged, and still belong, to the nature of Christ. The firstborn of all creation is the letters of the Word—the Hebrew Alphabet.

When our Heavenly Father, Messiah Yeshua, and the Holy Spirit created the alphanumeric Hebrew Alphabet out of the Messiah's essence, God ordered spiritual forces of creation through its twenty-two sacred letters before there was even one other act of creation. Therefore, we can say that God's Quantum 22 frame our reality by carrying the expression of the three unified spheres of light: The Heavenly Father, The Son, and the Holy Spirit. There are infinite possibilities and infinite combinations to our created reality. Each letter of the Hebrew Alphabet has a plethora of dimensions along with a unique framework of expression in Christ. They are infinitely creative.

THE INFINITY CROSS

Everything in creation starts and flows from the Head, which Scripture tells us is Messiah Yeshua: *"15 But, speaking the truth in love, may grow up in all things into Him who is the Head—Christ— 16 from whom the whole body, joined and knit together by what every joint supplies, according to the effective working by which every part does its share, causes growth of the body for the edifying of itself in love"* (Ephesians 4:15-16).

The Hebrew word *bereshit* (בְּרֵאשִׁית) opens Scripture and points to the start of Creation. God gave us this word first so we can understand that He is the beginning of everything created. Reb Jeff reveals that the very first verse, the very first word, and the very first letter contains what may be called a grammatical "mistake." The word *bereshit* is in the construct state and should be translated as "in the beginning of." The problem grammatically speaking is the word following *bereshit* in Genesis 1:1 is not a noun. The next word is *bara* (בָּרָא), which is a verb that means "He created." Obviously, this is not a mistake, but a divine statement with intentional significance. What doesn't make sense grammatically makes sense spiritually. Reb Jeff reveals that the world was created, but it never stopped being created. The Bible begins by telling us that it doesn't exist in time like other stories. It exists in a suspended moment that cannot be pinpointed on a timeline. The difference between two little dots (בְּ) and a little "T" (בָּ) shape under that great big BET is the difference between the Bible that tells a conventional story and one that tells a story that exists outside of time and within all time at the same time.[14]

The Word of God was in the midst of His Quantum 22 whirlwind during the fractal creative instance of Creation while at the same time (before the foundation of the world) a void was created in Messiah Yeshua through His pre-creation crucifixion (refer to the previous "Three Spheres and the Hebrew Living Letters" chapter): *"18 For you know that God paid a ransom to save you from the empty life you inherited from your ancestors. And it was not paid with mere gold or silver, which lose their value. 19 It was the precious blood of Christ, the sinless, spotless Lamb of God 20 God chose Him as your ransom long before the world began, but now in these last days He has been revealed for your sake"* (1 Peter 1:18-20 NLT).

This void in Christ allowed the creation of all worlds and realties, which means that Creation came from the Infinity Cross . . . from inside of Him. Everything, including us, is inside of Him *"in the beginning"* and even now. Christ hung in eternity on the Infinity Cross with a cube in its midst, which produced/produces infinite fractals (i.e., things that are infinity self-similar).

The void created in Christ is shown in the second verse of Genesis 1: "*¹ In the beginning God created the heavens and the earth. ² The earth was without form [tohu (תֹהוּ)], and void [bohu (בֹהוּ)]; and darkness was on the face of the deep. And the Spirit of God was hovering over the face of the waters*" (Genesis 1:1-2 *Additions mine*). "*Tohu bohu*" (תֹהוּ וָבֹהוּ) is the Hebrew phrase in Genesis 1:2, which means without form (empty) and void. *Tohu bohu* describes the earth immediately preceding the release of the divine light of God. *Tohu* (תֹהוּ) is also mentioned in *Job 26:7*— "*He stretches out the north over empty space* (תֹהוּ); *He hangs the earth on nothing.*" Talk about mind-blowing! How does God hang something on nothing? "Nothing" is so far beyond what we can think or imagine. To even grasp that the earth hangs on nothing, we must use something, like words.

The ALEF-TAV building blocks, the heavens, the earth (without form and void), darkness and water are all mentioned before the Father of Lights' impartation of His uncreated light in Genesis 1:3: "*³ Then God said, 'Let there be light'; and there was light. ⁴ And God saw the light, that it was good; and God divided the light from the darkness. ⁵ God called the light Day, and the darkness He called Night. So the evening and the morning were the first day*" (Genesis 1:3-5). Job 26 adds an interesting perspective to this when we consider the Three Spheres: "*He drew a circular horizon on the face of the waters, at the boundary of light and darkness*' (Job 26:10).

THREE SPHERES ACTIVE DURING CREATION

All Three Spheres that bear record in heaven—the Father, the Word, and the Spirit (1 John 5:7)—were active during creation. As the Holy Spirit hovered, the Heavenly Father's voice imprinted on the water of the Word. "*By the word of God the heavens existed long ago and the earth was formed out of water and by water*" (2 Peter 3:5 NASB).

Sometimes we try to separate the Three Spheres to understand them. However, we must comprehend that when we are learning about one of them, all Three Spheres are present due to their *echad* oneness (plurality in oneness). With that in mind, let us further explore how the Father, the Word (i.e., the Son), and the Holy Spirit were active during the first day of creation.

The Father of Lights is the giver of every perfect gift of Creation through His word of truth: "*¹⁷ Every good gift and every perfect gift is from above, and comes down from the Father of Lights, with whom there is no variation or shadow of turning. ¹⁸ Of His own will He brought us forth by the word of truth, that we might be a kind of firstfruits of His creatures*" (James 1:17-18). From the start, our Heavenly Father established "*the word of truth, which is the gospel of your salvation*" (Ephesians 1:13). As sons of the Living God who live according to His great grace, we are admonished to "*Be diligent to present yourself approved to God, a worker who does not need to be ashamed, rightly dividing the word of truth*" (2 Timothy 2:15). Lovers of God correctly hear, understand, and teach the living truth flowing from the fountain of life.

When the Father of Lights said, "*Let there be light!*" He established the spiritual foundation of the Universe. Many years ago, Yeshua shared with me: "Just as the spiritual realm is founded on The Father's love, so is the physical realm founded on the waters." "*¹ The earth is the* LORD*'s,* *and all its fullness, The world and those who dwell therein. ² For He has founded it upon the seas, And established it upon the waters" (Psalms 24:1-2).*

There is life in the Hebrew Living Letters. There's an exchange of love within. Behold, the Hebrew Living Letters are the elemental material of the Word of God in creation. Behold, the Hebrew Living Letters spiral into the DNA of Creation, which is the protoplasm of the universe. The spiritual foundation of the earth—the Father's love—is what holds all Creation together in Christ. Step into the middle of His Hebrew Living Letters DNA Strand where Abba's love is transformative. Intertwine with the burning heart of love in the Hebrew Living Letters.

The Creative Word of God and His Hebrew Living Letters are the Living Expression of the Heavenly Father's voice. We can also say that ALEF-TAV (את) and His letters are the Father's exact expression. The Living Expression of His reflection is the raw materials for all things created: *"¹ In the beginning the Living Expression was already there. And the Living Expression was with God, yet fully God. ² They were together—face to face, in the very beginning. ³ And through His creative inspiration this Living Expression made all things, for nothing has existence apart from Him! ⁴ A fountain of life was in Him, for His life is light for all humanity. ⁵ And this Light never fails to shine through darkness—Light that darkness could not overcome!" (John 1:1-5 $_{TPT}$)*. The Aramaic reads *"In Him were lives"* (plural) in John 1:4, which not only conveys many human lives but also spiritual life and eternal life. Simply put, in Messiah Yeshua is *zoe* life in every form.

The Holy Spirit is the nurturer and life-giving breath of Creation. Most languages designate the same term for breath and spirit. True and unadulterated life-giving breath does not have earthly origins. God blew the breath of life into Adam (Gen. 2:7). *"The Spirit of God has made me, and the breath of the Almighty gives me life" (Job 33:4)*. Through the Spirit of the Living God all living things are created.

Mankind has a special impartation of the Holy Spirit, which we will go over shortly in the "The Seven Spirits of God" section. Before the Spirit came down like a dove from heaven and rested on Yeshua, Scripture reveals that the Spirit can surround a person like a garment (Jud. 6:34). The Spirit can fall upon a person and hold their hand (Ezek. 11:5; 37:1). The Spirit resting upon one person can be divinely imparted to others. The Holy Spirit rested on the children of Israel when they crossed the Red Sea. [15] The Holy Spirit inspired the division of the country by lot among the tribes. [16] The Holy Spirit caused the soul of the widow woman's son to come back into him after Elijah cried out and YHVH heard his voice (1 Kings 17:22). The sweet psalmist of Israel spoke by the Spirit of YHVH s voice (2 Sam. 23:2). According to the Hebrews, "the visible results of the activity of the Holy Spirit are the books of the Bible, composed under its inspiration."[17]

When I asked our Beloved what were the waters that the Spirit of God hovered over in Genesis 1:2? He pointed me to Genesis 1:1. The key lies in the mysterious reference of ALEF-TAV (את). The thirteenth letter of the Hebrew Alphabet has a pictograph of water. It is the letter MEM (מ), from which we get the Hebrew word for water—*mayim* (מים). The Lord told me that the waters that the Holy Spirit was brooding over in Genesis 1:2 were the waters of MEM (מ). He added that all the other Hebrew letters were in the waters of MEM too, like some sort of watery alphabet soup. MEM (מ) was the carrier, but they were all there. Therefore,

MEM (מ) and company (ת ר ק צ פ ע ס נ ל כ י ט ח ז ו ה ד ג ב א) along with the brooding of the Spirit of the Living God were what received the imprint of the Father of Lights' voice to facilitate Creation.

This picture was re-iterated when Messiah Yeshua—the Word of God—was baptized in water and a dove representing the Holy Spirit rested upon Him. The Book of Luke especially reveals that it was a literal dove that rested upon Yeshua (Luke 3:22). The idea of a dove-like form is found in Jewish Literature "the voice of the dove" is translated as "the voice of the Holy Spirit" in Cant. ii.12. Ben Zoma interprets *Genesis 1:2 "And the Spirit of God moved on the face of the waters"* to mean "As a dove that hovers above her brood" (Hag. 15a). [18]

The frequency of the Holy Spirit in a believer's life is proportional to the measure of the stature of the fullness of Christ that's within (Eph. 4:13). When a person is born from above through the acceptance of Yeshua Ha Machiach (Jesus Christ) as their Lord and Savior, they receive a ten-percent deposit of the Holy Spirit which they are then tasked to steward and grow (Rom. 10:9-10; 2 Cor. 1:22; Eph. 1:13-14). The ultimate goal for mankind is the complete fullness of 100-percent in Christ where we accurately bear the exact image of the Righteous Messiah in the unique way that we were each created.

It is in the crossover place of heaven and earth that the Quantum 22 of ALEF-TAV (את) reside, sustain, and maintain the physical Universe through the word (letters) of His power: "*²[God] has in these last days spoken to us by His Son, whom He has appointed heir of all things, through whom also He made the worlds; ³ who being the brightness of His glory and the express image of His person, and upholding all things by the word* [Hebrew Living Letters] *of His power, when He had by Himself purged our sins, sat down at the right hand of the Majesty on high*" (Hebrews 1:2-3 *Addition mine*).

Wisdom was with YHVH—the Father, the Son, and the Holy Spirit—before there was ever an earth: "*²² "The LORD possessed me* [Wisdom] *at the beginning of His way, Before His works of old. ²³ I have been established from everlasting, From the beginning, before there was ever an earth. ²⁴ When there were no depths I was brought forth, When there were no fountains abounding with water. ²⁵ Before the mountains were settled, Before the hills, I was brought forth; ²⁶ While as yet He had not made the earth or the fields, or the primal dust* [quantum atoms] *of the world. ²⁷ When He prepared the heavens, I was there, When He drew a circle on the face of the deep, ²⁸ When He established the clouds above, When He strengthened the fountains of the deep, ²⁹ When He assigned to the sea its limit, So that the waters would not transgress His command, When He marked out the foundations of the earth, ³⁰ Then I was beside Him as a master craftsman; And I was daily His delight, Rejoicing always before Him, ³¹ Rejoicing in His inhabited world, And my delight was with the sons of men*" (Proverbs 8:22-31 *Additions mine*).

Notice the "primal dust" mentioned in verse 26, which I believe can be the primal dust of His Quantum 22. Also, notice the circle (sphere) reference in verse 27, which coincides with the phrase "*darkness was on the face of the deep*" in Genesis 1:2. The darkness of the "*darkness upon the face of the deep*" is not merely the absence of light. *The Stone Edition* of *The Chumash* says that it is "a specific creation, as is clearly stated in Isaiah 45:7 'He who forms the light and creates darkness.'" [19]

Primordial water was also an integral material of creation. Consider that the Hebrew word for "heavens" in Genesis 1:1 is *shamayim* (שָׁמַיִם), which can be separated into SHIN (שׁ) and *mayim* (מִים)—the SHIN waters. Recall *Mayim* is the Hebrew word for water. If we consider that the Hebrew Word Picture for water consists of three Hebrew letters—MEM, YUD, MEM—we see a hand between two waters because MEM stands for water and YUD stands for hand. Whose hand? Behold, ALEF-TAV's hand whose hidden reference just precedes *shamayim* (שָׁמַיִם) in the first verse of every Bible: (בְּרֵאשִׁית בָּרָא אֱלֹהִים אֵת הַשָּׁמַיִם וְאֵת הָאָרֶץ) We can see this reality during the second day of creation when God's hand separated the waters above from the waters below (Gen. 1:6-8).

Another clue is that it is common to use SHIN (שׁ) as shorthand for *El Shaddai* or Almighty God whose hidden name is ALEF-TAV (את). Therefore, we can behold the connection between the created heavens *shamayim* (שָׁמַיִם) with the original and most ancient waters *(mayim)* of creation in Genesis 1:2. Additionally, the Hebrew term for "earth" (אֶרֶץ) refers to a state of watery chaos, suggesting running water.

Before the waters were parted on the second day of creation, the Spirit of the Living God brooded over the primordial waters that were in the void created in Christ: *"¹In the beginning God created the heavens and the earth. ² The earth was without form, and void; and darkness was on the face of the deep. And the Spirit of God was hovering over the face of the waters" (Genesis 1:1-2).*

The Hebrew word for "hovering/brooding" is *rachaph* (רָחַף), which has three Hebrew letters. It is a primitive root that's an action word—a verb. It can be defined as: to move, to hover, to brood, to flutter, to grow soft, or to relax. *Gesenius Hebrew-Chaldee Lexicon* adds to be affected with the feeling of tender love; hence, to cherish as well as to tremble (Jer. 23:9). So, when the Holy Spirit brooded over the waters with all the Quantum 22 letters within, God's tender loving care hovered to make all things beautiful: *"He has made everything beautiful in its time. Also He has put eternity in their hearts, except that no one can find out the work that God does from beginning to end" (Ecclesiastes 3:11).* Additionally, the Holy Spirit infused the spark of life into the ancient waters of creation when the Spirit hovered; thus, adding a Spirit-infused life sparkle to the Hebrew Living Letters in the waters.

There are two things that received the *"it was very good"* designation in *Genesis 1:31* that didn't get the prior *"God saw that it was good" (Genesis 1:4,10,12,18,21,25)* stamp of approval. These two things are the waters and firmament of Day 2 as well as Day 6's creation of mankind —male and female. Are they connected? Behold, the waters above and the waters below with a firmament in between exists outside and inside each person made in God's own image. Quantumly-speaking water is the holder of all information. Its purpose is to enable and sustain life. Humans are two-thirds water, as is the organism that we exist on—Earth.

This is the incredible foundation of our world. Spiritually, it is the tender loving care of the beautiful three that bear witness in heaven (1 John 5:7). Physically, it is the water of the Word that is elementally made up of foundational letters—*otiyot yesod*. Recall that through the *otiyot yesod*, God formed substance out of what appeared to be chaos and brought forth existence from non-existence.

MIRACULOUS WATERS OF MEM (מ)

We could say that the SHIN waters remained the waters above the firmament—Heaven—while the waters of MEM were set apart as the waters below the firmament. The life sparkling waters of MEM are the Hebrew Living Letters being disbursed throughout creation.

It is not a coincidence that the miraculous waters of MEM (מ) show up as a supernatural sustenance in the Wilderness. In the Wilderness, there were three great divine gifts that assisted God's people—the manna, the cloud(s) of glory and the perpetual well of water, which the Hebrews say are connected to Moses, Aaron, and Miriam respectively. Additionally, we can map the manna to the Father of Lights who spoke *"Let there be light"* on Day One: *"So He humbled you, allowed you to hunger, and fed you with manna which you did not know nor did your fathers know, that He might make you know that man shall not live by bread alone; but man lives by every word that proceeds from the mouth of the LORD" (Deuteronomy 8:3)*. The cloud(s) of glory can be mapped to the One called the Word of God: *"They will see the Son of Man coming in a cloud with power and great glory" (Luke 21:27)*. And the perpetual well of water in the Wilderness can be mapped to the Holy Spirit: *"Jesus answered, 'Most assuredly, I say to you, unless one is born of water and the Spirit, he cannot enter the kingdom of God'" (John 3:5)*.

The rolling rock that accompanied the Children of Israel on their wanderings in the desert became known as Miriam's Well. Long before rock-n-roll, there was a rolling rock in the Wilderness; and that rock was Christ: *"For I do not want you to be unaware, brethren, that our fathers were all under the cloud and all passed through the sea; and all were baptized into Moses in the cloud and in the sea; and all ate the same spiritual food; and all drank the same spiritual drink, for they were drinking from a spiritual rock which followed them; and the rock was Christ"* (1 Corinthians 10:1-4 $_{NASB}$).

Historically, Miriam's Well provided fresh potable water not only for the people, but also for their livestock. It also provided fresh living water by which the dry places bloomed with green pastures and delightful flowers. Everywhere their foot tread, as they were led by the Pillar of Cloud by day and the Pillar of Fire by night, was blessed and abundantly full of life. No wonder God's people loved and respected the wise and God-fearing prophetess.

Personally, I think Miriam's connection to the rolling rock—the Messiah—that followed the Children of Israel pouring forth living water started with her Spirit-led abandoned worship at the Red Sea. After Moses led the people to miraculously cross the Red Sea, Moses and all the people broke into praise singing the "Song of the Sea" (Hebrew: שירת הים, *Shirat HaYam*), which is also known as the "Song of Moses" listed in Exodus 15:1-18. It is followed by a much shorter version sung by Miriam and other women, as they euphorically danced by the Spirit (Exo. 15:20-21).

Then, Exodus Chapter 15 goes right into how the Israelites went three days into the wilderness and found no water. They came to the bitter waters of Marah and grumbled to Moses. Moses cried out to the Lord who showed him a tree to throw into the water to make it sweet: *"So he cried out to the LORD, and the LORD showed him a tree. When he cast it into the waters, the waters were made sweet" (Exodus 15:25)*. After that the people grumbled again about a lack of water at Rephidim whereupon Moses was instructed by God to stand on the rock at Horeb and strike the rock, so water would come out of it (Exo. 17:1-7). This is how the precious living waters of Miriam's Well came to be in the Wilderness.

One of the ways that the Bible describes Yeshua is hanging on a tree (Gal. 3:13). When Moses threw the tree in the bitter waters to make the waters sweet, it was a prophetic foreshadow of the Crucified One purifying the bitter waters of the creation through the Cross in our spacetime continuum. *"[19] For it was the Father's good pleasure for all the fullness to dwell in Him, [20] and through Him to reconcile all things to Himself, whether things on earth or things in heaven, having made peace through the blood of His cross" (Colossians 1:19-20 [NASB]*). Behold, the redemptive wood comes from the ultimate eternal Tree of Life with the fruit of His Hebrew Living Letters.

WHERE THE WATER AND WINE MEET

The miraculous change of water into wine demonstrated Yeshua's power over all things down to the molecular level. Water conveys both temperature and pressure, which govern molecular structures. Water is a miracle. When it freezes, it expands. When it is heated, it super expands. Water is the only liquid known to naturally do this. When we contemplate water introspectively, we realize it buffers and transmits signals through our bodies. Water is the egg carton of our cells. It's the medium of life.

"Wine is a complex mixture of several hundred compounds, many of them are found at very low concentrations; however, they play an important role in its evolution and quality. In general, the average concentrations of the major components are water, 86%; ethanol, 12%; glycerol and polysaccharides or other trace elements, 1%; different type of acids, 0.5%; and volatile compounds, 0.5%." [20]

A trio of physicists at the University of Leicester have revealed that while the water into wine miracle could not occur spontaneously, it was still physically possible. The researchers "determined the extra energy required for turning water into wine, finding that Jesus would have had to instantaneously input more than 250,000 kJ—roughly half the amount of energy in an average lightning bolt—into the system for it to work. The conclusion: It must have been a miracle." [21]

John 2:11 tells us that the Messiah's first miracle at the wedding in Cana was to manifest His glory. During a *Mystic Mentoring* Group Ascension on 5 October 2023, God showed us the water into wine miracle in the Spirit. First, we saw Yeshua standing over the stone water jars; and then, we saw tiny whirlpools stirring the water at the molecular level. We understood that the hydrogen and oxygen atoms that made up the molecules of water were being stirred before Yeshua zapped them with electricity that looked like lightning.

Wine is 98-percent water and ethanol (alcohol). The chemical formula for ethanol is C_2H_6O, which means that four hydrogen atoms and two carbon atoms had to have been added to the H_2O water. Yeshua inserted His life and light into the wine. The molecular change of water into wine happened when Yeshua moved. He was the catalyst for the molecular change in the water. The best-saved-for-last new wine at Cana came from the carbon input from Yeshua's own DNA. The wedding guests literally tasted Yeshua. They tasted His wine. They tasted Him *"Oh, taste and see that the LORD is good; Blessed is the man who trusts in Him!" (Psalms 34:8).* What we saw in the Spirit was Yeshua's earthly body was enraptured at the same time His heavenly body was co-laboring.

Where the water and wine meet, miracles happen. The crossover place of heaven and earth is where the water turns to wine. You can literally see the miracle happen because the molecular change makes light. A contemporary of Nikola Tesla—Walter Russell—speaks of the spiral light structures that comprise the material world. These spiral (stirred) light structures may rightfully be considered to be the direct predecessors of the spiral light structures of elementary particles. [22]

The chemical elements needed to turn 120-150 gallons[23] of water into the finest of wine came from ALEF-TAV (את), just like "in the beginning." I like to think of ALEF-TAV (את) as Messiah Yeshua's elemental or quantum name because all the elements on the Periodic Table come from ALEF-TAV (את)'s own DNA via the outflow of His Hebrew letters. He is the eternal source for all the elements in the Universe with His delightful letters being the hyperquantum, subatomic foundation for them. It bears repeating that ALEF-TAV (את)'s Hebrew letters are the elemental substance of the Word that maintains and sustains our world and all others.

QUANTUM ENTANGLEMENT WITH THE QUANTUM ONE

ALEF-TAV (את) is the Quantum One. The one indivisible system of Christ, in Christ, and through Christ is the origins for all of creation: *"[15] He is the image of the invisible God, the firstborn over all creation* [the letters of the Hebrew Alphabet]. *[16] For by Him all things were created that are in heaven and that are on earth, visible and invisible, . . . things were created through Him and for Him. [17] And He is before all things, and in Him all things consist* [as one indivisible system]*" (Colossians 1:15-17* Additions mine*).*

Heinrich Päs writes in the *newscientist.com:* In the face of new evidence, physicists are starting to view the cosmos not as made up of disparate layers, but as a quantum whole, linked by entanglement. When we apply quantum mechanics to the entire cosmos, we realize that the universe isn't fundamentally made of separate parts at all but is instead a single quantum object. If quantum entanglement can be applied to the entire cosmos, then the universe is "a single indivisible unit," in the words of quantum pioneer David Bohm. [24]

Science is catching up with the Word of God: *"Then the seventh angel sounded: And there were loud voices in heaven, saying, 'The kingdoms of this world have become the kingdoms of our Lord and of His Christ and He shall reign forever and ever!'" (Revelation 11:15).* The Greek word for "world" in the phrase *"kingdoms of this world"* is literally *kosmos* (i.e., cosmos). [25]

According to *space.com*, "quantum entanglement is a bizarre, counterintuitive phenomenon that explains how two subatomic particles can be intimately linked to each other, even if separated by billions of light years of space. Despite their vast separation, a change induced in one will affect the other." [26]

Quantum entanglement with the Quantum One is key to the transformation of mankind back into the original image of God, like Adam and Eve before The Fall (Gen. 1:26-27). Mankind's change from corruptible to incorruptible and from mortality to immortality will come from the greatest quantum entanglement with Love Himself (1 Cor. 15:53). It is written: *"Nothing in all creation can separate us from God's love for us in Christ Jesus our Lord!" (Romans 8:39 $_{CEV}$)*.

Immortality is the new wine being poured out for God's precious people in this Kingdom Day. Behold, a large cluster of grapes. The Lord is peeling off the skin to reveal the glory within. It's time to focus on the flesh of the grapes, which is the heart of the matter. It is the essence of the new wine full of sweetness and divine glory.

The huge cluster of grapes that Joshua and Caleb carried out of the Promise Land speaks of the eternal reality of the immortality found in Messiah Yeshua (1 Cor. 15:52-54). It is the best saved for last wine spoken of in John 2:10. Isaiah 65:8 tells us that the new wine is found in the cluster. The new wine of immortality is poured out from a brand-new wineskin found only in the cluster of the overcomers. The fabric of the new wineskin is the cluster of those who are connecting in oneness to the Quantum One—ALEF-TAV (את).

Come! It's time to stand under His new wine fountain of glory next to the winepress. Behold, His Kingdom Wine is being poured out to and through the sons of light. This is the new wineskin of immortality where this mortality will take on immortality, as we become one, as His sons of light. This is the new wine being poured out on all creation. [27]

What Bridal Hearts give up will be minor compared to what we will gain. Perhaps, one of the greatest prizes of the high calling of God in Christ is immortality. The door to immortality manifesting here on earth is Yeshua Himself. It is not a coincidence that Messiah Yeshua declared twice three different I AM statements during the Feast of Dedication (Hanukkah): *"I am the Door" (John 10:7,9)*, *"I am the Good Shepherd" (John 10:11,14)*, and *"I am the Light of the World" (John 8:12; John 9:5)*. Once again, the Greek word for "world" in *"I am the Light of the World"* is cosmos. The Good Shepherd of our souls is dropping some major hints about the Door to returning to our primordial cherubic state before The Fall, as sons of light, but we have to be willing to surrender all for Jesus to be able to see it and be it.

When Yeshua says, *"As long as I am in the world [kosmos], I am the light of the world [kosmos]" (John 9:5 $_{Additions\ mine}$)*, He is connecting His disciples to His Sermon on the Mount: *"You are the light of the world [kosmos]" (Matthew 5:14 $_{Additions\ mine}$)*, which has a future time dimension when the fully mature Body of Christ on earth will literally manifest being the light of the world exactly like Yeshua did with all 12-strands of their DNA turned on. We will go into this more thoroughly in the next chapter.

To understand the fullness of what it means to be a Shining One, just like Yeshua, we need to study Solomon's Ten Copper Chariot Lavers as well as the entire passage that portrays what Yeshua did during Hanukkah—John 8:12-10:42.

The Ten Copper Chariot Lavers in Solomon's Temple shows His kings and priest made after the Righteous Order of Melchizedek the way to coming into His truth for becoming His living eternal temple, here on earth, as it is in heaven. In First Corinthians 6:19, we are told our bodies (both physical and spiritual) are the Temple of the Holy Spirit. When our DNA is transformed into the exact image of the DNA of God, our body, our soul, and our spirit will be filled with the Seven Spirits of God.

Technically, God has shown us[28] that when we are filled with the fullness of the Seven Spirits of God throughout our triune being (body, soul, and spirit), He will then seamlessly weave together our physical and our spiritual bodies. God loves our physical bodies too. Our temples of the Holy Spirit are so precious to Him. The Lord's seamless weaving together is a molecular integration of our physical man and our spiritual man when God deems that we are ready to go back to Eden and beyond.

"Integrate" indicates blending into a functioning or unified whole. It's about uniting mankind's triune being in oneness with the Holy Trinity of the Heavenly Father, the Son, and the Holy Spirit as well as being incorporated into a larger wholeness unit, like the Manchild Company, the fullness of the Bride of Christ, etc. Through God's gracious and awe-inspiring love, He makes a way to fully integrate us physically and spiritually. In this wholeness place and state, overcomers are given the morning star (Rev. 2:28), which ALEF-TAV is: *"I, Jesus, . . . I am the Root and the Offspring of David, the Bright and Morning Star" (Revelation 22:16).*

Let's circle back to some basics that involve the Seven Spirits of God. The Seven Spirits of God are the sevenfold nature of Christ that Yeshua walked in here on earth according to Isaiah 11:1-2. The seven branches of the Golden Menorah in God's Temple are connected to the Seven Spirits of God (Rev. 4:5). Yeshua tells us that He is the sprouting vine and His people are His branches in John 15:5. When we understand that Yeshua is also referred to as The Branch that has the Seven Spirits of God resting upon Him, we can also extrapolate that the Branch of the LORD in its fullness is Christ living fully in His people who are filled with the Seven Spirits of God in their body, in their soul, and in their spirit.

All that is given through the amazing gift of the Seven Spirits of God is designed to bring the Bride of Christ into a place where she is untainted, without spot or wrinkle (Eph. 5:27). The Seven Spirits of God are part of the government of God, and they reside in mature/maturing saints who are bringing an increase of His government of peace and righteousness within and without (Isa. 9:7). Believers who fully mature into their royal priesthood function are both His Manchild Company and His prepared Bride who will have Christ lived out through them by means of the Seven Spirits of God.

The Lord our God is coming IN His people before He comes FOR His people. The fullness of the One New Man in Christ is the Manchild Branch, which is the Melchizedek priesthood filled with the Seven Spirits of God. They are the finishing stone of the temple (Zech. 4:6-10). Operating as both kings and priests, they are a plummet in the earth, a standard of righteousness ruling authority as they execute His righteous judgments.[29] We could say the kings and priests made after the Righteous Order of Melchizedek operate in the Messiah Yeshua's Seven Spirits of God. And, at the same time, the multimembered Bride of Christ with full lamps are those filled with the Seven Spirits of God as well.

The Ten Copper Chariot Lavers are key to the Bridal Restoration of our DNA because the infilling of the Seven Spirits of God in a person's body, soul, and spirit is a progressive perfection process. Forget about onerous perfectionism. The being "made perfect" concept in John 17:23 has nothing to do with perfectionism: *"I in them, and You in Me, that they may be MADE PERFECT IN ONE; and that the world may know that You have sent Me, and has loved them, as You have loved Me" (John 17:23).*

The word "perfect" in the Greek is an action verb *teleioō*, which communicates the action of completing, finishing, accomplishing, consummating in character, fulfilling an event, or simply making something perfect. It comes from the Greek adjective *teleios*, which speaks about a perfect full age man—one who is complete in labor, complete in growth, complete in mental character, and complete in moral character. If we dig deeper into *teleioō*'s very root, we get the Greek noun *telos*, which describes setting out for a definite point or goal. Think of a telescope when you think of *telos*, because it's a point aimed at as a limit. It is a conclusion of an act or state to the utmost. Messiah Yeshua in us and the Father in Yeshua that we may become the finishing generation.

Every king and priest who have entered the Kingdom of God are leading their own inner fire bride forth (John 3:5). The three components that are present in the perfection of those being made after the Righteous Order of Melchizedek are the same transformative agents that we must do to press into the Bridal Restoration of our DNA: [1] Daily Communion, [2] Daily Crucifixion, and [3] Daily Bread, God's Word. Know that if you unlock the secrets of the Ten Copper Chariot Lavers, you turn the key to unlocking immortality in your DNA. Only you and Messiah Yeshua have access to this immortality key within you. Though men, beasts, and technology may try to break this code, understand that your transformation cannot happen without Messiah Yeshua's permission and guidance.

It is not a coincidence that the Ten Copper Chariot Lavers used in Solomon's Temple were dismantled and taken captive by Babylon. The Ten Copper Chariot Lavers used to wash (cleanse) the blood from the voluntary burnt offering sacrifices, which is a picture of offering ourselves daily as a living sacrifice to the Good Shepherd, so He can choose what is the best way to burn up our carnal nature. Just like the Ten Copper Chariot Lavers were dismissed and treated like junk, so has our 10-strands of shadow DNA.

Before we go further into the mysteries of the Ten Copper Chariot Lavers, let's lay the foundation for living sacrifices—you and me—becoming a burnt offering. *"¹ Beloved friends, what should be our proper response to God's marvelous mercies? To surrender yourselves to God to be his sacred living sacrifices. And live in holiness, experiencing all that delights His heart. For this becomes your genuine expression of worship. ² Stop imitating the ideals and opinions of the culture around you, but be inwardly transformed by the Holy Spirit through a total reformation of how you think. This will empower you to discern God's will as you live a beautiful life, satisfying and perfect in His eyes" (Romans 12:1-2 $_{TPT}$).*

Leviticus 1 speaks of the burnt offerings for His Dwelling Place. The following are the most important principles behind the Burnt Offering:

[1] First and foremost, it is a voluntary offering given exclusively to the Lord.

[2] It is also called the Elevation Offering or Ascension Offering because it raises one's spiritual level when done right.

[3] It is considered superior to other offerings because it is voluntary and offered in its entirety by fire on God's Altar.

[4] The Burnt Offering is called the *olah* (עֹלָה) in Hebrew. The ancient word picture (pictograph) for its three Hebrew letters tells us that an *olah* is what the shepherd brings forth. This means that the Good Shepherd of our souls selects what is the best sacrifice. He selects the best thing that will crucify your carnal nature (Matt. 16:24-26). In this Kingdom Day, the Great Shepherd of our soul will guide us to the best path that we each need to take to become just like Him.

[5] We bypass animal sacrifices because Yeshua offered up the perfect sacrifice of His earthly body once for all (Heb. 9:12). Yeshua became our Burnt Offering on the Cross, and His Bridal Body will be like Him in His death, burial, and resurrection for the purpose of the redemption of our own body and His Corporate Body (Rom. 8:23; Col. 1:24).

[6] The ultimate spiritual principle behind the Burnt Offering, which remains steadfast forever is: *"Behold, I have come to do Your will, O God"* (Hebrews 10:7).

[7] The fulfillment of the picture of the Burnt Offering in our fresh dispensation is the Baptism of Fire that the mature sons of God must choose to voluntarily go through to become part of the Lamb's Pure and Spotless Bride. Remember, we don't get to choose what we will offer as a Burnt Offering. It's the Great Shepherd who is in the midst of God's Kingdom that chooses for us what our Elevation Offering will be, so every member of His Bride can resonate at the exact same frequency as our heavenly Bridegroom.

The Ten Copper Chariot Lavers ultimately point to the *Merkabah* (God's Chariot Throne) process. Ezekiel Chapter 1 reveals the *Merkabah*, which portrays a fourfold divine communion with the Father, the Son, the Holy Spirit, and yourself (a person). Please refer to Book 3— *Taking on Common Union*—in the *Understanding the Order of Melchizedek: Complete Series*.

True communion requires you to examine yourself, so you can move into deeper dimensions within yourself. So does the *Merkabah* require a person to go deeper into the Kingdom of God within (Luke 17:21). When a person focuses inward and descends to the place where they currently operate through Christ's Cross, it should cause introspection, as they gaze into the reflecting pool to see the face of Yeshua. There they can ask for assistance examining themselves according to the righteous and true plumb line of Messiah Yeshua. If done properly with sincerity and humility, it will cause an inner transformation step by crucial step.

There are many mysteries of God embedded within you. Not the least of which is the Kingdom of God, which serves as a springboard for believers to ascend legally "in Christ" by the Blood of Yeshua (Heb. 10:19).

Messiah Yeshua has given us the best pattern for Ascension in Christ: "*⁹ (Now this, 'He ascended'—what does it mean but that He also first descended into the lower parts of the earth? ¹⁰ He who descended is also the One who ascended far above all the heavens, that He might fill all things.)" (Ephesians 4:9-10).* We actually descend to the platform within the heart of our own earth (the place in which we currently operate) before we ascend in Christ. Recall that this descension platform within a person acts as a springboard for their ascension.

Christ's death caused Him to descend, and His resurrection caused Him to ascend to be seated in heavenly places at the right hand of the Father (Eph. 1:20). We must press on to continually live in a descended state on earth as a continual burnt offering as well as live in an ascended state continually living in heavenly places in Christ. For more information, check out Book 4—*Hitting the Bull's Eye of Righteousness* in *Understanding the Order of Melchizedek: Complete Series.*

The Apostle Paul speaks of both: [1] Continually living in a descended state: *"I have been crucified with Christ [in Him I have shared His crucifixion]; it is no longer I who live, but Christ (the Messiah) lives in me; and the life I now live in the body I live by faith in (by adherence to and reliance on and complete trust in) the Son of God, who loved me and gave Himself up for me" (Galatians 2:20 $_{AMP}$).* [2] Continually living in an ascended state: *"⁴ But God, being rich in mercy, because of His great love with which He loved us, ⁵ even when we were dead in our wrongdoings, made us alive together with Christ (by grace you have been saved), ⁶ and raised us up with Him, and seated us with Him in the heavenly places in Christ Jesus, ⁷ so that in the ages to come He might show the boundless riches of His grace in kindness toward us in Christ Jesus. ⁸ For by grace you have been saved through faith; and this is not of yourselves, it is the gift of God" (Ephesians 2:4-8 $_{NASB}$).*

THE SEVEN SPIRITS OF GOD

Let us dive a little deeper into the Seven Spirits of God through the richness of the Word. Why does the Apostle John add grace and peace to you from the Seven Spirits before the throne *after* blessing the seven churches with grace and peace from Him who is and who was and who is to come?

"¹ The Revelation of Jesus Christ, which God gave Him to show His servants—things which must shortly take place. And He sent and signified it by His angel to His servant John, ² who bore witness to the Word of God, and to the testimony of Jesus Christ, to all things that he saw. ³ Blessed is he who reads and those who hear the words of this prophecy, and keep those things which are written in it; for the time is near. ⁴ John, to the seven churches which are in Asia: Grace to you and peace from Him who is and who was and who is to come, and from the seven Spirits who are before His throne, ⁵ and from Jesus Christ, the faithful witness, the firstborn from the dead, and the ruler over the kings of the earth. To Him who loved us and washed us from our sins in His own blood, ⁶ and has made us kings and priests to His God and Father, to Him be glory and dominion forever and ever. Amen" (Revelation 1:1-6).

The precious blood of Yeshua (Jesus), the Hebrew Living Letters, and the Seven Spirits of God were components of creating our physical world, which are all essences of the exact re-presentation in the Son. Messiah Yeshua is pure white light while the Seven Spirits of God are the rainbow colors that make up that spotless white light. These seven make up the Nature of Christ:

[1] Red—Spirit of the Lord

[2] Orange—Spirit of Wisdom

[3] Yellow—Spirit of Understanding

[4] Green—Spirit of Counsel

[5] Light Blue—Spirit of Might/Power/Strength

[6] Indigo—Spirit of Knowledge/Revelation

[7] Purple—Spirit of the Fear of the Lord

The Lord has shown me that the Seven Spirits of God are unique to man. When God formed man of the dust into a clay vessel of the ground, He breathed into his nostrils the breath of life. *"And the Lord God formed man of the dust of the ground, and breathed into his nostrils the breath of life; and man became a living being" (Genesis 2:7).* When God breathed into man's nostrils the breath of life, man became a living being made in God's own image through the impartation of the Seven Spirits of God. *"The Spirit of God has made me, And the breath of the Almighty gives me life" (Job 33:4).*

Just as the Hebrew Living Letters are to the Word of God, so are the Seven Spirits of God to the Holy Spirit. Additionally, the Seven Spirits of God are the breath of the Almighty, which created man. The Seven Spirits of God are solely reserved for humanity. When God breathed the breath of life into Adam's nostrils, the Seven Spirits of God were woven throughout man's body, soul, and spirit; thus, making each person a temple of the Holy Spirit. *"17 But he who is joined to the Lord is one spirit with Him. ... 19 Or do you not know that your body is the temple of the Holy Spirit who is in you, whom you have from God, and you are not your own? 20 For you were bought at a price; therefore glorify God in your body and in your spirit, which are God's" (1 Corinthians 6:17, 19-20).*

How does one engage the Seven Spirits of God who are also governors and tutors to help the sons of the Living God mature to inherit all things (Gal. 4:2) as well as return to our primordial pre-fallen state? From the depths of your being desperately seek for your soul to be filled with the Seven Spirits of God, like they did the Son of Man when He walked the earth and like mankind was originally created in the beginning.

The Seven Spirits of God not only make up the nature of the Messiah and the original nature of man, but they're also the chief princes of highest praise that are inextricably connected to God's Dwelling Presence Glory—His *Shekinah*. Hebrew mystics tell us the *Shekinah* is a divine rainbow that radiates colors in two directions. Yeshua walked in this crossover place of heaven and earth at the same time, and so shall the righteous kings and priests made after the Order of Melchizedek.

Scripture tells us that when we abide in Christ, we walk in the same manner as He did (1 John 2.6). One of the ultimate realities for God's Devout Ones is we are supposed to be made perfect and complete like the High Priest of the Order of Melchizedek. This means we are supposed to be completely filled with the Seven Spirits of God here on earth, just like Yeshua was. We can glean clues on how to be completely filled with the Seven Spirits of God by studying all the references to the "Son of Man" and "Son of God" in the Bible. We must walk like the Pattern Son—Son of Man—to become like unto the Son of God (Heb. 7:3b).

The Book of the Revelation of Jesus Christ also reveals a significant pattern for overcomers becoming the seven lamps before the throne. The great multitude who come out of the tribulation are shown to be *"before the throne" (Revelation 7:9,15)*, just as the Seven Spirits of God burn *"before the throne" (Revelation 4:5)*. This signifies the fullness of the Seven Spirits of God in the peoples, tribes, tongues, and nations whose garments/robes are made white by the Blood of the Lamb (Rev. 7:14), which means that their physical and spiritual temples (bodies) are filled with the fullness of the Seven Spirits (Holy Spirit) and have also been integrated with the Heavenly Father who sits on the throne along with the Lamb who is in the midst of the throne (Rev. 7:15-16).

Pay attention to how the Apostle John heard a loud voice, like a trumpet, saying, I am the Alpha and Omega, the First and the Last. When John turned to see the voice of the Almighty, he saw seven golden lampstands and One in the midst of them like the Son of Man (Rev. 1:13). This One like the Son of Man has a voice of the sound of many waters. Not only is this a picture of Messiah Yeshua, but also the peoples, multitudes, tongues, and nations who are one with Almighty God (Ezek. 1:24; Rev. 1:15; Rev. 17:15). They have His sharp two-edged sword coming out of their mouth (Rev. 1:16), and they have the keys to Hades and Death—immortality, (Rev. 1:18).

We are told that the seven lampstands that have the One like the Son of Man in their midst are the seven churches. The Greek word for "churches" here is *ekklēsia*, which is a gathering, an assembly, a legislative body, a community gathered for worship, but more importantly, being united in one body.

To become the Messiah bodily (Col. 2:9), His Corporate Unified Body must listen to what the Spirit says to the churches (Rev. 2:7,11,17,29; Rev. 3:6,13,22) via the voice of the Almighty, so we can overcome.

> [1] Overcomers eat from the Tree of Life (Rev. 2:9). The loveless church of Ephesus overcomes by flowing in the fullness of the red-hot passionate love inherent in the red flame of the Spirit of the Lord.

> [2] Overcomers are not hurt by the second death (Rev. 2:11). The persecuted church of Symrna overcomes by flowing in the fullness of the orange flame of the Spirit of Wisdom.

> [3] Overcomers receive the hidden manna and white stone with a new name (Rev. 2:17). The compromised church of Pergamos overcomes by flowing in the fullness of the yellow flame of the Spirit of Understanding.

[4] Overcomers receive power over the nations and receive the morning star (Rev. 2:26-29). The corrupt church of Thyatira overcomes by flowing in the fullness of the green flame of the Spirit of Counsel.

[5] Overcomers are clothed in white garments, are not blotted out from the Book of Life, and Yeshua confesses your name before Heavenly Father and His angels (Rev. 3:5). The dead church of Sardis overcomes by flowing in the fullness of the light blue flame of the Spirit of Might.

[6] Overcomers become a pillar in God's Temple as well as the Name of God, the New Jerusalem, and Yeshua's new name are written on you (Rev. 3:12) The faithful church of Philadelphia overcomes by flowing in the fullness of the indigo flame of the Spirit of Knowledge.

[7] Overcomers are granted to sit with the Messiah in His throne, as He overcame and sat with His Father (Rev. 3:21). The lukewarm church of Laodicea overcomes by flowing in the fullness of the purple/violet flame of the Spirit of Fear of the Lord.

The Book of Revelation shows Messiah Yeshua holding seven stars, which are the angels of the seven churches, and He is walking in the midst of the seven lampstands—His unified corporate body. Once His unified body overcomes these seven challenges, immediately, we are in the Spirit and there's a door standing open in heaven (Rev. 4:1-2). Behold, a throne in heaven and One sitting on the throne (Rev. 4:2). Then, we see an emerald rainbow around the throne, which represents new life . . . the new heaven and new earth of the New Jerusalem. Then, we are told that there are seven lamps burning before the throne, which are the Seven Spirits of God (Rev. 4:5).

These seven lamps represent a person who has overcome to the utmost; and therefore, completely filled to the brim and overflowing with the Seven Spirit of God in their body, soul, and spirit, as originally designed.

Don't stop there, because the four living creatures full of eyes in the middle and around the throne with the Sea of Glass are before the throne too along with the seven lamps of fire (Rev. 4:6). Please check out my *Understanding the Order of Melchizedek: Complete Series* book for more revelation on the four living creatures. [30]

There are six wings mentioned in Revelation 4:8 where each set of wings represents the heavenly redemption of a person's body, soul, and spirit. This is the revelation that the Lord gave me when I meditated upon: *"Thus were their faces. Their wings stretched upward; two wings of each one touched one another, and two covered their bodies" (Ezekiel 1:11).*

Finally, we are privileged to witness the Lion of the Tribe of Judah, the Root of David, being worthy to open the scroll (a person's scroll) and loose the Seven Seals of the Human Body,[31] which are connected to the Seven Spirits of God (Rev. 5:1; Isa. 11:1). *"⁵ But one of the elders said to me, 'Do not weep. Behold, the Lion of the Tribe of Judah, the Root of David, has prevailed to open the scroll and to loose its seven seals.' ⁶ And I looked, and behold, in the midst of the throne and of the four living creatures, and in the midst of the elders, stood a Lamb as though it had been slain, having seven horns and seven eyes, which are the Seven Spirits of God sent out into all the earth" (Revelation 5:5-6).*

In a righteous sense, these seven seals are loosed to enable the Seven Spirits of God to weave together a person's spiritual body and natural body. Thus, fusing the triune nature of a person into an integrated whole with the Holy Three in One. Notice that when the Apostle John looked at the Lion of the Tribe of Judah, he saw a Lamb as though slain, having seven horns and seven eyes, which significantly are the Seven Spirits of God sent out into all the earth—my earth and yours. The unsealing of the fullness of the Seven Spirits of God in a person's body, soul, and spirit can only be done by the Lamb slain before the foundation of the world. *"^{18}Knowing that you were not redeemed with corruptible things, like silver or gold, from your aimless conduct received by tradition from your fathers, ^{19} but with the precious blood of Christ, as of a lamb without blemish and without spot. ^{20} He indeed was foreordained before the foundation of the world, but was manifest in these last times for you ^{21} who through Him believe in God, who raised Him from the dead and gave Him glory, so that your faith and hope are in God" (1 Peter 1:18-21).*

Did you notice that the Lamb with the Seven Spirits of God (seven horns and seven eyes) was in the midst of three things? The throne, the four living creatures, and twenty-four elders. I will mostly leave it to you to do your own research; but notice how the four living creatures and 24 elders both fall down before the Lamb when He takes the scroll (Rev. 5:8). They all have harps—worship—and golden bowls—prayers.

The new song that they sing gives us a hint of who the four living creatures and twenty-four elders represent: *"^{8} Now when He had taken the scroll, the four living creatures and the twenty-four elders fell down before the Lamb, each having a harp, and golden bowls full of incense, which are the prayers of the saints. ^{9} And they sang a new song, saying: 'You are worthy to take the scroll, And to open its seals; For You were slain, And have redeemed us to God by Your blood out of every tribe and tongue and people and nation, ^{10} And have made us kings and priests to our God; And we shall reign on the earth'" (Revelation 5:8-10).*

Behold a kingdom mystery concealed within the first five books of the Book of Revelation now revealed. It's all about kings and priests made after the Righteous Order of Melchizedek going back to Eden and beyond for the total redemption of man's body, soul, and spirit. *"Now may the God of peace Himself sanctify you completely; and may your whole spirit, soul, and body be preserved blameless at the coming of our Lord Jesus Christ" (1 Thessalonians 5:23).* Please note that I continually emphasize the "Righteous" Order of Melchizedek because there is a Satanic Order of Melchizedek whose goal is to usurp the righteous rule of the kingdoms of this earth (and they have succeeded for way too long). We as His Crucified Ones don't throw away the righteous ruling in Christ of YHVH's kings and priests, we take greater dominion in Christ, as we rectify anything unrighteous within and without.

Notice that the fruits of the Spirit for His rainbow kids are the antithesis of the works of the flesh: *"^{16} I say then: Walk in the Spirit, and you shall not fulfill the lust of the flesh. ^{17} For the flesh lusts against the Spirit, and the Spirit against the flesh; and these are contrary to one another, so that you do not do the things that you wish. ^{18} But if you are led by the Spirit, you are not under the law. ^{19} Now the works of the flesh are evident, which are: adultery, fornication, uncleanness,* *lewdness,* ²⁰ *idolatry, sorcery, hatred, contentions, jealousies, outbursts of wrath, selfish ambitions, dissensions, heresies,* ²¹ *envy, murders, drunkenness, revelries, and the like; of which I tell you beforehand, just as I also told you in time past, that those who practice such things will not inherit the kingdom of God.* ²² *But the fruit of the Spirit is love, joy, peace, longsuffering, kindness, goodness, faithfulness,* ²³ *gentleness, self-control. Against such there is no law.* ²⁴ *And those who are Christ's have crucified the flesh with its passions and desires.* ²⁵ *If we live in the Spirit, let us also walk in the Spirit"* (Galatians 5:16-25).

Transformed Ones are crowned with the Seven Spirits of God and clothed with seven-fold light. Father God illustrated this via a Group Ascension in Christ. ³²

> We see storm clouds rolling in. We hear Abba's voice in the lightning and thunder. We are pulled into the middle of infinity, which was shown as an infinity sign. It feels like a portal where we are hurled toward the Heavenly Father. We are being summoned/called. We are being sucked in by the Father's breath. We land on His lips. We hear: "You're on My lips." Abba Father is announcing His children of His kingdom. We see a lightning rainbow. We hear Abba say that the children of His Kingdom: "They are My rainbow kids."
>
> This rainbow [His rainbow of the Seven Spirits of God] is contained and protected by His lightning: *"And out of the throne proceeded lightnings and thunderings [voices] and noises: and there were seven lamps of fire burning before the throne, which are the Seven Spirits of God"* (Revelation 4:5).
>
> We feel full, equipped, empowered, excited, and on top of things. The air around the throne is clean, refreshing, and has an ionizing smell, like after the rain. The throne has a bright, clean, honest, and pure scent. People are going to smell the Transformed Ones. Those caught up to the throne are going to smell like the Throne Room (Rev. 12). We get seated in heavenly places (Eph. 2:6). He is seated in us: *"God reigns over the nations; God sits on His holy throne"* (Psalms 47:8). *"For God is the King of all the earth"* (Psalms 47:7).
>
> *"*¹⁴ *Now thanks be to God who always leads us in triumph in Christ, and through us diffuses the fragrance of His knowledge in every place.* ¹⁵ *For we are to God the fragrance of Christ among those who are being saved and among those who are perishing.* ¹⁶ *To the one we are the aroma of death leading to death, and to the other the aroma of life leading to life. And who is sufficient for these things?* ¹⁷ *For we are not, as so many, peddling the Word of God; but as of sincerity, but as from God, we speak [on His lips] in the sight of God in Christ"* (2 Corinthians 2:14-17).
>
> It is the intimacy with God that carries the fragrance of Christ. We see purple fireworks coming from the heart . . . coming from the earth and exploding. God's rainbow kids are the kings and priests made after the Righteous Order of Melchizedek who walk in the fear of the Lord. The purple fireworks have to do with *"the shields of the earth belong to God"* (Psalms 47:9).

Let's further explore the kingdom mystery about the kings and priests going back to Eden and beyond for the total redemption of mankind's body, soul, and spirit. Remember that the Lord told me that the Seven Spirits of God are unique to man. When God formed man from the dust into a clay vessel of the ground, He breathed into man's nostrils the breath of life (Gen. 2:7); and thus, man became a living being made in God's own image through the impartations of the Seven Spirits of God. When Almighty God breathed the breath of life into man's nostrils, the Seven Spirits of God were woven throughout a person's body, soul, and spirit, making each a temple of the Holy Spirit (Job 33:4; 1 Cor. 6:19-20).

The Book of Revelation of Jesus Christ reveals a significant blueprint for overcomers literally becoming the seven lamps before the throne. Recall both the seven lamps and a great multitude of victorious people are positioned before the throne:

"And pulsing from the throne were blinding flashes of lightning, crashes of thunder, and voices. And burning BEFORE THE THRONE were seven blazing torches, which represent the seven Spirits of God" (Revelation 4:5 $_{TPT}$).

"⁹ After this I looked, and behold, right in front of me I saw a vast multitude of people—an enormous multitude so huge that no one could count—made up of victorious ones from every nation, tribe, people group, and language. They were all in glistening white robes, standing BEFORE THE THRONE and before the Lamb with palm branches in their hands. ¹⁰ And they shouted out with a passionate voice: 'Salvation belongs to our God seated on the throne and to the Lamb!' . . . ¹³ Then one of the elders asked me, 'Who are these in glistening white robes, and where have they come from?' ¹⁴ I answered, 'My lord—you must know.' Then he said to me, 'They are ones who have washed their robes and made them white in the blood of the Lamb and have emerged from the midst of great pressure and ordeal. ¹⁵ For this reason they are BEFORE THE THRONE of God, ministering to Him as priests day and night, within His cloud-filled sanctuary. And the Enthroned One spreads over them His tabernacle-shelter'" (Revelation 7:9-10,13-15 $_{TPT}$).

The seven lampstands with the One like the Son of Man in their midst are the seven churches, which are people united in one Body with Christ as the Head. Each member of Christ's Corporate Body must listen and do what the Spirit says to the seven churches in Revelation chapters 2 and 3.

Yeshua holds mankind's transformation in His hands. *"He had in His right hand seven stars, out of His mouth went a sharp two-edged sword, and His countenance was like the sun shining in its strength" (Revelation 1:16).* The seven stars in the One like the Son of Man's right hand are the angels of the seven churches. The Greek word for angels is *angelos* from *aggellos,* which means a messenger "angel," or "sent one." Hebrews 1:7 also speaks of *angelos:* *"Who makes His angels spirits and His ministers a flame of fire."* Both the Hebrew Scriptures and Greek Septuagint say that the "sent one" or "messenger" can either be a divine-angelic or a divine-human messenger.

Resurrection life for the sons of God who have come into the fullness of being kings and priests of the Righteous Order of Melchizedek is a type of *theosis,* which is a transformation of a human being back into our original created state. We need to shift our mindsets about "resurrection" from a post-mortem experience to a living and active transformation of God's True Church.

Enoch models this reality. In 2 Enoch 22, we see Enoch transformed into an angel, resurrected, and then sent back to earth after his mortality took on immortality (1 Cor. 15:53). This is also our journey back to the future where His devout ones will be transformed back into the very image of God.

The "angels spirits" in "who makes His angels spirits and His ministers a flame of fire" is *angelos pneuma* in Greek. These are the human angelic messengers whose greatest potential within is the fullness of the Sevenfold Spirit of Christ. The same measure of the Holy Spirit in Messiah Yeshua is our goal. Every thought, decision, and action matters. Never forget His great grace and kindness that leads us to repentance, redemption, and resurrection.

The "seven stars" in the One like the Son of Man's hand can be equated to the sevenfold light and energy of the Spirits of God. The Greek word for "stars" is *aster*. *Aster* is connected both to the One born king of the Jews and the wise men who follow His star.

"The mystery of the seven stars" (Revelation 1:20) is connected to *"the angel of the church of Ephesus" (Revelation 2:1)*, which reveals: *"write, these things says He who holds the seven stars in His right hand, who walks in the midst of the seven golden candlesticks: 'I know your works, your labor, your patience, and that you cannot bear those who are evil. And you have tested those who say they are apostles and are not, and have found them liars; and you have preserved and have patience, and have labored for My name's sake and have not become weary. Nevertheless I have this against you, that you have left your first love* [more focused and concerned with doing things for the Lord than being with Him, loving Him]. *Remember therefore from where you have fallen; repent and do the first works* [according to the perfect will of the Father], *or else I will come to you quickly and remove your lampstand* [includes your transformation potential] *from its place—unless you repent. But this you have, that you hate the deeds of the Nicolaitans, which I hate. He who has an ear, let him hear what the Spirit says to the churches* [potential transformation candidates]. *To him who overcomes I will give to eat from the Tree of Life, which is in the midst of the Paradise of God* [immortality, eternal life]' " *(Revelation 2:1-7* Additions mine*)*.

It is not a coincidence that the fourth church in Thyatira connects repentance to overcomers being given "the morning star [*aster*]" (Rev. 2:28). *"And to the angel* [messenger connected to the fourth Spirit of Counsel] *of the church in Thyatira write, 'These things says the Son of God, who has eyes like a flame of fire* [seven eyes, which are the 7 Spirits of God sent out into all the earth (Rev. 5:6), which are the 7 lamps of fire burning before the throne, which are the 7 Spirits of God (Rev. 4:5)], *and His feet like fine brass : "I know your works, love, service, faith, and your patience; and as for your works, the last are more than the first. Nevertheless I have a few things against you* [which hinder your own transformation], *because you allow your wife* [you are married to the unfaithful and conniving] *Jezebel, who calls herself a prophetess, to teach and seduce My servants to commit sexual immorality and eat things sacrificed to idols* [violating two of the four Essential Commandments for all believers (Acts 15:20)]. *And I gave her time to repent of her sexual immorality, and she did not repent. Indeed I will cast her into a sickbed, and those who commit adultery with her into great tribulation, unless they repent of their deeds. I will kill her children with death, and all the churches shall know that I am He who searches the minds and hearts. And I will give to each one of you according to your works. Now to you I say, and to the rest in Thyatira* [contending for the fullness of the Spirit of Counsel], *as many as do not have this doctrine, who have not known the depths of Satan* [Jezebel knows the depths of Satan], *as they say, I will put on you no other burden. But hold fast what you have till I come. And he who overcomes, and keeps My works until the end, to him I will give power over the nations—He shall rule them with a rod of iron; They shall be dashed to pieces like the potter's vessels'—as I also have received from My Father; and I will give him the morning star* [Yeshua - Rev. 22:16]. *He who has an ear, let him hear what the Spirit says to the churches" ' " (Revelation 2:19-29 _{Additions mine}).*

Then, we have the fifth church in Sardus being given instruction from He who has the Seven Spirits of God and the seven stars [*asters*]: *"And to the angel* [messenger connected to the fifth Spirit of Might] *in Sardus write, 'These things say He who has the Seven Spirits of God and the seven stars: "I know your works, that you have a name that you are alive, but you are dead. Be watchful, and strengthen the things which remain, that are ready to die, for I have not found your works perfect before God, Remember therefore how you have received and heard; hold fast and repent. Therefore if you will not watch, I will come upon you as a thief, and you will not know what hour I will come upon you. You have a few names even in Sardus who have not defiled their garments; and they shall walk with Me in white, for they are worthy. He who overcomes shall be clothed in white garments* [become integrated Shining Ones], *and I will not blot his name from the Book of Life; but I will confess his name before My Father And before His angels. He who has an ear, let him hear what the Spirit says to the churches" ' " (Revelation 3:1-6 _{Additions mine}).*

Behold, Transformed Ones overcome the seven multi-faceted challenges of the seven churches.

TRANSFORMED ONES

Yeshua and the Integrated Ones' bodies will be identical—His mature resurrected body. *"*^{19} *For the earnest expectation of the creature waiteth for the manifestation of the sons of God.* ^{20} *For the creature was made subject to vanity, not willingly, but by reason of him who hath subjected the same in hope,* ^{21} *Because the creature itself also shall be delivered from the bondage of corruption into the glorious liberty of the children of God.* ^{22} *For we know that the whole creation groaneth and travaileth in pain together until now.* ^{23} *And not only they, but ourselves also, which have the firstfruits of the Spirit, even we ourselves groan within ourselves, waiting for the adoption, to wit, the redemption of our body" (Romans 8:19-23 _{KJV}).*

"The creature" that waits for the manifestation of the sons of God is the "Living Creature" of Ezekiel 1, which is made up of the mature/maturing kings and priests made after the Righteous Order of Melchizedek.

When Romans 8:23 refers to "the firstfruits of the Spirit," it can be talking about the ten-percent deposit of the Holy Spirit a person receives when they receive Messiah Yeshua (Jesus Christ) as their Lord and Savior. *"*^{13} *And you also were included in Christ when you heard the message of truth, the gospel of your salvation. When you believed, you were marked in him with a seal, the promised Holy Spirit,* ^{14} *who is a deposit guaranteeing our inheritance until the redemption of those who are God's possession—to the praise of his glory" (Ephesians 1:13-14 _{NIV}).*

"Firstfruits" implies more fruit than the initial ten-percent deposit of the Holy Spirit for several reasons; but especially due to the phrase that follows *"ourselves also, which have the firstfruits of the Spirit"* which is *"waiting for the adoption . . . the redemption of our body" (Romans 1:23).*

Once a person receives the deposit/guarantee of the Holy Spirit, Messiah Yeshua will integrate the physical and spiritual bodies of those who *"press on toward the goal to win the [supreme and heavenly] prize to which God in Christ Jesus is calling us upward" (Philippians 3:14 $_{AMP}$)*.

"But our labors are in heavenly things, from whence we look for our Savior, our Lord Jesus Christ, who shall transform our poor body to the likeness of His glorious body, according to His mighty power, whereby He is able even to subdue all things to Himself" (Philippians 3:20-21 $_{Lamsa's\ Aramaic}$).

Transformation is a heavenly inheritance. *"³ Blessed be the God and Father of our Lord Jesus Christ, who according to His abundant mercy has begotten us again to a living hope through the resurrection of Jesus Christ from the dead, ⁴ to an inheritance incorruptible and undefiled and that does not fade away, reserved in heaven for you* [waiting to be woven together with your physical body] *⁵ who are kept by the power of God through faith for salvation ready to be revealed in the last time" (1 Peter 1:3-5 $_{Additions\ mine}$)*. Add to your understanding that a person's soul is the crux for their transformation: *"receiving the end of your faith—the salvation of our souls" (1 Peter 1:9)*. This theme continues: *"²² Since you have purified your souls in obeying the truth through the Spirit in sincere love of the brethren, love one another fervently with a pure heart, ²³ having been born again, not of corruptible seed but incorruptible, through the word of God which lives and abides forever" (1 Peter 1:22-23)*.

Transformation happens to those who inherit of the Kingdom of God (Gal. 5:19-21; 1 Cor. 6:9-10). These are the ones who have incorruption and immortality put on them.

"⁴⁵ And so it is written, The first man Adam was made a living soul; the last Adam was made a quickening spirit. ⁴⁶ Howbeit that was not first which is spiritual, but that which is natural; and afterward that which is spiritual. ⁴⁷ The first man is of the earth, earthy; the second man is the Lord from heaven. ⁴⁸ As is the earthy, such are they also that are earthy: and as is the heavenly, such are they also that are heavenly. ⁴⁹ And as we have borne the image of the earthy, we shall also bear the image of the heavenly. ⁵⁰ Now this I say, brethren, that flesh and blood cannot inherit the kingdom of God [those who do works of the flesh will not inherit the Kingdom of God]; *neither doth corruption inherit incorruption. ⁵¹ Behold, I shew you a mystery; we shall not all sleep, but we shall all be changed* [transformed], *⁵² In a moment, in the twinkling of an eye, at the last trump: for the trumpet shall sound, and the dead* [those who have died completely to themselves] *shall be raised incorruptible, and we shall be changed. ⁵³ For this corruptible must put on incorruption, and this mortal must put on immortality. ⁵⁴ So when this corruptible shall have put on incorruption, and this mortal shall have put on immortality, then shall be brought to pass the saying that is written, Death is swallowed up in victory" (1 Corinthians 15:45-54 $_{KJV\ Additions\ mine}$)*.

For better clarity, let's look at the definitions of words like corruption, incorruption, mortal, immortality, transform, transformation, transfigure, and transfiguration in my *Merriam-Webster's Collegiate Dictionary (10ᵗʰ Edition)*:

corruption ~*noun* 1a: impairment of integrity, virtue, or moral principle : DEPRAVITY 1b: DECAY, DECOMPOSITION 1c: inducement to wrong by improper or unlawful means 1d: a departure from the original or from what is pure or correct 2 *archaic:* an agency or influence that corrupts[33]

incorrupt ~*adj* : free from corruption : as a *obs* : not affected by decay b: not defiled or depraved : UPRIGHT c: free from error[34]

incorruptible ~*adj* : incapable of corruption as a: not subject to decay or dissolution b: incapable of being bribed or morally corrupted[35]

incorruption ~*noun archaic* : the quality or state of being free from physical decay[36]

immortal ~*adj* 1: exempt from death 2: exempt from oblivion : IMPERISHABLE 3: connected with or relating to immortality ~*noun* : one exempt from death[37]

immortality ~*noun* : the quality or state on being immortal a: unending existence b: lasting fame[38]

mortal ~*adj* 1: causing or having caused death : FATAL 2a: subject to death 2b: POSSIBLE, CONCEIVABLE 2c: DEADLY 3: marked by unrelenting hostility: IMPLACABLE 4: marked by great intensity and severity : EXTREME 5: HUMAN 6: of, relating to, or connected with death[39]

transfiguration ~*noun* 1a: a change in form or appearance: METAMORPHOSIS 1b: an exalting, glorifying, or spiritual change[40]

transfigure ~*vt* to give a new and typically exalted or spiritual appearance to : transform outwardly and usually for the better *syn* TRANSFORM[41]

transform ~*adj* 1a: to change in composition or structure. 1b: to change the outward form or appearance of 1c: to change in character or condition : CONVERT 2: to subject to mathematical transformation 3: to cause (a cell) to undergo genetic transformation ~*vi* : to become transformed : CHANGE *syn* TRANSFORM implies a major change in form, nature, or function. METAMORPHOSE suggests an abrupt or startling change induced by or as if by supernatural power. TRANSMUTE implies transforming into a higher element or thing. CONVERT implies a change fitting something for a new or different use or function. TRANSMOGRIFY suggests a strange or preposterous metamorphosis <frog into prince> TRANSFIGURE implies a change that exalts or glorifies. [42]

transformation ~*noun* 1: an act, process, or instance of transforming or being transformed 4b: genetic modification of a cell [physical body] by the uptake and incorporation of exogenous DNA [caused by factors or an agent from outside the organism or system : introduced from or produced outside the organism or system; specifically : not synthesized within the organism or system] [43]

Just as Yeshua's body did not see corruption, so it is for His Identical Integrated Ones. "*³⁰ For he [David] was a prophet, and he knew that God had sworn by an oath to him that of the flesh, He would raise up One to sit on His throne, ³¹ So he foresaw and spoke concerning the resurrection of Christ, that His soul was not left in the grave, neither did His body see corruption. ³² This very Jesus, God has raised up, and we are all His witnesses*" (Acts 2:30-32 *Lamsa's Aramaic – Additions mine*).

The bodies of the Transformed Ones are raised in incorruption, raised in glory, raised in power, raised a spiritual body, and made a quickening spirit. "*⁴² So also is the resurrection of the dead. It is sown in corruption; it is raised in incorruption: ⁴³ It is sown in dishonor; it is raised in glory: it is sown in weakness; it is raised in power: ⁴⁴ It is sown a natural body; it is raised a spiritual* body. There is a natural body, and there is a spiritual body. *⁴⁵ And so it is written, The first man Adam was made a living soul; the last Adam was made a quickening spirit" (1 Corinthians 15:42-45 ₖⱼᵥ).*

The Transformation of His Bride/Remnant/144000 Firstfruits cannot be duplicated by man or Fallen Angels. Yeshua has kept this secret to Himself and the Father. Even though Fallen Angels and fallen mankind have taken matters into their own hands, the King of kings and the Lord of lords will overturn their plans. Out of this, transformation life will spring anew where they will glow like the midday sun. Watch for the glass ceiling created by man/demons, meant to contain us to two realms, to open up to the unlimited where heaven and earth are connected. Watch for the firstfruits transformation of the 144-thousand that will crack the glass ceiling and pave the way for others. Watch the glass ceiling crack and disappear. Watch the glass ceiling shatter into millions of crystals (shards). Yeshua and Abba Father are making all things beautiful in its time. [44]

In Scripture, transfiguration is pictured as a temporary condition, yet it gives us a glimpse into the permanent transformation of His Crucified Ones. *"¹ Now after six days Jesus took Peter, James, and John his brother, led them up on a high mountain by themselves; ² and He was transfigured before them. His face shone like the sun, and His clothes became as white as the light. ³ And behold, Moses and Elijah appeared to them, talking with Him. ⁴ Then Peter answered and said to Jesus, 'Lord, it is good for us to be here; if You wish, let us make here three tabernacles: one for You, one for Moses, and one for Elijah.' ⁵ While he was still speaking, behold, a bright cloud overshadowed them; and suddenly a voice came out of the cloud, saying, 'This is My beloved Son, in whom I am well pleased. Hear Him!' ⁶ And when the disciples heard it, they fell on their faces and were greatly afraid. ⁷ But Jesus came and touched them and said, 'Arise, and do not be afraid.' ⁸ When they had lifted up their eyes, they saw no one but Jesus only"* (Matthew 17:1-8).

Transformation is an instantaneous change, but also part of the process that comes from loving our Beloved. It is a natural flow, which the enemy cannot duplicate because it only flows from being connected in oneness to the True Head of Messiah Yeshua.

QUANTUM LEVELS

The mature resurrected Body of Christ—Yeshua and His Transformed Ones—is multidimensional with quantum levels. Another Ascension in Christ experience revealed a different perspective of creation, which included the creation of quantum levels. [45]

> We first saw a spiral staircase that we ascended with stars all around us. As we focused on the perfect heart of the Heavenly Father, we approached something that was above us that was covered in light, which pulled us into itself, like a tractor beam. We could feel the light come through us from our feet upward. It swirled and filled us internally, then we radiated light. We understood that there is oneness in this light. The Messiah's light merged with our DNA; and then, the light cascaded down. There was a circle to our left going clockwise. The light with the substance of the Messiah is in eternity (circle), which releases something that cascades down.

When we looked through the light of the Messiah, we saw that the light was taking shape. It's creation taking shape when the words "Let there be light" were uttered by the Father of Lights.

Eternity rains down from the Messiah. Eternity is being released from Yeshua. In the streams of light, there are little Hebrew letters. It's the particles of the Word. It is part of Yeshua's essence, yet the Hebrew Living Letters are separate sentient beings that have a consciousness. Just as the Seven Spirits of God are to the Holy Spirit, so are the Hebrew Living Letters to Messiah Yeshua—the Word of God.

The light is composed of the Father's perfect will. It originates from the Heavenly Father's heart, which is the Messiah. We sense the Father's Presence in union. The longer that we are in the light, the chunkier it gets. Chunky Creation is home. The light is Messiah Yeshua emanating from the Father's will.

We are drawn right into the midst of Yeshua to become one with Him and His creative flow. We are literally pulled right into the Messiah's heart. There is a space there. We feel a little whirlwind around us. There is warmth, love, and passion of the Father in the midst of Yeshua's heart. It's so beautiful. You can feel that love behind creation (at the intersection place). His love overwhelms all of our senses. We feel a pulsating heartbeat around and within. It's the rhythm of life. This very secret place is most holy. It is warm. It is complete love. There are numerous highways (pathways) in this very secret place that lead to countless places . . . the veins and arteries of the Messiah.

We can feel that we are in a broad, infinite place, yet it feels cozy. Around us are spheres with Hebrew Letters and other things in them. Each sphere has a specific purpose. There is attention to detail in the truth in love. Swirling around the spheres are different colored jewels with light radiating from each one of them. The jewels are expressions of specific attributes and revelations of the Heavenly Father. It is in all of us. We all have built-in time-released revelations of the Father. We are coming into this time now.

As the Body of the Messiah, we are one with His heart. All of us are in a semi-circle. Each of us has different gems that have a time and purpose for its release. We can hear the various jewels' frequencies that combine to form a symphony. The highest tone is a sapphire. Amethyst is next highest, then a yellow gem. The red ruby has a middle tone while the emerald is a baritone and an orangish-red precious stone has a frequency that's really low.

Its appearance is like the Northern Lights (Aurora Borealis). The colors and sounds are swirling. Their frequencies are dancing with each other. Each color has its own unique dance. When various colors swirl together, it makes something totally new. We see all the colors dance above us. The colors all merge, then they explode with a massive amount of power. This massive explosion goes in all directions with a tremendous amount of speed. It may seem out-of-control, but it is in divine order. Everything goes where it is supposed to go according to His divine purpose. It is synchronized to His Divine Pattern.

Each has a specific place in Creation's design that it is being thrust into. It's in numerical order. The Hebrew Living Letters take their number form. We can see complicated mathematical equations.

The explosion goes into several dimensions. There are threads between these dimensions. Behold, the eternal realm affects different dimensions at different times. Even though these dimensions are distinct and separate, they are connected. The common thread is the creative force of light and frequency. The thread is being pulled and the dimensions are being pulled together. The Blood of Yeshua is the light that runs through it all. It is His veins and arteries of light that are connecting the dimensions.

We shrink to the size of an elemental particle, so we can travel with light. We are instructed to step into the first photon, which is a *Merkabah*—as one. We understand that each and every atom has a direct connection to the light—the heart of the Father—the Messiah. Every atom's direct tie to the Father's heart is a pipeline of love, provision, health, healing, etc. The blueprint of Creation is that everything is connected to the Father and the Son. Each thing released during Creation came from the original One Love. It is the umbilical cord for life flowing and bringing His substance to all things.

We see the various shapes of the five Platonic solids, which are connected to the shape of the gems. The different colors, geometric shapes, and frequencies are revelatory expressions of the Father. We find ourselves inside one of the geometric shapes with our hands up in a V-shape. We are part of the shape. We get Phi, which contains the perfect design of mankind.

There are "things" coming out of the first photon (*Merkabah*), but the first photon is not changing. While we are inside the first photon, the Holy Spirit draws near vibrating. The Holy Spirit's hovering vibration is all around us. There is multiplication flowing from it.

We see graph paper waving while we are still inside the original photon. There are cubes flowing through the Creator's graph paper. These cubes go to the coordinates where His blueprint shows they are supposed to be. When the graph paper goes through us, we move through the Creator's blueprint and see vibration coming off the cubes hitting the graph. Geometry is simply vibration seen. The solids that we experience are dimensional frequencies.

When we look at the wavy graph, we can see the thread that connects to other dimensions. The path of redemption is Yeshua's Blood, which is the connecting thread. When we follow the redemptive path to different dimensions, we can look in the distance and see that it goes on forever. Yet, it circles back to the Source—Messiah Yeshua.

You can traverse dimensions where one dimension intersects another via a thread. We understand that a fifth-dimension person can be in the fifth, fourth, or third dimension. Basically, they can be in any dimension that is below or equal to their resonance. Adam and Eve could go into all dimensions until they fell in the Garden.

And now to unlock the understanding . . . Unlocking is traveling in Him. He resides in all creation. We hear: "Quantum Levels."

DNA—THE MULTIDIMENSIONAL CODE OF LIFE

With a spiritual transformation comes a mental and physical change. Deoxyribonucleic acid (DNA) is a nucleic acid that contains all the genetic instructions used in the development and functioning of all known living things. [46] It's the multidimensional code of life. DNA is such an astounding creation that its complexities can be difficult to ascertain. Just think about how mind-blowing is the reality that there is enough information in one DNA molecule to fill 1000 books. [47] "If all the DNA in your body were placed end-to-end, it would stretch from here to the Moon more than 500,000 times. In a human, it takes three billion letters to represent a copy of you." [48]

A person's DNA is their personal blueprint, which contains all their physical, mental, emotional, and spiritual information. But don't stop there, because each person also has the DNA record of all their previous generations, included in their own DNA.

DNA IS A LANGUAGE

DNA is a language that functions much like Hebrew. We can trace each Hebrew word back to a two letter "parent root," just as a human being has two parents. In a woman's egg cell, there are 22 autosomes and 1 sex chromosome. In a man's sperm cell, there are also 22 autosomes and 1 sex chromosome. For the creation of a healthy human, you need a total of 46 chromosomes. It is not happenstance that the Bible was originally written in both the Hebrew and Greek languages. When we add the 24 letters of the Greek Alphabet to the 22 letters of the Hebrew Alphabet, we get a total of 46.

It is significant that chromosomes are in the shape of an "X" or a cross. Just as the chromosomes represent the gathering of two people to form a body, so does the Cross *"that He might reconcile them both to God in one body through the cross" (Ephesians 2:16).*

For the most part, the Hebrew "parent root" can't be used in a sentence without the addition of a third letter. Hebrew verbs require a three-letter consonant root. This corresponds to their Three Spheres Source. This three letter Hebrew primary root is called the "child root." The DNA helix is formed through the combination of two nucleotides, which are Adenine and

Thymine (A-T) or Cytosine and Guanine (C-G) combinations. For amino acids of the genetic code to be formed into proteins, the nucleotides must combine into three letter combinations to make codons.

The genetic code consists of 64 codons. Amino acids are specified by these codons. When DNA is in the nucleus, the DNA contains two letter pairs held together by hydrogen bonds. These two letter pairs are unable to form a human body without leaving the nucleus and taking on a third letter, just like the Hebrew language. To take on a third letter, the DNA helix must "unwind" to copy itself onto messenger RNA (mRNA), which makes one long string of letters, just like an ancient Torah scroll. Therefore, a DNA Strand is made of letters, letters make words, and words make sentences, which are called genes. Genes tell the cell to make proteins. These proteins allow a cell to perform special functions, such as working with other groups of cells to make sight possible.

Just as Scribes transcribed one Torah scroll onto another, the DNA's letters are transcribed by mRNA. Messenger RNA is also called the Servant RNA because they can go outside the nucleus of a cell while the DNA must stay inside. DNA doesn't leave the nucleus to preserve the two parent's original copy. When the mRNA leaves the nucleus of a cell, it is translated through a hole called the gatekeeper.

The three letters represent amino acids. There are 22 amino acids that are read by the ribosomes, so that protein can be produced. We can basically say that "a DNA molecule is the instruction manual for the forming of amino acids."[49] Significantly, there are 22 proteins formed from the DNA molecule's amino acids, which corresponds to the Quantum 22 Hebrew Living Letters. As we have seen, the 22 letters of the Hebrew Alphabet point to their source Messiah Yeshua—the Almighty ALEF-TAV (את).

The genetic information that's read by the ribosomes determines what will be formed. Scientists have discovered that the genetic material for our eyes, ears, nose, hands, feet, liver, heart, stomach, etc. are the same. It is the epigenome that translates the information to cause each part of our human body to exist. "The epigenome is the collection of all the epigenetic marks on the DNA in a single cell."[50] The epigenome does not alter the DNA sequence, it turns on or off different parts of the sequence.

The epigenome translation of the genetic code is like the Holy Spirit teaching and leading a person how to interpret the Word of God correctly, so that they can be formed into Messiah Yeshua's exact image in their own unique presentation: *"until Christ is completely and permanently formed (molded) within you"* (Galatians 4:19 AMP). It's interesting that DNA dwells in the midst of a matrix of water. Please recall that the Holy Spirit brooded over the waters of MEM before the Father of Lights spoke: *"Let there be light!"* (Genesis 1:3). In fact, DNA's double-helix ladder shape is caused by the two letter roots wrapping themselves around each other to avoid contact with their wet environment.[51]

Psalms 139 speaks of a person being formed in the womb: *"My bones weren't hidden from you when I was being put together in a secret place, when I was being woven together in the deep parts of the earth"* (Psalms 139:15 CEB). The word "woven" is the Hebrew word *râqam* (רָקַם), which means to fabricate, embroider, needlework, or the twisted threads of DNA. Not only is the womb likened to the earth—the holder of the treasures of the dust of the earth—but in Hebrew thought, the womb is likened to waters.

DNA IN THE DARKEST PART OF A CELL

Bio-chemists Francis Crick and James Watson discovered DNA in 1953 when they investigated the nucleus of a cell—the darkest part. [52] Please recall that darkness was on the face of the deep before the Father of Lights spoke the release of uncreated light to initiate creation: *"The earth was without form and void, and darkness was upon the face of the deep; and the Spirit of God was moving over the face of the waters" (Genesis 1:2 $_{RSV}$)*. It was in the thick darkness of our three-part cell structure that light was discovered in the form of DNA—the scrolled Book of Life.

There are over 60 trillion cells [some say 37.2 trillion cells] within the human body.[53] Incredible as it may seem, each of these cells are intelligent. They have the ability to communicate with each other. Each of these cells have memory. They can transfer sequences that can cause the body and mind to react in various ways. Each of these cells carries and regulates many levels of information. They carry the ancestral blueprint of a person's past generations, including the first couple before the Fall. [54]

The three main parts of a cell are the cell membrane, the cytoplasm, and the nucleus. The cell membrane is like the skin of the cell, which is a thin but tough wall surrounding each cell. A cell membrane can also be called a plasma membrane. The cell's membrane has intelligence, allowing useful substances to enter the cell while blocking harmful substances. It also forces out waste products. The cytoplasm is the interior of the cell that performs most of the constant work that keeps it alive. It is mainly composed of water, plus some solids. The specialized components within the cytoplasm that perform specific functions are called organelles. Each organelle is surrounded by a separate membrane. Mitochondria is an organelle responsible for energy transactions that are necessary for the cell's survival. Lysosomes are organelles that digest undesirable materials within a cell. [55] Basically, some organelles make proteins. Some organelles change food molecules into materials used for energy and growth. Interestingly, a network of tubes transport material within the cytoplasm of each cell. The main organelle is the nucleus, which has the necessary information for cell growth and reproduction. While there are multiple organelles, there is only one nucleus per cell. The nucleus is located near the center of a cell and is its control center. [56] In addition to DNA, the nucleus contains chromosomes that hold the cell's genes, which determines the characteristics of the cell. When a cell makes copies of itself, the genes pass on the cell's traits to the new cell.

The three main parts of a cell can correspond to the three main components of God's Temple: Outer Court, Inner Court (Holy Place), and the Holy of Holies. The darkest part of a cell, where DNA molecule resides, is the Holy of Holies. It is the Torah Scroll—the Word of God—within the Ark of the Presence. Moses testifies: *"3 So I made an ark of acacia wood, hewed two tablets of stone like the first, and went up the mountain, having the two tablets in my hand. 4 And He wrote on the tablets according to the first writing, the Ten Commandments, which the*

LORD had spoken to you in the mountain from the midst of the fire in the day of the assembly; and the LORD gave them to me. ⁵ Then I turned and came down from the mountain, and put the tablets in the ark which I had made; and there they are, just as the LORD commanded me" (Deuteronomy 10:3-5).

Remember when Jesus said to Pilate: *"My Kingdom is not of this world" (John 18:36)*? The *Dead Sea Scrolls* gives us some clues when it describes the Holy of Holies as the *raz nihyeh*—the mystery of being or the mystery of becoming. The mystery of humanity becoming what we were originally designed to be is connected to the kingdoms of our earth becoming the kingdoms of our God at the cellular level (Rev. 11:15). Wise people were exhorted to gaze on the *raz nihyeh* to understand the birth time of salvation and to know who will inherit glory.

"Cells are largely composed of compounds that contain carbon. The study of how carbon atoms interact with other atoms in molecular compounds forms the basis of the field of organic chemistry and plays a large role in understanding the basic functions of cells. Because carbon atoms can form stable bonds with four other atoms, they are uniquely suited for the construction of complex molecules. These complex molecules are typically made up of chains and rings of hydrogen, oxygen, and nitrogen, as well as carbon atoms." [57]

It's not a fluke that relatively few elements compose most of the Earth's outer crust. Silicon, oxygen, hydrogen, and aluminum amount to more than 90 percent of the Earth's crust. Of greater significance, is that two of these four elements—hydrogen and oxygen—also account for more than 99 percent of our human bodies as well. [58] Please note that our next chapter will explore the elements of the Periodic Table.

Not only do quantum elements link man to Earth, but so does the Hebrew word for man—*Adam* (אדם)– and the Hebrew word for ground or earth is *adamah* (אדמה). When we break down the name of the first man created, we understand that Adam's name is derived from the first letter of the Hebrew Alphabet ALEF (א) and the word *dam* (דם). The primitive two letter word *dam* means blood, as in life is in the blood (Lev. 17:11). *"Therefore, brethren, having boldness to enter the Holiest by the blood of Jesus" (Hebrews 10:19).*

When mankind continually strives to do the perfect will of the Heavenly Father (John 5:19), the mechanics of eternal life are released line upon line, precept upon precept. *"Sacrifice and offering You did not desire, But a body You have prepared for Me... Then I said, 'Behold, I have come—In the volume of the book it is written of Me—To do Your will, O God'"* (Hebrews 10:5,7). *"And this is eternal life, that they may know You, the only true God, and Jesus Christ whom You have sent" (John 17:3).*

SHADOW DNA

Eternal life here on earth necessitates a transformation of mankind's DNA, which takes us back to the Garden of Eden before the Fall. The restoration and redemption of mankind's DNA is humanity returning to their original creation path. Back to who we really are. Everything originally encoded in Adam's DNA when God made him still resides within all of mankind

(Gen. 1:27). However, this time redeemed mankind will be quickening spirits instead of living souls: *"And so it is written, The first Adam was made a living soul; the last Adam was made a quickening spirit" (1 Corinthians 15:45 $_{KJV}$)*. The last Adam is Messiah Yeshua (Jesus Christ). He brought the Original DNA Template for man back to earth in His Body of cells.

97% of humanity's DNA, that has not been activated yet, has been waiting for mankind to rightly position themselves in oneness with the Father, the Son, and the Holy Spirit through connecting in love as well as doing only what they see and hear the Father doing (John 5:19,30). This will fully stimulate the process for this mortality taking on immortality and this corruption taking on incorruption (1 Cor. 15:53). I cannot overemphasize that the restoration of mankind's DNA can only be fully gained through being "in Christ"—being in ALEF-TAV (את): *"For as the Father raises up the dead and quickens them, even so the Son quickenth whom He will" (John 5:21 $_{KJV}$)*.

An eternal divine template exists behind mankind's DNA, which is an energetic duplicate blueprint "in Christ" that has an additional ten strands of DNA. Humanity has been quantumly entangled with God since the beginning of Creation and even before the foundation of the cosmos. Quantum entanglement is a phenomenon that occurs when one or more objects are connected in such a way that they can be thought of as a single system. All of creation is in the Three Spheres System due to the entirety of its essence existing in ALEF-TAV (את).

Scientists have labeled these additional ten strands junk DNA. Junk DNA contains fragmented DNA with unplugged information, which makes them dormant. My guess is that humanity's dormant DNA contains both non-physical as well as physical components that is waiting to be tuned into so that it can be turned on.

According to Scripture, the term "10 Shadow DNA Strands" is more accurate than Junk DNA. *Tzelem* is the Hebrew word for the divine image of man that God created: *"Then God said, 'Let Us make man in Our image [tzelem], according to Our likeness'" (Genesis 1:26 $_{Addition\ mine}$)*. *Tzelem* is derived from *tzel*, which means "shadow." Mankind is a shadow of the Holy Three in One—the Us in *"Let Us make man in Our image."* Colossians 2:17 speaks of Biblical practices connected to our physical existence on earth, *"which are a shadow of things to come; but the BODY IS OF CHRIST" (Colossians 2:17 $_{KJV}$)*. One of the interpretations of a *"shadow of things to come"* is the manifestation of Christ's Body coming into its fullness as a perfect One New Man in the Messiah (Eph. 2:15; Eph. 4:13).

When 97% of our DNA gets activated, our 2-strand DNA-helix will become 12-strands of DNA. This will come with the capacity of being more multidimensional, like Adam and Eve before their encounter with the serpent. Eating the forbidden fruit caused Adam, Eve, and their progeny to disconnect from their higher *tzel* image. They forfeited their higher levels of consciousness, forgetting its multidimensional nature. The sin that caused the breakage of our DNA has locked humanity in a third-dimension spacetime continuum loop of mortality and death; but thanks be to God that Messiah Yeshua holds the keys to both death and eternal life: *"^{54} So when this corruptible has put on incorruption, and this mortal has put on immortality, then shall be brought to pass the saying that is written: 'Death is swallowed up in victory.' ^{55} 'O Death, where is your sting? O Hades, where is your victory?' ^{56} The sting of death is sin, and the strength of sin is the law. ^{57} But thanks be to God, who gives us the victory through our Lord Jesus Christ.*

⁵⁸ *Therefore, my beloved brethren, be steadfast, immovable, always abounding in the work of the Lord, knowing that your labor is not in vain in the Lord" (1 Corinthians 15:54-58).*

The serpent is sometimes used as a metaphor for the structure of DNA in the world. Take for example that medical establishments display the Caduceus image of two serpents forming the double helix on a pole. Genesis 3:15 is the first prophecy of the coming Seed of the Original DNA Template, which will restore all creation. This prophecy basically announces the serpent (i.e., damaged-fallen DNA) will be usurped by the Seed of the Original DNA Template for humanity: *"And I will put enmity between you and the woman, And between your seed and her Seed; He shall bruise your head, And you shall bruise His heel" (Genesis 3:15).*

The Book of Genesis could be called the Book of Genetics. If we reverse the Scriptural address of Genesis 3:15 to Genesis 15:3, we are privy to an encoded conversation between God and Abram about the coming Seed: *"And Abram said, Behold, to me thou hast given no seed; and, lo, one born in my house is mine heir" (Genesis 15:3 ₖⱼᵥ).* "The genetically fallen one (old nature) born in your house into sin that is missing the genetic marker [for the immortal light body] is not the heir. The Original Template DNA Strand is reprogramming/recoding"⁵⁹ more than 37.2 trillion cells in your body⁶⁰ "to restore you to the Light Being you are." ⁶¹

BURNT OFFERINGS AND TEN COPPER CHARIOT LAVERS

Currently, our less-than-capacity 10-Strands of Shadow DNA needs to be redeemed through the 10 Copper Chariot Lavers process, which we spoke of in the previous chapter. Please allow me to further clarify. Even though all of humanity has the same root, each person is unique. Their experiences are different. Their ancestors are different. Their DNA is different. Therefore, the redemption of everyone's DNA will be different, even though there are basic principles at work. God knows the way each of us needs to take, so we can become just like Him.

Let's go over how Burnt Offerings worked before and during King Solomon's day before we bring these principles forward into this Kingdom Day. Modern Hebrew says that a Burnt Offering is an *olah* (עֹלָה), but it's technically a *qorban olah* (קָרְבַּן עוֹלָה). *Olah* means "what is brought up" while *qorban* means "to come near." Therefore, a *qorban olah* was an elevation offering that caused a person to come near and go up when done properly. This concept is emphasized when we understand that the Hebrew noun *olah* (עֹלָה) is formed from the active participle of the verb *alah* (עָלָה), which means "to cause to ascend." ⁶²

The name *olah* (עֹלָה) is based on the descriptive phrase *"an offering made by fire unto the Lord" (Leviticus 1:9).* ⁶³ When an *olah* was a voluntary sacrifice, it could be a young bull, a ram, a year-old goat, a turtle dove, or a pigeon. The first use of the word *olah* in Scripture refers to the sacrifices of Noah *"of every clean beast, and of every clean fowl, and offered burnt offerings on the altar" (Genesis 8:20).* It's significant that the first burnt offerings were required to be offered on an "altar of earth" (Exo. 20:24). The next noteworthy mention of *olah* is the near-sacrifice of Isaac by Abraham on Mount Moriah: *"Take now your son, your only son Isaac, whom you love, and go to the land of Moriah, and offer him there as a burnt offering [olah] on one of the mountains of which I shall tell you" (Genesis 22:2 ₐddition mine).*

After Moses' Tabernacle was built, the place of sacrifice shifted from an "altar of earth" to the Brazen Altar, which was by the corporate door of the Tabernacle of Meeting. The first chapter of the Book of Leviticus describes the types of Burnt Offering sacrifices as well as their purposes, and circumstances. The sacrifices were required to be "unblemished" (Num. 1:10). The list of blemishes included animals that were blind, broken, maimed, ulcer, eczema, or scabs (Lev. 22:21-22). Once the sacrifices were determined to be unblemished, they were brought to the north side of the Brazen Altar to be offered up. The blood from the sacrifice was carefully collected by a priest, so it could be sprinkled on the Altar's four outside corners (Num. 1:11). Unless the offering was a bird *(olat haof)*, its flesh was divided and placed on the wood of the Altar, which had a continual fire from heaven burning on it. The Burnt Offering slowly burned until it was reduced to ashes.

Burnt Offerings could either be the Daily Burnt Offering, the Sabbath Burnt Offering, the Biblical Festival Burnt Offering, or a voluntary one. The Daily Burnt Offering was presented during the morning and evening prayer (third and ninth hour), which were always a year-old lamb accompanied with grain (bread) and wine (Exo. 29:38-42). This is symbolic of communion with the Lamb of God. The Sabbath Burnt Offering doubled the elements of the Daily Burnt Offering (Num. 28:9-10). This is symbolic of a double-portion of grace flowing into time from eternity for entering His rest. His rest is His Head resting on His Body. The Biblical Festival Burnt Offerings were part of the corporate gatherings where God's people met Him at the specific times set out in Leviticus 23: Passover, Pentecost, Feast of Trumpets, Day of Atonement, and the Feast of Tabernacles (included sacrifices for all nations). This is symbolic of partaking of the Lord's Table and His glorious celebrations. The voluntary Burnt Offerings could be offered by anyone. In fact, the *olah* was the only offering in God's Temple that could be offered by non-Jews. All that was required was for a sincere seeker to lay their hands on the head of their sacrifice with all their strength to confess their sins and ask for forgiveness from the Most Holy God. *"If we confess our sins, He is faithful and just to forgive us our sins and to cleanse us from all unrighteousness" (1 John 1:9).*

Not only did the water in the 10 Copper Chariot Lavers in Solomon's Temple cleanse the internal organs and feet of the sacrifices according to Josephus,[64] but they were also used to wash the hands and feet of the priests before they performed their Temple service (Exo. 30:18-21). [65] Therefore, the 10 Copper Chariot Lavers in Solomon's Temple (1 Kings 7:38) washed both the innermost parts of the sacrifice—you and me—as well as our hands and feet, symbolizing sanctifying what we do and where we go. There is a double emphasis on the sanctified path due to the priests washing their hands and feet at the same time in the copper lavers, which was a feat that required coordination and concentration. A priest had to cleanse his right hand and right foot at the exact same time while keeping his balance, then switch to his left hand and left foot.

What most people miss about a voluntary Burnt Offering is that a person must first come to the Shepherd of their souls to ask: What is the best Burnt Offering (*olah*) for me? Once our Good Shepherd chooses what we need to die to in this world, then we must count the cost and voluntarily choose to deny ourselves, take up our cross, and follow Him.

Sometimes, we don't initiate the asking. Many times, I have found that difficult circumstances in my life revealed the *olah* that God had chosen for me. Instead of arguing with God (or others) when your hidden sin gets revealed, may I suggest that you start by running the situation by God. First, lay the "perceived" issue on His Altar, so you won't hear the idols of your own heart; then, earnestly seek God. If He confirms needing to get rid of a particular sin, that's when we must do the inner work of denying ourselves of that particular sin, taking up our cross, and following Christ by joining Him on His cross (Matt. 16:24). Two verses later in Matthew 16:26, we are asked the question: "What will a man give in exchange for his soul?" Know that whatever sin that we refuse to deny ourselves will be the very issue that we will exchange for the total redemption of our soul.

Everything in life is controlled by frequency. The frequency that you or I oscillate is determined by our DNA activation. Our DNA activation is determined by the percentage of oneness with the Word of God, which upholds everything. Know that our mental or verbal assent to a "truth" is not enough to change our DNA. We must all perform *The Shema* by hearing, understanding, and doing something to become just like Abraham where *"by works was faith made perfect. And the Scripture was fulfilled which says, Abraham believed God, and it was imputed to him for righteousness: and he was called the Friend of God" (James 2:22-23).* Know that Messiah Yeshua's Bride perfectly resonates at the frequency of the Hebrew Living Letters of the Word of God.

REDEMPTION OF DNA

Persistent sin or generational sin is also called iniquity, which Christ paid the price for on the Cross: *"All we like sheep have gone astray; We have turned, every one, to his own way; And the LORD has laid on Him the iniquity of us all" (Isaiah 53:6).* After we accept the free gift of Jesus Christ as our Lord and Savior, we still have to work out our salvation with fear and trembling to shine as lights in the world in ever increasing brightness (Phil. 2:12-15). Like most works of the Lord in our lives, we must partner with Him to the saving of our soul (Heb. 10:39) unto the total redemption of our bodies (Rom 8:23).

The Lord has revealed that the iniquity of a person and their generational bloodline is captured in their 10-Strands of Shadow DNA. [66] It remains captured there until His Pillar of Righteousness is planted in your midst and you are washed by the waves of His glory, which means that the lights in the 10-Strands of Shadow DNA are turned on after you perfectly tune into Him.

During a Group Ascension,[67] the Lord showed us the process a person goes through for the redemption of their DNA. From the inception, the Lord was telling us that He was buttoning up the redemption of DNA for a person's body, soul, and spirit. Though our sins be as scarlet, they will be white as wool (Isa. 1:18), as we press toward the goal of the upward calling of God in Christ Jesus (Phil. 3:14).

First, we encountered an orange atmosphere—the land of wisdom. Then, the backdrop changed to a yellow atmosphere of understanding before it went all red, relating to the Spirit of the Lord. Next, the atmosphere changed to a medium to light blue color of the Spirit of Might; and then, to the purple hues of the Spirit of the Awesome Reverential Fear of the Lord. The green atmosphere of the Spirit of Counsel followed before everything changed to the indigo hue of the Spirit of Knowledge where the redemption of our DNA kicked in and turned on. We felt the indigo knowledge of the Lord in our mouth, redeemed the eating from the Tree of Knowledge of Good and Evil.

In our journey back to Eden and beyond the Spirit of Knowledge comes last, not first, in the restoration process of the Seven Spirits of God (Isa. 11:2) where the God of Peace sanctifies our spirit, soul, and body completely, fully, and wholly (1 Thess. 5:23). It is the kind of knowledge of truly being known by Him and truly knowing Him. *"When Christ, who is your life, appears, then you also will appear with him in glory" (Colossians 3:4 $_{NCB}$)*.

The Lord led us to receive the planting of Messiah Yeshua's Pillar of Righteousness, as we exhaled the iniquity in our DNA. As forerunners, we were led according to the perfect heart of the Heavenly Father to exhale three times while His Pillar of Righteousness was fixed in our body, our soul, and our spirit.

During the atmospheric changes of the Seven Spirits of God, we experienced picking twelve bunches of lilies of the valley, which we knew confirmed the redemption of humanity's 12-Stands of DNA. Picking lilies in the Bride's low places caused her sweet fragrance to grow. Then, we saw another solitary Lily of the Valley that was red, drenched in the Blood of the Lamb. *"I am the Rose of Sharon, And the Lily of the valleys. Like a lily among thorns, So is My love among the daughters" (Song of Solomon 2:1-2)*. Notice how Yeshua declares, *"I am the Lily of the valleys;"* and then, He compares His Bride to Himself: Like a lily among thorns [prickly people], so is the one My heart adores. This single Lily of the valleys was doing the redeeming by His Blood, as we volunteered to surrender all and lift up holy hands as an offering by fire—Burnt Offerings. All that was not of Him burned up. The Shepherd and Lover of our souls knows who we truly are.

ANTENNA TO GOD

A group led by Leonard Horowitz—an internationally known public health authority and award-winning author—took three years of multidisciplinary study of health science, mathematics, genetics, and physics to understand DNA. These experts in their fields concluded that DNA appears to be a spiritual antenna to God.

"DNA's coiled design, vibrating action, and 'electrogenetic' function makes spiritual as well as physical evolution possible. Life's genomes are empowered by waves and particles of energized sound and light which, more than chemicals or drugs, switch genes 'on' or 'off.' Likewise, genetic inheritance is energetically transmitted 'bioacoustically and electronically' through special water molecules that form the electrogenetic matrix of the 'Sacred Spiral.' These hydroelectric geometric structures—most shaped like pyramids, hexagons, and pentagons—direct physical as well as spiritual development, according to researchers." [68] In Dr. Horowitz's book *DNA: Pirates of the Sacred Spiral*, he writes about the uses and abuses of DNA—"The Sacred Spiral". The book's description says: "Evidence proves DNA is nature's bioacoustic and electromagnetic energy receiver, signal transformer, and quantum sound and light transmitter." [69]

JACOB'S LADDER AND THE PILLAR OF RIGHTEOUSNESS

Genesis 28 tells of Jacob encountering the gate of heaven when he stayed at Bethel for a night. He took a stone and put it at his head before laying down to sleep. Whereupon he dreamt of a ladder that was set up on earth with its top reaching to heaven with holy angels ascending and descending on it. At the top of the ladder, YHVH showed up: *"*^{13} *And behold, the L*ORD *stood above it and said: 'I am the L*ORD *God of Abraham your father and the God of Isaac; the land on which you lie I will give to you and your descendants...* ^{15} *Behold, I am with you and will keep you wherever you go, and will bring you back to this land; for I will not leave you until I have done what I have spoken to you'" (Genesis 28:13,15).*

The ladder set up on earth with its top reaching to heaven points to the Messiah and the sons of man that are being made like unto the Son of God: *"And He said to him, 'Most assuredly, I say to you, hereafter you shall see heaven open, and the angels of God ascending and descending upon the Son of Man'" (John 1:51).* This ladder is connected to humanity's spiral staircase DNA that has been created and is sustained through the Word of God, which is literally *"the gate of heaven" (Genesis 28:17).*

After Jacob woke up, he took the stone that was under his head and set it up as a pillar ... a Pillar of Righteousness (Gen. 28:18). The Hebrew word for the "pillar" that Jacob anointed is *matstsêbâh* (מַצֵּבָה). Not only does *matstsêbâh* mean a memorial stone, but also a pillar, an image, or a stock of a tree—the Tree of Life Messiah Yeshua. Jacob called his anointed Pillar of Righteousness the House of God—*Bethel* (Gen. 28:17). Hebrews 1:3 speaks of the "express image of His essence and upholding all things by the Word of His power." The Greek word for "express image" is *charaktḗr* (χαρακτήρ), which is defined to be the exact expression (exact image) of anything, a marked likeness, or a precise reproduction in every respect.

DNA IS LIGHT

The word "light" is found 264 times in the Bible. The number 12 represents divine government or divine authority. When 264 is divided by 12, we get 22. Therefore, 22 not only represents the Word of God and its Quantum 22 Hebrew Living Letters, but also "light." Scripture confirms the Word of God is light: *"Your word is a lamp to my feet and a light to my path" (Psalms 119:105);* and this light is the light of life: *"Then Jesus spoke to them again, saying, 'I am the Light of the World. He who follows Me shall not walk in darkness, but have the light of life'" (John 8:12).*

TAV (ת) is the 22nd letter of the Hebrew Alphabet. When we layer TAV on top of the light of the Word of God and the Hebrew Living Letters, we behold the Light of the World—Messiah Yeshua—dying on the cross to save the world (John 3:17) because the ancient mark for TAV is a cross.

Before Yeshua was crucified, He gave us a preview of the Light Bodies mankind can attain. When Yeshua was transfigured before His closest inner circle—Peter, James, and John—He showed us the prize of the high calling of God in Christ Jesus (Matt. 17:1-8; Phil. 3:14). When Yeshua died on 22—the Cross—He pierced the third dimension's restrictive veil, tearing it from top to bottom, heaven to earth (Matt. 27:51); thus, making a way for mankind's fifth or fourth dimensional glorious bodies:

> *"For our citizenship is in heaven, from which we also eagerly wait for the Savior, the Lord Jesus Christ, who will transform our lowly body that may be conformed to His glorious body, according to the working by which He is able even to subdue all things to Himself" (Philippians 3:20-21).*

> *"For it pleased the Father that in Him all the fullness should dwell, and by Him, to reconcile all things to Himself, by Him. Whether things on earth or things in heaven, having made peace through the blood of the cross. And you, who once were alienated and enemies in your mind by wicked works, yet now He has reconciled in the body of His flesh through death, to present you holy and blameless, and above reproach in His sight—if indeed you continue in the faith, grounded and steadfast, and are not moved away from the hope of the gospel which you have heard, which was preached to every creature under heaven" (Colossians 1:19-23).*

Recall that the double helix DNA structure contains 22 also. The number 22 can be regarded as a code for one of the building blocks for a divine human in the guise of a body's 22 amino acids. When we understand that 22 denotes the word, the quantum letters, and "light," the scientific studies about DNA communicating with light and is created from light starts making more sense. Scientists are saying that light appears to be a fundamental part of our being. The original design of the human body is the most complex expression of light interacting with water, which interacts with the expression of the building blocks of life itself—protein.

Just as the Word of God is made up of letters, the four molecules called bases—Adenine, Thymine, Cytosine, and Guanine (ATCG) make up the alphabet or the language of DNA. It's no coincidence that four letters make up the most sacred name of the Creator—YHVH—who is The Bridge back to Eden and beyond for all of humanity.

Let's further explore DNA being made of four ATCG building blocks called nucleotides. The nucleotides attach to each other in pairs. Adenines attach to the Thymines and Cytosines attach to the Guanines to form chemical bonds called base pairs. These chemical bonds are what connect two DNA strands.

Nucleotides combine with a third letter to make codons. Codons literally means "code on." Recall that humans have 64 codons that deliver specific instructions to cellular chemistry to manufacture proteins. Even though humans have 64 codons, most of us only have 20 turned on. This means that mankind is operating way below our potential.

Codons are powerful molecules that work in groups of three to produce light within the DNA of mankind. Think of the three nucleotides as a metaphysical trinity that's connected to the Father, the Son (the Word), and the Holy Spirit. These are the Three Spheres in charge of Jacob's Ladder—all the DNA in Creation. They are in control of the cellular living word of 64 codons within our bodies that are longing to be completely turned on.

When Adam and Eve fell through ingesting forbidden fruit, they fell in consciousness from the elevated fifth dimensional reality to the limited third dimensional experience. They not only fell in consciousness, but also genetically. The Original DNA Template that produced immortality and pure light began to malfunction, sending faulty codes throughout their corporeal bodies.

Adam and Eve's bodies began to resonate with a frequency less than the premium state of eternal life. After the faux pas of partaking of the Tree of Knowledge of Good and Evil, their bandwidth lowered and descended to the third dimensional realm of illusions and lies where their cellular intelligence began to receive commands that eventually produced death. We could say that a death and destruction protocol was initiated: *"And the LORD God commanded the man, saying, 'Of every tree of the garden you may freely eat; but the Tree of the Knowledge of Good and Evil you shall not eat, for in the day that you eat of it you shall surely die'" (Genesis 2:16-17)*.

As their carnal minds hooked into the third dimension became more active, they and their prodigy were filled with worldly desires: the lust of the flesh, the lust of the eyes, and the pride of life (1 John 2:16). Their minds produced thoughts, and their bodies produced actions conducive to corruption due to the forbidden fruit hack exploiting mankind's DNA. Wars, conflict, strife, sickness, disease, deformities, religion, slavery, and death were its results.[70]

Always remember that the last Adam—Messiah Yeshua—brought back the Original DNA Template when He came to earth. Upon His death, burial, and resurrection, the override for the death and destruction protocol was initiated. Yeshua's Body of Cells are the Promised Seed, the Living Word, made flesh: *"And the Word was made flesh and dwelt among us, (and we beheld His glory), the glory as of the only begotten of the Father), full of grace and truth" (John 1:14$_{KJV}$)*. With Yeshua's incarnation, His original DNA Template chromosomes were composed of DNA structured by four bases: A-T and C-G, which were transcribed by His perfect RNA. His 64 combinations of three letters (codons) perfectly expressed 22 amino acids. It is truly awe-inspiring that there are abbreviations or full versions of the fourteen most common names of God (most found in the Hebrew Scriptures) that appear about 500,000 times in each human cell. This means that God's Names appear trillions of times (500,000 × 10 trillion) in the human body. We appear to be saturated with the names of God. One of the most common four-letter words in human genes is YHVH. It appeared 500 times in each cell; and its common abbreviation—YH—appears 35,000 times.[71]

Behold, the unblemished Lamb of God is the Pattern Son Yeshua who is the exact image of the perfect man with all 64 codes on. 64 = 6 + 4 = 10. Ten is God's divine law at the molecular level to produce life. There are famously Ten Commandments, which are fulfilled by truly loving God and people (Matt. 22:37-40). This is the internal Living Word where a person must demonstrate the reality of: *"If you love Me, keep My commandments" (John 14:15)*.

We see the commandment number of 10 again when we decode another mystery of mankind's physical body, which is built of 46 chromosomes. The "46" in 46 chromosomes is 46 = 4 + 6 = 10. Humans have 23 pairs of chromosomes (46 in total). One set comes from each biological parent. One full set of 46 chromosomes is called a genome.[72]

Adam and Eve's story reveals that mankind started with two people. Amazing as it may seem, this Biblical story mirrors what modern science sees in human genetics today. If the Bible is correct and Adam is really the ancestor for all mankind, there should be only one original male ancestor for humanity. "In the mitochondrial DNA (mtDNA), there's a little piece of DNA that's only inherited from your mother. Since all people get their mtDNA from their mother, we can build a family tree of all the [genetic] females in the world. There should only be one female ancestor for everyone on earth today. Most genes come in two versions; because Eve was taken from Adam, she probably got Adam's genome, except the Y chromosome. Adam only had two codes of each gene, so you can only have two versions of each gene. Turns out. This is exactly what we found in modern human genetics." [73]

It appears that science is proving the veracity of Scripture at every turn. "A Northwestern University team recently caught DNA doing something that has never been seen before: it blinked. For decades, textbooks have stated that macromolecules within living cells, such as DNA, RNA, and proteins, do not fluoresce on their own... But now Professors Vadim Backman, Hao Zhang, and Cheng Sun have discovered that macromolecule structures in living cells do, in fact, naturally fluoresce. 'Everybody has overlooked this effect because nobody asked the right question,' said Backman, Walter Dill Scott Professor of Biomedical Engineering in Northwestern's McCormick School of Engineering... The reason why no one spotted the fluorescence before? The molecules were in the 'dark state,' a condition in which they do not absorb or emit light... Backman, Zhang, and Sun discovered that when illuminated with visible light, the molecules get excited and light up well enough to be imaged without fluorescent stains. When excited with the right wavelength, they even light up better than they would with the best, most powerful fluorescent labels." [74]

Biophotons are the bits of light called photons that are spontaneously generated by most living cells. Research suggests that these biophotons are created in the 98% of each DNA molecule that is not used for genetic coding of behavior—in our shadow DNA. [75] "We now know that the biphotons emitted from our cells are highly coherent energy that may be responsible for the operation of our biological systems." [76] A "Russian scientist named Peter Gariaev placed DNA inside a quartz container and zapped it with a laser (a very coherent form of light). The DNA absorbed the light and stored it for up to thirty days inside a corkscrew shaped spiral in an exact blueprint of light reflecting the DNA pattern. Even more interesting is that the spiral of light stayed in the same place even after the quartz container and the DNA had been removed. Gariaev's work suggests some unknown force is holding the light in place. One possible explanation is that DNA is responding to an external energy field. This energy field is exchanging information with your cells in the form of light. In essence, our bodies are working as a giant antenna that is constantly sending and receiving signals from the field. This research even suggests that all living biological systems, including us, have the exact blueprint of our physical bodies stored in a field of light." [77]

Humans are more than the atoms and molecules that make up our bodies. We are beings of light. When a human body emits biophotons, they are part of the visible electromagnetic spectrum but are 1000 times lower than the sensitivity of our naked eyes. [78]

Light is order. Light is the Heavenly Father's commanding order of all elements into something that we can touch and feel. There is a force holding us together—the Father of

Lights. CERN has been trying to their utmost to figure out the force that holds everything together. They need to get back to the Biblical basics of life. God's command is the force, which is the basis of light and order. This is the light and the order of the multidimensional reality of humanity's DNA that's been made in God's own image.

∞

CREATION'S QUANTUM ELEMENTS

"Let there be . . ." (Genesis 1:3,6,14).

Quantum physics studies elements of nature that are divided into discrete units or energy packets called "quanta." This is where we get the word "quantum." The quantum world is the world of subatomic particles and photons that interact with one another on the smallest scale of the universe. When you think "quantum," always remember that it is a world much smaller than microscopic ingredients[79] because it is believed that these invisible quanta of energy are the smallest units into which something in our physical realm can be partitioned. In 1918, Max Plank was awarded the Nobel Prize for his discovery that all of nature is made up of these invisible quanta of energy. [80]

I have a hypothesis (born out of my own Ascension in Christ experiences and Biblical studies) that there are infinite levels to the Creator's Hebrew letters, which includes and goes beyond Planck's constant. Planck's constant is currently believed to be "a fundamental universal constant that defines the quantum nature of energy and relates the energy of a photon to its frequency." [81] I believe there are infinite levels and dimensions to letters of the Hebrew Alphabet because they are sustained and flow as part of the eternal essence of ALEF-TAV (את). For more about Messiah Yeshua's name ALEF-TAV (את), please refer to *ALEF-TAV's Hebrew Living™ Letters: 24 Wisdoms Deeper Kingdom Bible Study*. Could the Circle of Life be a "biofeedback loop" from the mind of Christ to the Body of Christ (in the "all creation" context) back to Messiah Yeshua?

The quantum model of the atom is the basis of every material thing as well as the first success of quantum mechanics. Quantum mechanics is a fundamental theory of physics, which is the foundation of all quantum physics. Quantum mechanics is a science dealing with the behavior of matter and light on the atomic and subatomic scale. [82]

The quantum model of the atom consists of protons, neutrons, and electrons. Protons and neutrons are bound together by nuclear forces to form a nucleus while tiny electrons orbit the core of the atom (i.e., the nucleus). In general, we can say that atoms are electrically neutral because they have a net electrical charge of zero. For instance, the simplest and lightest hydrogen atom has only one proton and one electron while deuterium is a hydrogen atom that has had a neutron added, which forms a heavier nucleus. [83]

Each atomic substance has a different configuration of protons and neutrons in its nucleus, plus a unique mass. In an electrically neutral atom, the number of negatively charged electrons always matches the number of positively charged protons.

Atoms with more neutrons than protons retain the same chemical properties, but some of these isotopes aren't stable. Unstable atoms are radioactive, which means that they emit one or more particles from the nucleus in a random fashion by a process called quantum tunneling. During fission in nuclear reactors, tunneling takes place within the fuel of the radioactive atoms. Electrical components, making modern electronics possible, also rely on the phenomena of quantum tunneling of electrons to function. [84] Quantum mechanics made modern chemistry possible too. Thanks to quantum mechanics, our modern society has things like plastics, nanotechnology, microbiology, telecommunications, cell phones, atomic energy, space crafts, etc.

Quantum mechanics is complex and extremely mathematical, yet its overarching concepts can be grasped when they are taken one simple step at a time. I highly recommend Michael O'Connell's book: *Finding God in Science* for an excellent and understandable breakdown of complex quantum subjects.

No one has actually seen an atom, but we know they exist based on repeatable physics measurements. *Proverbs 8:26* calls atoms *"the primal dust of the world"* or *"the beginning of the dust"* that cannot be seen. All physical matter is made up of molecules and molecules are made up of atoms.

Quantum mechanics is an insult to the strictly material view of the world. More than any other scientific theory, quantum mechanics reveals the mind of God. [85] According to Genesis 1:26-27 man was made in the image of God, so let's contemplate that a human mind possibly functions by the rules of quantum mechanics too. Since it is impossible to predict what a quantum particle will do at any given moment, then the behavior a person chooses can't be totally predicted. Yet, a person's behavior is not random. Sane people are aware of their choices, and they are completely free to act without being constrained by what has taken place before. God is the originator of freedom. "The root of freedom is inherent in the physics of the universe through chance events. Freewill and liberty must be of paramount importance to God since it also brings a great deal of trouble to Him and to us." [86] Most significantly, freedom makes true love possible. God controls the universe through physical laws that include random chance at the smallest scales. God does not have a need to control every random occurrence in the stream of time, He designed the universe to do that for Him.

Scripture speaks of God giving us the freedom to choose life or death, blessings or curses (Deut. 30). Our choices and behaviors decide whether good or evil will occur in the world—our world and others. God is not responsible for the evil deeds of men. God has an eternal point of view outside time. Chance events and God's sovereignty are compatible because God made that our reality.

By the quantum uncertainty principle, God may select the outcomes of otherwise random events without violating His own physics. Such divine tampering would be undetectable to science, except for people witnessing its results. This is how miracles happen. God has access to every physical law that He created, and He has the power to perform whatever He wishes. He knows all the natural laws—His Word in nature—and He has access to all of them. [87]

Is your understanding of God too small? Miracles like the restoration of the sight of one born blind are simply a different class than the "extremely likely" event. God has the power, the ability, and the knowledge to recreate or rebuild biological optical structures at the molecular level by changing individual quantum states. We just don't understand how; hence, we call them miracles. God's verifiable miracles result in real physical outcomes. For God to do miracles, He does not have to suspend natural law. However, miracles do require an execution of natural law in accordance to the perfect will of the Heavenly Father: " *19 Then Jesus answered and said to them, 'Most assuredly, I say to you, the Son can do nothing of Himself, but what He sees the Father do; for whatever He does, the Son also does in like manner.... 30 I can of Myself do nothing. As I hear, I judge; and My judgment is righteous, because I do not seek My own will but the will of the Father who sent Me' " (John 5:19,30).*

The righteous Metatron image of Messiah Yeshua (Jesus Christ) is a multidimensional, multi-membered eternal reality. The righteous Metatron can do the miraculous works of God; because everyone in the righteous Metatron Matrix is one in Christ, tapping into the Head—Yeshua. Since "Metatron" is a controversial topic, please allow me to speak about it for a bit. Metatron has been branded as New Age Occult for many years, which causes most believers in Christ to simply stay away. However, God's people need to understand that if there is a counterfeit whose origins are righteous, there is a godly counterpart that can and will be redeemed. [88]

God Himself introduced me to Metatron as the Great Golden Scribe, which means I did not learn about the righteous Metatron from a teacher, a book, or a video. While writing the book *SANTA-TIZING: What's wrong with Christmas and how to clean it up,*[89] I would desperately seek the Lord for His assistance. I truly needed divine aid to fulfill the vow that I made to Messiah Yeshua's face three times to tell the Christian Church: "Christmas will be (is) the Golden Calf of America (and the entire world)."[90,91] As I researched the vast topics of "mixture," "Babylon," "Christmas," and the "Golden Calf" both historically and scripturally full-time for ten years, I would diligently seek His face and pray for a move of God. The deeper I dove, the more I knew that God's people needed divine assistance in getting rid of their sacred cow. I know I did.

Many times, when I desperately cried out, the Lord would send a golden scribe angel who was extremely helpful! I was so grateful and so focused, as I pressed into fulfilling this divine mandate, that I didn't even ask who the golden scribe angel was until after the book was published in 2008. When I mentioned getting assistance writing *SANTA-TIZING* from a golden scribe angel, someone asked me who he was. I had never thought of asking until that moment. [92]

What shocked me is that the Lord told me it was the Great Scribe Metatron, Enoch as Metatron as well as Melchizedek Enoch in Merkabah Form. All of these, He told me are different descriptors of the same reality of that golden scribe angel that practically co-authored *SANTA-TIZING*.[93] Some basic realities about the glorious and righteous Metatron: [94]

[1] Metatron is Biblical. It's in the Pseudo-Jonathan Targum (translation) of *Genesis 5:24*: *"Enoch worshiped in truth before the Lord, and behold he was not with the inhabitants of the*

earth because he was taken away and he ascended to the firmament at the command of the Lord, and he was called Metatron, the Great Scribe." [95]

[2] Metatron is the undifferentiated state of the Messiah. "Undifferentiated" means no difference; therefore, the Righteous Metatron consists of the fully mature Head Messiah Yeshua and His 100-percent fully mature members of His Body who have pressed into being made into His exact image. [96]

[3] The Righteous Metatron is the fully mature Body of Christ that can be classified as a oneness matrix with the Father the Son, and the Holy Spirit. [97]

[4] The Righteous Metatron is an eternal reality of the fullness of the Heavenly One New Man in Christ which is a multi-membered entity that's under the headship of Messiah Yeshua (Jesus Christ). [98]

[5] There is a false Metatron whose unrighteous head is a pagan sun god. Take your pick. Apollo is one. Another is "Metatron Mithra" who is connected to Christmas and the demonic control of spacetime. This means that no one connected to Christmas can be part of the righteous Oneness Metatron Matrix. [99]

[6] The Righteous Metatron is a multidimensional reality.

[7] We are still discovering who the Righteous Metatron is and what he does. During various *Mystic Mentoring* Group Ascensions in Christ, it is not uncommon to encounter the Righteous Metatron in Enoch or other forms.[100] [Note: Please be careful on Facebook and other platforms. A couple other groups are trying to steal our "Mystic Mentoring" and "Mystic Mentoring in Christ" image, identity, and anointing, so please be discerning. You can find us at https://www.mysticmentoring.com/.]

[8] At least 90 percent of "Metatron" on Google is off.

If you need further Biblical insight and clarification about "Ascension in Christ" [101] and the righteous "Order of Melchizedek," please refer to two other books I wrote:

- *Ascension Manual* [102]

- *Understanding the Order of Melchizedek: Complete Series* [103]

Let's get back to the Creation's Quantum Elements topic. Quantum mechanics include paradoxical concepts like the dual nature of light. Light can behave like a wave or a particle. Light is pure energy. Quantum physics reveals that light behaves differently depending on how it is measured. Light energy acts like a particle when a position measurement is taken while displaying wave properties otherwise. The particle and wave dual nature of light are called complementary properties; because they both exist and complement one another, even though they seem to contradict each other. [104] All isolated quantum particles—electrons, protons, and neutrons—have the same dual-natured behavior. [105] Recall that the components of an atom are the building blocks of matter, just as the Hebrew Living Letters are building blocks of the Word.

Quantum particles seem to interact with their environment in a knowing way. The physicist Freeman Dyson expresses that "it appears that mind, as manifested by the capacity to make choices, is to some extent inherent in every atom."[106] Other scientists agree. [107]

A measurement performed by an observer in a quantum experiment causes all the possibilities of the wave function to "collapse" to a single outcome. [108] When we hear, we are observing. When we see, we are observing. When we perceive, we are observing. When we observe, we are collapsing the wave function.

As a king and priest of the Righteous Order of Melchizedek, you are called to bring heaven to earth: *"Your kingdom come. Your will be done, On earth as it is in heaven" (Matthew 6:10 $_{NASB}$)*. This happens when we follow the Pattern Son Yeshua by only doing what we see the Father doing. Obvious to the most causal observer, the key concept here is sons of the Living God seeing what the Father is doing. "Seeing" in John 5:19 not only means to see and discern bodily with your physical eyes; but also to perceive by one's senses, to discover or know through experience, and to have the power of understanding. Let's all taste and see that our Heavenly Father is so good, as the One New Man in the Messiah made after the Righteous Order of Melchizedek brings heaven to earth.

Quantum measurement has many spiritual counterparts. One deals with the matter of faith. As with quantum particle-wave measurements, we will get what we measure for. Yeshua tells us: *"With the same measure you use, it will be measured to you" (Mark 4:24)* and *"According to your faith let it be to you" (Matthew 9:29)*. Deciding to believe God, or not, is something like the result of a measurement for a wave or a particle in quantum physics.

Measurement and consciousness are deep parts of the fabric of our reality. Strong objective reality seems to disappear in the quantum realm. Surely, one of the most amazing implications in all science is the role of the observer in quantum physics. Quantum physics is the basis for everyone's reality of their own experiences. However, it is erroneous to say that things don't exist until they are measured, or that reality is whatever we want it to be; because those concepts leave out the ultimate observer—the Creator Himself.

Proverbs 23:1 asks us to *"observe carefully what is before you,"* because a prudent person should mindfully observe before they act with wisdom, understanding, and knowledge. In fact, everything a person does follows the pattern of observing, choosing, and then acting. God wants us based in truth in love where we observe the golden rule: Do unto others what you'd like done to yourself.

In 1989, Astrophysicist George Smoot and other scientists of the Cosmic Background Explorer (COBE) satellite program basically discovered that the universe started as a single point in the distant past.[109] This was YUD (׳). Their discoveries essentially forced cosmology into agreement with the Book of Genesis. George Smoot remarked, "If you're religious, it's like looking at God."

Michael O'Connell relates: "The Initial Singularity was not just the beginning of matter and energy, but it was also the beginning of time and space. According to quantum mechanics theory, the smallest division of time is Planck time, an unimaginably short duration of 10^{-43} seconds. For all practical purposes, events separated by a duration of a Planck time are instantaneous. Scientists believe they worked out how the universe unfolded from one Planck unit of time forward. Yet, no one knows what happened before this because our scientific equations and observations break down at the singularity, the beginning."[110] Please refer to the "Three Spheres and the Hebrew Living Letters" chapter.

Essentially, it was truly void with no disorder at the moment Creation began. Creation was a blank canvas if you will. Everything was perfect because the Lord God Almighty is. The Initial Singularity is the One from whom all light—all energy and matter—came. More accurately, the Initial Singularity is the place where God intersected the physical universe causing the universe to exist. Our *Echad* God (plurality in one) is the outside agent who was the Creator that started Creation, and who used their joyous Hebrew Living Letters to be the intersecting elements of Creation between the spiritual and material realms. This is why I call the letters of the Hebrew Alphabet "hyperquantum."

In the infant universe, there was no material thing at first, just energy in the form of pure uncreated light. By the way, the Blood of Christ is congealed light. There was also the information encoded in the sacred letters of the Word of God. When you read about the Word in the beginning, always keep in mind that the Word is made up of Hebrew letters. *"In the beginning was the Word, and the Word was with God, and the Word was God" (John 1:1). "In the beginning God created* [את] *the heavens and the earth." (Genesis 1:1)*. Additionally, the infant universe had magnificent and unending possibilities.

"All possible states of elementary particles in our universe are 10^{90}.[111] That is a mind-boggling 10 followed by 90 zeroes. It would be just like God to create a universe with so much possibility, with all the possible quantum states for the 10^{90} elementary particles in the universe. Only He could formulate the wave equation for such a universe. It would be just like the God of Abraham and Moses to create so many creatures on whom to bestow His love, to create a universe that is more than mere matter, one that is full of potential—potential that can be quantified and realized."[112]

The description of Genesis 1 matches the scientific evidence for what actually happened. Science says that matter came into existence in a few moments, condensing out of the light energy field as the infant universe expanded. Remember that Einstein proved mathematically that energy and matter are different forms of the same thing, which was powerfully demonstrated by the explosion of the atomic bomb in 1945. Matter and energy being the same substance is a similar concept to God's Word being the same substance as God.

Scientists know that the earliest matter to appear in the primordial universe was a super-hot and super-dense liquid, identified as quark-gluon plasma. Quarks are the fundamental particles of matter that combine to make protons and neutrons.[113] Quark-gluon plasma has been observed in high-energy physics laboratories.

Listen to Mark Peplow's description of quark-gluon plasma as you keep in mind Genesis 1:2's depiction of creation in terms of liquid: *"And the Spirit of God was hovering [brooding] over the face of the waters."* Mark Peplow reports: "The resulting liquid is almost 'perfect': it has a very low viscosity and is so uniform that it looks the same from any angle."[114] Waves propagated through the quark liquid of the early universe like sound waves rippling through water. These ripples from the waves produced very slight variations, which led to slightly higher and slightly lower densities in various regions of space.

The fresh new universe grew enormously during a period of rapid expansion called "inflation." The inflation expansion of space was driven by energy released in a kind of state transition as the universe cooled, similar to the way energy is released when ice condenses out of chilled water. When the state transition was complete, the universe slowed to a smaller rate of expansion. Keep in mind that scientists know that the universe is still expanding. The inflation theory matches the COBE deep space observations exactly.[115] It also matches the account in Genesis 1— *"God made the expanse, and separated the waters that were below the expanse from the waters that were above the expanse; and it was so"* (Genesis 1:7 [NASB]).

As the inflationary universe expanded 1000 times, the quark-gluon plasma cooled, which caused protons and neutrons to form. When the universe cooled another 1000 times, light atomic nuclei formed.[116] As space further expanded, free electrons were captured by atomic nuclei for the first time. At that moment, the soupy opaque universe transitioned to a transparent form. This is when light propagated through the newborn universe unimpeded.[117]

At this point, Genesis 1:4-5 comes into play, because the light of the cosmic background radiation became visible across the universe, and light parted from darkness. *"⁴ God saw that the light was good; and God separated the light from the darkness. ⁵ God called the light 'day,' and the darkness He called 'night.' And there was evening and there was morning, one day"* (Genesis 1:4-5 [NASB]).

Mario Livio tells us that the earliest complete atoms were mostly the lighter ones, like hydrogen, helium, and deuterium with only small amounts of lithium.[118]

During this early epoch, there were no stars in the heavens according to science, as the background radiation dimmed and darkness prevailed again for a time until the first stars coalesced and ignited, which ushered in the current era of sustained starlight in the universe. Billions of stars formed clusters that comprised the earliest galaxies.[119] *Genesis 1:8* marks the completion of the expanse of heaven: *"God called the expanse 'heaven.'"*

Michael O'Connell shares that many of the first stars were massive, like a hundred times greater in size than our solar system's Sun. These first stars burned hot and fast. They went through their nuclear fuel and formed some of the heavier atomic elements. Then, because these first stars went Supernova, they exploded away their outer shells of hydrogen and helium, along with other essential elements, like oxygen, carbon, and iron. The resultant resource-rich dust clouds blew through space, making it possible for more modern stars and planets like ours to form. Only the later generation of stars possessed the atomic elements necessary to support life.[120]

The Milky Way's Earth formed in a long-gone atomic dust cloud, which is described in Genesis 1:9-10: *"⁹ Then God said, 'Let the waters below the heavens be gathered into one place, and let the dry land appear'; and it was so. ¹⁰ And God called the dry land 'earth,' and the gathering of the waters He called 'seas'; and God saw that it was good"* (Genesis 1:9-10 [NASB]).

I find it interesting that Genesis details the creation of the Earth with its plants and fruit-bearing trees before the creation of its time-keeping elements of the Sun, the Moon, and the Stars (See Genesis 1:9-19). The lights in the sky of Creation are a different light from the pure uncreated light that started it all.

Recall that the Spirit of God, darkness, and the waters are mentioned before the pristine uncreated light in the Book of Genesis: "*¹ In the beginning God created [את] the heavens and the earth. ² The earth was without form, and void; and darkness was on the face of the deep. And the Spirit of God was hovering over the face of the waters. ³ Then God said, 'Let there be light'; and there was light" (Genesis 1:1-3* _{Addition mine}*).* Remember that Psalms 24:2 reveals that the earth and all its fullness have been founded on the waters.

A simple water molecule consists of two hydrogen atoms and one oxygen atom bound together to form a molecule—H_2O. Water may be the most remarkable substance in the universe. It exists as a liquid across a remarkably broad range of temperatures, which enables a thriving biosphere on Earth. Both hydrogen and oxygen are gases at ambient temperature and pressure. Yet, when they combine explosively, they make a molecule of water. Chemists can calculate exactly how much water is produced and the amount of energy released in this chemical reaction. The same is true for all chemical compounds.

Every kind of atom that mankind has so far discovered appears on the Periodic Table of Elements where each atom has its own unique properties. For example, a lead atom has 82 protons, 82 electrons, and 125 neutrons, which is a heavy element. Versus helium, which causes balloons to float in the air.

Chemistry happens everywhere around us and within us. The human body is made up of 63% hydrogen, 25.5% oxygen, 9.5% carbon, 1.4% nitrogen, plus some other chemical elements. In fact, the chemical formula for a hypothetical "human molecule" is: $H_{375,000,000} O_{132,000,000} C_{85,700,000} N_{6,430,000} Ca_{1,500,000} P_{1,020,000} S_{206,000} Na_{183,000} K_{177,000} Cl_{127,000} Mg_{40,000} Si_{38,600} Fe_{2,680} Zn_{2,110} Cu_{76} Mn_{13} F_{13} Cr_7 Se_4 Mo_3 Co_1$.[121]

Quantum physicist, inventor, musicologist, and visionary David Van Koevering takes this a bit further: "I am here to announce that the frequencies of your body include 31 fundamental frequencies. Eleven of these frequencies are the dominate frequencies—the active frequencies—and 20 of these frequencies are trace elements in your body."[122] John Tussey collaborated with the late David Van Koevering to record and release these frequencies of the Periodic Table of Elements.[123]

The letters of the Hebrew Alphabet are frequencies and fractals of the Word of God—Yeshua. YHVH created our natural world by first creating 22 Hebrew letters; and then, the Lord combined them to create the physical reality of everything we know. Each letter is a living creature that has a particular spiritual energy formed from the substance of ALEF-TAV (את).

A fractal is a natural phenomenon that exhibits a repeating pattern displayed on every scale (infinitely self-similar). The fractal codes within the Hebrew Living Letters contain the laws of Creation because they come from the Lawgiver and Creator Himself. The foundational Hebrew letters are the links to the voice of God between the physical realm and spiritual realm, which sustains and maintains Creation, as we know it: *"upholding all things by the word of His power" (Hebrews 1:3).* The essence of the Messiah in His delightful letters has always been in Creation's mix.

Most of what mankind has known has been on the material side of the veil—the Periodic Table of Elements. However, when we marry these chemical elements to their Hebrew letter counterparts, we behold a profound and incredible key, which links the spiritual side of the veil to the natural. This handshake, if you will, is sovereignly causing our big beautiful world to be.

This book charts the Quantum 22 letters to Creation's elements. We will map the Hebrew Living Letters more thoroughly to their corresponding elements on the Periodic Table in each chapter that spotlights an individual Hebrew letter. There we will delve into more specifics about each element's atomic number, atomic symbol, atomic weight, atomic structure as well as its properties and uses. The atomic weight is the average weight of an element with respect to all its isotopes (two or more forms of the same element that contains the same number of protons but different numbers of neutrons in their nuclei. For example, hydrogen has three isotopes.).

When it comes to the energy of a Hebrew letter, the atomic mass is key. The atomic mass represents the mass of a single atom. If we want to understand what mass is, we need to show where energy comes from, as Einstein said, $E = mc^2$. The deepest and most fundamental energetic source comes from ALEF-TAV (את) through His Hebrew Living Letters.

The following is a chart that gives you a sneak preview of where we are going in the redemption of the Quantum 22.

HLL #	Hebrew Letter	Element Name	Atomic Symbol	Atomic #	Atomic Weight	Protons, Electrons & Neutrons
1	Alef (א)	Hydrogen	H	1	1.0080	1, 1, 0
1000	Alef (א)					
2	Bet (ב)	Helium	He	2	4.0026	2, 2, 2 (usually)

3	Gimel (ג)	Lithium	Li	3	6.94	3, 3, 4
4	Dalet (ד)	Beryllium	Be	4	9.0122	4, 4, 5
5	Hei (ה)	Boron	B	5	10.81	5, 5, 5
6	Vav (ו)	Carbon	C	6	12.011	6, 6, 6
7	Zayin (ז)	Nitrogen	N	7	14.007	7, 7, 7
8	Chet (ח)	Oxygen	O	8	15.999	8, 8, 8
9	Tet (ט)	Fluorine	Fl	9	18.998	9, 9, 10
10	Yud (י)	Neon	Ne	10	20.180	10, 10, 10
11	Kaf (כ)	Sodium	Na	11	22.990	11, 11, 11
20	Kaf (כ)	Calcium	Ca	20	40.078	20, 20, 20
12	Lamed (ל)	Magnesium	Mg	12	24.305	12, 12, 12
30	Lamed (ל)	Zinc	Zn	30	65.38	30, 30, 34 (^{64}Zn)
13	Mem (מ)	Aluminum	Al	13	26.982	13, 13, 14
40	Mem (מ)	Zirconium	Zr	40	91.224	40, 40, 51
14	Nun (נ)	Silicon	Si	14	28.085	14, 14, 14
50	Nun (נ)	Tin	Sn	50	118.71	50, 50, 69
15	Samech (ס)	Phosphorus	P	15	30.974	15, 15, 16
60	Samech (ס)	Neodymium	Nd	60	144.24	60, 60, 84
16	Ayin (ע)	Sulfur	S	16	32.06	16, 16, 16
70	Ayin (ע)	Ytterbium	Yb	70	173.05	70, 70, 103
17	Pey (פ)	Chlorine	Cl	17	35.45	17, 17, 17 (^{37}Cl)
80	Pey (פ)	Mercury	Hg	80	200.59	80, 80, 120
18	Tzadik (צ)	Argon	Ar	18	39.95	18, 18, 22
90	Tzadik (צ)	Thorium	Th	90	232.04	90, 90, 142
19	Koof (ק)	Potassium	K	19	39.098	19, 19, 20
100	Koof (ק)	Fermium	Fm	100	[257]	100, 100, 157
20	Resh (ר)	Calcium	Ca	20	40.078	20, 20, 20
200	Resh (ר)					
21	Shin (ש)	Scandium	Sc	21	44.956	21, 21, 24
300	Shin (ש)					
22	Tav (ת)	Titanium	Ti	22	47.867	22, 22, 26
400	Tav (ת)					

When you think of the redemption of the Quantum 22 Hebrew Living Letters, contemplate how everything had a higher bandwidth before The Fall. When God spoke causing matter to be, our cosmos contained the full range of the glory of God. All the frequencies and colors of His beautiful voice and His letters were living and active on Earth, the Milky Way, and the entire Universe.

Creation is God's sound. The brilliant David Van Koevering proclaims: "The entire electromagnetic spectrum was sung into existence by Jesus Christ who sustains all things."[124] In six days, God created all the frequencies that He called very good: *"Then God saw everything that He had made, and indeed it was very good. So the evening and the morning were the sixth day" (Genesis 1:31).*

Everything that we see, touch, and experience in the natural realm is a result of cymatics, which is the study of sound and vibration made visible. The multiple forms of universal geometry, symmetry, and beauty that emerge through resonance in various mediums allow us to ponder the wonder of the nature of sound, vibration, and form itself.[125] Johannes Kepler puts it this way: "Where there is matter, there is geometry."

Geometry is vibrations seen. Geometric shapes, some call sacred geometry, are simply sounds made visible. We interact with geometric shapes every day from the spherical shape of Earth to the hexagonal shapes of snowflakes and honeycombs to the fractal golden spiral whose growth factor is PHI (ϕ)—the golden ratio 1:618—that we see in flowers, crystals, and all of creation, including galaxies. And don't forget about the fabulous Fibonacci numbers of 1-1-2-3-5-8-13-21-34-etc. whose successive divisions reveal the prized golden ratio of 1:1.618033988, which surprisingly is not a number but a universal phenomenon of successive Fibonacci wavelengths.[126]

Dr. Robert Moon was a physicist, chemist, and engineer who worked on the original Manhattan Project. He gave us the Moon Model where he applied the five Platonic solids as a configuration of the atomic nucleus. Basically, Dr. Moon was saying space can be quantized, which means space is not empty. It has structure.[127] Dr. Moon came up with the geometric basis for the periodicity of the elements.[128] He solved quantum physics by the concept of a geometric nucleus where the nucleus isn't just some big ball of protons, but the protons are shells. He taught that protons aren't particles, but waves, and waves are geometric. Therefore, each proton is a corner of a geometric object. For instance, oxygen has 8 protons; therefore, it's a cube with 8 corners. We are totally dependent on oxygen to breathe. As we inhale oxygen, we inhale a bunch of little cubes. There is power in your breath. We either take in black cubes of mortality, blue cubes of life, or golden cubes of immortality. Since oxygen is 62.55% of all matter on Earth, it causes all matter to want to go into a cube pattern. It's our home frequency. It's what feels good to the people on earth and is most useful to life. The cube is simply a fractal of the original image of the Universe. Paint Metatron's Cube blue by fulfilling His commandments and they (you) will become gold. For more on this subject, please refer to "A Tale of Three Cubes" video.[129]

Behold, the Hebrew Living Letter fractals are literally the Word of God "made flesh." They are the dancing hyperquantum substance of His breath. May the gentle and glorious touch of the DNA Blueprint of His sacred letters awaken you to the full measure of the Messiah's peace, joy, love, hope, and health. May you become entrained in the pure light and laughter included in His endless love.

80

HEBREW LIVING LETTER PRINCIPLES

As we have seen, the Hebrew letters are the building blocks for all creation. They are the individual spiritual forces that are/were inherent in the Word of God—Messiah Yeshua Himself. When God spoke "Let there be," the individual spiritual forces of the Hebrew letters that originally belonged, and still belong to the nature of the Messiah (Christ), were first created out of nothing but God's desire; then mobilized as the protoplasm of all created light and life. Hence, we can say these letters are the protoplasm of creation, which significantly holds many of its secrets.

Not only are the letters of the Hebrew Alphabet the protoplasm of creation, but so is the Blood of the Lamb slain before the foundation of the world along with water and the Holy Spirit mentioned in Genesis 1. *"⁶ This is He who came by water and blood—Jesus Christ; not only by water, but by water and blood. And it is the Spirit who bears witness, because the Spirit is truth. ⁷ For there are three that bear witness in heaven: the Father, the Word, and the Holy Spirit; and these three are one. ⁸ And there are three that bear witness on earth: the Spirit, the water, and the blood; and these three agree as one" (1 John 5:6-8)*. The Hebrew Living Letters flowed out of the Word who is the Lamb of God slain before the foundation of the world (1 Pet. 1:18-20).

His pre-creation crucifixion created an empty space for Creation to come into existence while Messiah Yeshua's crucifixion in Creation at *"the Place of the Skull, which is called in Hebrew Golgotha"* [130] [131] made a way for its redemption: you, me, the whole wide world, and the entire universe. *"¹³ He has delivered us from the power of darkness and conveyed us into the kingdom of the Son of His love, ¹⁴ in whom we have redemption through His blood, the forgiveness of sins. ¹⁵ He is the image of the invisible God, the firstborn over all creation.¹⁶ For by Him all things were created that are in heaven and that are on earth, visible and invisible, whether thrones or dominions or principalities or powers. All things were created through Him and for Him. ¹⁷ And He is before all things, and in Him all things consist. ¹⁸ And He is the head of the body, the church, who is the beginning, the firstborn from the dead, that in all things He may have the preeminence" (Colossians 1:13-18)*.

The Hebrew letters are the elemental units of the frequencies of the Father of Light's voice and the Messiah's Blood, which contains their intrinsic essence. These are the essential powers of the Word that holds all things together (Heb. 1:3). The Most High God's delightful righteous letters provide the crystalline lattice of life where His sound is made visible (and continues to be made visible) in geometric form.

We now know through quantum physics that everything is energy. That means everything has a unique frequency in vibratory motion. This is something that science has only started to take seriously since the turn of the last century. However, this was something the founder of the Hermetic teachings taught as early as the first century AD.

Atoms are in a state of rapid motion, which manifests a very rapid state and mode of vibration. This means that all forms of physical matter manifest as vibration, energy, or frequencies at the quantum level.

When the Hebrew mystics say that the letters of the Hebrew Alphabet were created first of all out of nothing but God's desire—His divine will—they are saying one of the fundamental purposes of the Hebrew Living Letters is that they facilitate the quantum reality of all created things.

So, how do we know what a Hebrew letter means? Let's begin by identifying three things. We start out with understanding that each letter is a picture, so what a letter looks like is important. These various pictures of the Hebrew letters are called pictographs, which is the earliest known form of writing. A pictograph is a way to represent information using images, like a picture of a person walking with a slash through it communicating: "Do not walk." Next, we keep in mind that there are numerical values to every letter, sometimes more than one. For example, the Hebrew letter MEM (מ) has an ordinal value of 13, a numerical value of 40, and an additional value of 600 for its end letter the Final MEM (ם). Then, we consider that the first time a letter appears in the Torah (i.e., the Word of God) or that letter begins a root of a word, we need to stop, take note, and learn because that word will teach us the meaning of that letter.

Before we investigate the Messiah's Hebrew Living Letters one at a time, let's go over some fundamentals of the Hebrew Alphabet as a whole.

ALPHABETIC ORDER OF THE HEBREW ALPHABET

The first written reference to the alphabetic order of the Alef-Bet comes three thousand years after the creation account in Genesis 1. It was transcribed by the man after God's own heart—King David *(David ha Melech)*. [132]

The entire Hebrew Alef-Bet in its correct order is only found in three books of the Bible: Psalms *(Tehillim)* composed by King David, Proverbs *(Mishlei)* composed by King Solomon, and Lamentations *(Megillas Eichah)* written by the Prophet Jeremiah. It's extremely significant that the acronym for these three books spells out the Hebrew word for "truth." [133]

When King David and King Solomon wrote the correct alphabetic order of the Hebrew Living Letters in ascending order, they were reminding all future generations of the proper priority of the King of King's supremacy over any human monarchy. [134] King David affirmed this when he personally vowed to not enjoy his palace until he prepared a Dwelling Place—Temple

—for God's *Shekinah* Glory. David states this most clearly in Psalms 132, which is one of the Song of Ascents *(Shir ha Ma'alos)*. *David ha Melech* composed Psalms 132 while the foundations for Solomon's Temple were being laid. Significantly, Psalms 132 is the only psalm of the fifteen Song of Ascents that contains all twenty-two letters of the Hebrew Alphabet. [135]

> "*¹ Lord, remember David and all his afflictions; ² How he swore to the Lord, and vowed to the Mighty One of Jacob: ³ 'Surely I will not go into the chamber of my house, or go up to the comfort of my bed; ⁴ I will not give sleep to my eyes or slumber to my eyelids, ⁵ Until I find a place for the Lord, A dwelling place for the Mighty One of Jacob.' ⁶ Behold, we heard of it in Ephrathah; We found it in the fields of the woods. ⁷ Let us go into His tabernacle; Let us worship at His footstool. ⁸ Arise, O Lord, to Your resting place, You and the ark of Your strength. ⁹ Let Your priests be clothed with righteousness, And let Your saints shout for joy. ¹⁰ For Your servant David's sake, Do not turn away the face of Your Anointed. ¹¹ The Lord has sworn in truth to David; He will not turn from it: 'I will set upon your throne the fruit of your body. ¹² If your sons will keep My covenant and My testimony which I shall teach them, Their sons also shall sit upon your throne forevermore.' ¹³ For the Lord has chosen Zion; He has desired it for His dwelling place: ¹⁴ 'This is My resting place forever; Here I will dwell, for I have desired it. ¹⁵ I will abundantly bless her provision; I will satisfy her poor with bread. ¹⁶ I will also clothe her priests with salvation, And her saints shall shout aloud for joy. ¹⁷ There I will make the horn of David grow; I will prepare a lamp for My Anointed. ¹⁸ His enemies I will clothe with shame, But upon himself His crown shall flourish'"* (Psalms 132:1-18).

King David illustrated through the pictures of the Hebrew Living Letters that God's Kingdom on earth should have been established first by writing His Holy Name YHVH (יהוה). By superimposing The Name over that of *David Ha Melech*, it demonstrated that the monarchy of King David remained in the shadow of the Kingdom of the Most High God: *"¹ He who dwells in the secret place of the Most High shall abide under the shadow of the Almighty. ² I will say of the Lord, 'He is my refuge and my fortress; My God, in Him I will trust'"* (Psalms 91:1-2). [136]

Judah *(Yehudah)* received the blessing from his father Jacob *(Yaakov)* that royalty should never depart from him: *"The scepter shall not depart from Judah"* (Genesis 49:10). Please notice that the spelling of the name *Yehudah* (יהודה) includes the four letters that comprise the Tetragrammaton YHVH (יהוה) with the addition of the letter DALET (ד). The DALET in the midst of YHVH represents King David who used his royal powers to champion the Kingdom of YHVH. [137]

Through much refining, King David learned to die to his own personal glorification. He died to his own Kingdom of Self to enhance and promote the Kingdom of God. As a direct reward for King David's self-sacrifice of his own personal comfort and desires in lieu of exalting the Most High God, he received the honor of Messiah Yeshua increasing the government of God and peace on earth upon the throne of David: *"Of the increase of His government and peace There will be no end, Upon the throne of David and over His kingdom, To order it and establish it with judgment and justice From that time forward, even forever. The zeal of the Lord of hosts will perform this"* (Isaiah 9:7).

The proclamation *"HASHEM will rule forever and ever" (Exodus 15:18)* comes at the end of Moses' Song by the Sea. The Song by the Sea contains the praises that the nation of Israel sang to their Most High and Holy Deliverer after they had miraculously crossed the Red Sea. Moses and all of God's people witnessed many miracles when they crossed the Red Sea, not the least was witnessing His Divine Deliverance and Power. God's people are coming to another Red Sea moment where there will be no doubt that God's mighty right hand delivers His own, as the enemy of their souls bears down to destroy them.

Let's take a look at all the Hebrew Living Letters in ascending order in honor of the King of Kings and Lord of Lords (Rev. 19:11-16). Know that when the letters of the Hebrew Alef-Bet appear in ascending order, they represent the attribute of mercy. When they appear in descending order, they represent the attribute of strict judgment.

THE SACRED LETTERS

Hebrew is the holy language of the Bible. The Hebrew Alef-Bet consists of twenty-two letters, which are all consonants. Each Hebrew letter has its own picture, sound, and numerical value. The presence of a dagesh (a dot placed within a letter) modifies the sound of a letter. It indicates a hard pronunciation, and essentially makes one Hebrew Living Letter into two.

The quantum letters of the Hebrew Alphabet are:

Hebrew Letter	Picture	Sound	Numerical Value
Alef/Aleph	א	(silent)	1
Bet/Beis	בּ	b as in boy	2
Vet/Veis	ב	v as in vine	2
Gimel/Gimmel	ג	g as in God	3
Dalet	ד	d as in day	4
Hei/Hay/He	ה	h as in hay	5
Waw	ו	w as in way	6
Vav	ו	v as in violin	6
Zayin	ז	z as in Zion	7
Chet/Het	ח	ch as in Bach	8
Tet/Tes	ט	t as in toy	9
Yud/Yod/Yood	י	y as in yes	10
Kaf	כ	k as in king	20
Lamed	ל	l as in lion	30
Mem	מ	m as in mother	40
Nun/Noon	נ	n as in now	50
Samech/Samekh	ס	s as in sin	60
Ayin	ע	(silent)	70

Pey/Pay/Pe	פ	p as in pastor	80
Fay	פ	ph as in alphabet	80
Tzadik/Tzadi/Tsade	צ	ts as in boots	90
Koof/Kuf/Qof	ק	k as in king	100
Resh/Reysh	ר	r as in run	200
Shin	שׁ	sh as in ship	300
Sin	שׂ	s as in sin	300
Tav	תּ	t as in toy	400
Sav	ת	th as in thin	400

The final two letters—TAV and SAV—were differentiated by *Ashkenazi* (European) Jews. However, in Modern Hebrew, they are both pronounced as *tav*, even when there is no dagesh (point) within the letter.

NUMERICAL VALUE

Every letter in the Hebrew Alphabet has a numerical value. The first ten letters (ALEF to YUD) each correspond to a number from 1 through 10. The next nine letters (KAF through KOOF) represent 20 through 100. The final three letters (RESH, SHIN, and TAV) are from 200 to 400. Letters of a Hebrew word can be added together to equal a given number, which is a similar practice to Roman numerals. These Hebrew letters retain their essential worth no matter where they are placed in a sequence. With this system, any Hebrew word or phrase has a specific numerical value.

Never forget that Hebrew is alphanumeric. Looking at the numerical values of Hebrew letters and words is called gematria. Some people don't appreciate, and even demonize, gematria. Looking at the numerical value of anything should not be the main meal, but a seasoning that adds taste. The Word of God is infinite. When the numerical value of a word or phrase is used properly, it can bring a greater understanding and appreciation for God's Truth. Looking at the numerical value of anything in the Word of God is more than wordplay or random happenstance. If we are to properly study the Scriptures, we should treat the text as living words, which includes the interplay of content with context with each letter, word, or phrase.

Take for example one of the most famous interplays of words and numbers bringing clarity to Jacob leaving his father-in-law's house. As Jacob leaves Laban's home to return to Israel, he sends his brother a message: *"Im Lavan garti"* (i.e., I have lived with Laban). The Hebrew Sage Rashi points out that the numerical value for *garti* is 613, which corresponds to the number of commandments in the Torah (the first five books of the Bible). Thus, Rashi interprets Jacob's message to Esau as: "Throughout the years that I lived with the evil Laban, I kept the 613 commandments." Why would Rashi conclude that Jacob was revealing more than Jacob's living arrangements for the past twenty years? The Hebrew word *garti* is from the root *ger*. *"Ger"* communicates the concept of a stranger or convert. Jacob could have used a more straightforward and appropriate verb to say: "I lived with Laban." *Garti* carries the connotation that "I was a stranger. I was not like them. I was different. I never converted to their wicked ways. I never fit in; because I lived differently." It was Jacob's hidden coded message that the whole time I was away from home, I stayed true to the lessons I learned from home (the household of faith). This is the context for Rashi and others making the supposition that Jacob kept all the commandments of God, even though he was in a household that did not follow the ways of the Most High God; because Jacob remained a stranger to their way of life. [138]

SIMPLE AND COMPOUND LETTERS

The Quantum 22 Hebrew Living Letters can be divided into two groups: simple letters and compound letters. The simple letters have shapes (pictures) that are unique. There are seven simple letters whose shapes are unique. They are ZAYIN (ז), KAF (כ), RESH (ר), NUN (נ), YUD (י), VAV (ו), and DALET (ד). The simple letters are written in this order to prompt people to remember them through the mnemonic *zikaron yud*. *Zikaron* means "memory" in Hebrew. [139]

Significantly, if we take the total number of letters in the Hebrew Alphabet (22) and divide it by the number of simple letters (7), we get the ratio of the circumference of a circle to its diameter—*pi* (π). The spherical shape of a circle points to the complete cycle of creation with the coming of the Messiah, which is displayed in the glorious letters of the Hebrew Alphabet. [140]

The remaining fifteen Hebrew Living Letters are formed by a combination of simple letters. They are called compound letters. For example, the letter ALEF is made up of the letter VAV and two YUDs. The composite letters for a compound letter can provide further depth to the meaning of a word or phrase as well as using the combined numerical value of a composite letter. [141]

The compound letters of the Hebrew Alphabet are:

Name	Picture	Combination	Composite Value
Alef/Aleph	א	י + ו + י (Yud+Vav+Yud)	10+6+10=26
Bet/Beis	ב	ו + ו + ו (Vav+Vav+Vav)	6+6+6=18
Gimel/Gimmel	ג	י+ ז (Yud+Zayin)	10+7=17
Hei/Hay/He	ה	ד + י (Dalet+Yud)	4+10=14
Chet/Het	ח	ד+ ו (Dalet+Vav)	4+6=10
Tet/Tes	ט	נ + ו (Nun+Vav) or כ + ו (Kaf+Vav)	50+6=56 20+6=26
Lamed	ל	כ + ו (Kaf+Vav)	20+6=26
Mem	מ	כ + ו (Kaf+Vav)	20+6=26
Samech/Samekh	ס	כ + ו (Kaf+Vav)	20+6=26
Ayin	ע	ז + ו + נ (Zayin+Vav+Nun)	7+6+50=63
Pey/Pay/Pe	פ	כ +י (Kaf+Yud)	20+10=30

Tzadik/Tzadi/Tsade	צ	י + נ (Nun+Yud)	50+10=60
Koof/Kuf/Qof	ק	ו + ר (Reysh+Vav)	200+6=206
Shin	ש	ו + ו + ו (Vav+Vav+Vav)	6+6+6=18
Tav	ת	ו + ר (Reysh+Vav)	200+6=206

HEBREW LETTERS AT THE END OF WORDS

There are five Hebrew Living Letters that change form when they appear at the end of a word. Even though they look different, they are still pronounced the same. Notice how these final five letters continue the progression of the numerical values from TAV (400). Thus, the complete numeric cycle begins at 1 with ALEF (א) and continues until the final TSADIK letter (ץ), which is equal to 900. [142]

Each of the five final letters can also be interpreted as having the same value as its normal counterpart. For instance, the Final KAF (ך) has a dual value of 20 and 500, the Final Mem (ם) has a dual value of 40 and 600, and so on. [143]

The final Hebrew letters at the end of a word are:

Default Name	Default Form	Final Name	Final Form	Number
Kaf	כ	Final Kaf	ך	500
Mem	מ	Final Mem	ם	600
Nun	נ	Final Nun	ן	700
Fay	פ	Final Fay	ף	800
Tzadi(k)	צ	Final Tzadi(k)	ץ	900

The twenty-seven letter Alphabet (including Final Kaf, Final Mem, Final Nun, Final Fay, Final Tsadik) appears only once in Scripture: *"Therefore, wait for Me, says God, for the day that I will rise to meet with you. For it is My judgment to assemble the nations, to gather kingdoms, to pour out My fury on them, by all the kindling of My wrath, for with the fire of My jealousy all the earth shall be consumed"* (Zephaniah 3:8). [144]

Due to the twenty-seven letter Alef-Bet proclaiming God's oneness, it is significant that Zephaniah 3:8 is followed by *"For them I will convert the peoples to a pure language that all of them call in the Name of YHVH, to worship Him of one accord"* (Zephaniah 3:9). This "pure language" was the original language universally used from the time of creation until the confusion of the one language at the Tower of Babel. This means that for approximately two thousand years the holy language unified the world: *"At this time the whole world spoke one language. Everyone used the same words" (Genesis 11:1 $_{ICB}$)*. Also, notice that the phrase "one language" has the same numerical value as the words "the language of creation." [145]

If we were to sum the higher value of the five final letters, we would get a total of 3500 = 500 + 600 + 700 + 800 + 900. Rabbi Yitzchak ben R' Yehudah haLevi explains that 3500 represents the full splendor of God's Glory throughout the entire universe, which is His revelation and manifestation with respect to three-dimensional space. [146]

The Hebrew Sage Ibn Ezra uses a dimensional explanation as to why the final letters have the same pronunciation as their corresponding regular letters. Ibn Ezra points out that the final letters shape represents a different dimension than their normal letter. The normal (regular) letters represent the dimension of breadth, as indicated by having a base that runs right to left. The final five letters represent depth due to them being bent vertically. Mathematically speaking, these two formats can represent the x and y-axis. [147]

The final letters of the Hebrew Alef-Bet have been called the "Alphabet of Redemption" by Pirkel d'Rabbi Eliezer. Since these letters appear at the end of a word, it illustrates that redemption is the final stage at the end of an exile. [148] Re-establishing a righteous foundation and the righteous operation of the Hebrew Living Letters are critical for the redemption of our world.

METHODS FOR INTERPRETING

Some say that there are at least seventy-three methods for interpreting the Torah (the first five books of the Bible). [149] If you want to dive deeper into various Hebrew methods for studying Scripture, I highly recommend Dovid Leitner's amazing work entitled *Understanding the ALEF-BEIS: Insights into the Hebrew Letters and Methods for Interpreting Them*. In *Quantum 22*, we will only focus on seven of these methods in addition to a letter's picture and numerical value. Please be led of the Holy Spirit by the perfect will of the Father when you study various elements of Scripture with these methods.

[1] Ordinal Value—The ordinal value is the numerical value of each Hebrew letter that is assigned to them based on the ascending order that they appear in the Hebrew Alphabet. The following are all twenty-two Hebrew Living Letters with their ordinal value as well as the basic picture symbology for each and its meaning: [150]

Ordinal Value	Letter	Picture	Symbolic Meaning
1	א - Alef	ox, bull	strength, leader, first
2	ב - Bet	tent, house	household, family, "in"
3	ג - Gimel	camel	to lift up, pride, provision
4	ד - Dalet	door	pathway, to enter
5	ה - Hei	behold	to reveal, to show, "the"
6	ו - Vav	nail, peg	to add, to secure, "and"
7	ז - Zayin	sword, weapon	to cut, to pierce
8	ח - Chet	fence, inner room	to protect, to separate
9	ט - Tet	snake, coiling	to surround, to twist
10	י - Yud	hand (closed)	to make, to work, deed done

11	כ - Kaf	palm (open hand)	to open, to cover, to allow
12	ל - Lamed	rod or staff	to prod, to go forward, to control
13	מ - Mem	water	liquid, mighty & massive (ocean), chaos, to come from (like water down a stream)
14	נ - Nun	fish	action, activity, life
15	ס - Samech	prop	to support, to turn aside, twist slowly (like plant being changed)
16	ע - Ayin	eye	to see, to understand, to know, to experience, to be seen
17	פ - Pey	mouth	to speak, to open, the beginning (like a river)
18	צ - Tsade	fish hook	to pull toward, desire, something inescapable, trouble, harvest
19	ק - Koof	back of the head	what is behind, last, final, or least
20	ר - Resh	head	a person, what is highest, what is most important
21	ש - Shin	tooth	to devour, to destroy, to consume, something sharp
22	ת - Tav	sign (cross)	ownership, to seal, to make a covenant, to join two things together, to make a sign

[2] Composite Value—The composite value of a Hebrew letter is the total numerical value of the letters that compose it. This only works for the fifteen composite letters. For example, we already saw that the letter ALEF (א) is composed of three letters YUD (י), VAV (ו), YUD (י) whose composite value is 10+6+10 = 26. [151]

[3] Mispar Katan (Sum of Digits)—In addition to the numerical value, ordinal value, and composite value is the *"mispar katan"* value. *Mispar katan* deals with dropping all the zeros for the numerical value of a letter so that only a basic integer from one (1) to nine (9) remains. For example, the letter DALET (ד) has a numerical value of four (4), which is also its *mispar katan*. The letter MEM (מ) with a numerical value of 40 and TAV (ת) with a numerical value of 400 also has a *mispar katan* of four. Even though these letters have different numerical values, they are of the same essence due to the concept of contraction in creation. It is its basic original essence. [152]

[4] Millui Fulfillment (Articulated Form)—Each letter of the Hebrew Alphabet can be written out in its expanded articulated form, or in other words, it is the written name of the letter itself. For example, the first letter ALEF (א) is spelled out as ALEF-LAMED-FAY (אָלֶף). This is the letter's "fulfillment." The following are the *millui* fulfillment of all twenty-two Hebrew Living Letters with some alternate meanings: [153]

Letter	Pronunciation	Fulfillment	Alternative Meaning
א – Alef	ah-lef	אָלֶף	prince, teaching, thousand
ב – Bet	bet or bayt	בֵּית	house, tent

ג – Gimel	geeh-mel	גימל	camel - nourish, benevolence, bridge
ד – Dalet	dah-let	דלת	door - poor man, elevation
ה – Hei	hey	הא	behold - to take seed, to be broken
ו – Vav	vahv	ואו	hook
ז – Zayin	ZAH-yeen	זין	weapon - crown, species, gender
ח – Chet	rhymes with "met," sound of "ch" as in Bach	חית	grace of life, fear
ט – Tet	rhymes with "mate," sound of "t" as in tall	טית	goodness
י – Yud	yood, rhymes with "mode"	יוד	hand - possession
כ – Kaf	kaf with sound of "k" as in kite	כף	palm, wing - spoon, power to suppress. kingdom life and action
ל – Lamed	lah-med	למד	to learn, to teach, royalty, heart
מ – Mem	mem with sound of "m" as in mom	מם	water - blemish
נ – Nun	noon	נון	fish - continue, offspring, heir
ס – Samech	sah-mekh	סמך	support, shield
ע – Ayin	ah-yeen	עַיִן	eye - color, fountain
פ – Pey	pay	פה or פא	mouth - here, present (tense)
צ – Tsade	tsah-dee	צָדִי	righteous, hunting
ק – Koof	kof with sound of "q" as in queen	קוף	monkey, surround, great strength
ר – Resh	raysh	רֵישׁ	head - beginning, destitute
ש – Shin	sheen	שִׁין	tooth - ivory, sleep, change
ת – Tav	tav	תָיו	sign - cross, musical notes

[5] Roshei Teivos (First Letters)—*Roshei Teivos* looks at the first letters in a phrase or verse for a deeper meaning. [154] This is what happened when the chief priests protested the sign that hung over Yeshua's head on the Cross: "*[19] Now Pilate wrote a title and put it on the cross. And the writing was: JESUS OF NAZARETH, THE KING OF THE JEWS. [20] Then many of the Jews read this title, for the place where Jesus was crucified was near the city; and it was written in Hebrew, Greek, and Latin. [21] Therefore the chief priests of the Jews said to Pilate, 'Do not write, "The King of the Jews," but, "He said, 'I am the King of the Jews.'"' [22] Pilate answered, 'What I have written, I have written'*" (John 19:19-22).

In Greek, Latin, and Hebrew, this may have looked like:

Greek: ΙΗΣΟΥΣ Ο* ΝΑΖΩΡΑΙΟΣ Ο* ΒΑΣΙΛΕΥΣ ΤΩΝ ΙΟΥΔΑΙΩΝ

Latin: Iēsus Nazarēnus, Rēx Iūdaeōrum

Hebrew: ישוע הנצרי ומלך היהודים

When a person looked at the first Hebrew letters of the words written on the sign, they saw it spelled YHVH and they received the message from their associated pictographs that "Salvation comes by the nailed hand of God." NOTE: These first letters appear on the right side of a Hebrew word; therefore, they represent the attribute of mercy.

[6] Sofei Teivos (Last Letters)—*Sofei Teivos* looks at the last letters in a phrase or verse, which is on the left side of a Hebrew word. These last letters represent the strict attribute of justice. [155]

[7] Mispar Merubah (Multiplication)—There are a variety of methods that are classified by the general term *mispar merubah* (multiplication). We won't be going into these methods much beyond an introduction. One multiplication approach is to multiply the numerical value of a letter by itself, which produces a squared numerical value for that word. [156]

Another approach is to multiply the numerical values of each letter by the other letters of a word. This is the most common multiplication method. An example of this comes from *Deuteronomy 20:19 - "Is the tree of the field a man?"* This analogy between a tree and a man can be easily understood by looking at the product of the numerical values of the letters of the Hebrew word for "man" $160 = 40 \times 4 \times 1$ (אדם), which is equal to the numerical value of the Hebrew word for "tree" $160 = 70 + 90$ (עֵץ).

DECIPHERING PICTOGRAPHS

One of the ways to receive deeper revelation for any Hebrew word is to look at the pictures and symbolism of the sacred letters in a word through the eyes of the Holy Spirit. This can be done for any Hebrew word. Recall that a Hebrew Word Picture can also be called a Pictograph.

When a person deciphers a Hebrew word through contemplating the pictures of its letters, it is a coded message, which is on the *remez* level of studying Scripture. Please recall that *remez* (רֶמֶז) means a "sign or hint" in Hebrew. *Remez* is the hidden or secretive meaning beyond the literal sense. It is the deeper esoteric, allegoric, or symbolic meaning of Scripture. Deciphering pictographs can be on the mysterious *sod* (סוֹד) level as well.

Let's consider an example for deciphering a Hebrew word pictograph. The first two verses in Psalms 119 begins with the Hebrew root word *'ešer* (אַשְׁרֵי), which is pronounced eh'-sher. *'Ešer* (אַשְׁרֵי) can be translated as "Blessed are."

" ¹ (אַשְׁרֵי תְמִימֵי־דָרֶךְ הַהֹלְכִים בְּתוֹרַת יְהוָה)

Blessed are the undefiled in the way,
Who walk in the law of the Lord!

² (אַשְׁרֵי נֹצְרֵי עֵדֹתָיו בְּכָל־לֵב יִדְרְשׁוּהוּ)

Blessed are those who keep His testimonies,
Who seek Him with the whole heart!
(Psalms 119:1-2).

Ešer (אֲשֶׁר) consists of three Hebrew letters: ALEF (א), SHIN (שׁ) and RESH (ר). If we look at the pictures for these letters and their basic meaning, we dive deeper into the Hebrew Word Picture for *ešer* (אֲשֶׁר):

Ordinal Value	Letter	Picture	Symbolic Meaning
1	א – Alef	ox, bull	strength, leader, first, Father
20	ר – Resh	Head	a person, what is highest, what is most important
21	שׁ – Shin	Tooth	to devour, to destroy, to consume, something sharp, El Shaddai, Son

As you gaze at the pictures and the basic symbolic meaning of the pictures for the magnificent letters for *'ešer* (אֲשֶׁר), check in with the Father, the Son, and the Holy Spirit to see what the One True God is personally speaking to you. When I did this, I received a revelatory message for *'ešer* (אֲשֶׁר) "blessed are." Blessed are those who know that Almighty God *(El Shaddai)* is what is highest and most important.

Please note: There can be multiple revelatory word picture messages for each Hebrew word. With my initial Spirit-led Hebrew Word Picture interpretation for *'ešer* (אֲשֶׁר), the first Scripture that popped into my mind was: *"12 Blessed is the man who endures temptation; for when he has been approved, he will receive the crown of life which the Lord has promised to those who love Him. . . . 16 Do not be deceived, my beloved brethren. 17 Every good gift and every perfect gift is from above, and comes down from the Father of lights, with whom there is no variation or shadow of turning. 18 Of His own will He brought us forth by the word of truth, that we might be a kind of firstfruits of His creatures" (James 1:12,16-18).*

To dig a little deeper, recall the pictures of these Hebrew letters and their alternate meanings:

Letter	Pronunciation	Fulfillment	Alternate Meaning
א – Alef	*ah-lef*	אָלֶף	prince, teaching, thousand
ר – Resh	*Raysh*	רֵישׁ	head – beginning, destitute
שׁ – Shin	*Sheen*	שִׁין	tooth – ivory, sleep, change, El Shaddai, Son

As you gaze at the alternate meaning of the pictures for the Hebrew Living Letters for *'ešer* (אֲשֶׁר), check in with the One True God again. When I did this, I received another revelatory message for *'ešer* (אֲשֶׁר): Blessed are those whom El Shaddai teaches how to change into the image of the Head—Christ.

With this alternate interpretation of the Hebrew Word Picture for *'ešer* (אֲשֶׁר), the following Scripture came to mind:

"15 but, speaking the truth in love, may grow up in all things into Him who is the head— Christ— 16 from whom the whole body, joined and knit together by what every joint supplies, according to the effective working by which every part does its share, causes growth of the body for the edifying of itself in love" (Ephesians 4:15-16).

HEBREW VOWELS

We will not be going into vowels because this book is not meant to be a Hebrew language course. We are simply focusing on some basics to assist in our understanding of various truths and mysteries of the Hebrew Living Letters.

Substance of Creation

ALEF (א)—1
MYSTERY OF ONENESS

REDEMPTION OF ALEF (א)

The redemption of God's Quantum 22 is solely done in oneness with the Creator. All creation waits for the manifestation of the sons of God to resonate at His perfect pitch, because of it (Rom. 8:19). The most fundamental part of creation that the Lord of All Creation—ALEF-TAV (את)—and His perfectly mature sons are redeeming to their pristine state is His exquisite Hebrew Living Letters.

To come into agreement with this righteous, holy, and truth-filled redemption of the Quantum 22, we must see as our Heavenly Father sees, perfectly (John 5:19). With this in mind, let us ruminate about an Ascension in Christ where a group of precious people sought the perfect will and perfect heart of our incredible Heavenly Father. We started with no agenda and looked solely to Him. It pleased the Father to show us (and you) a specific illustration about the Redemption of ALEF's story. We were led to entitle it: "The AA Meeting":

> Hear the song "You Are My All in All." It's an AA Meeting with Abba with two As—ALEFs. We hear: "Be still and know that I am God. I am the first Alpha—ALEF (א)." It is so peaceful here where "I am that I am."
>
> Hear His majestic voice. Feel the Heavenly Father's love in your heart. It's a perfect place. The first commandment starts with an ALEF (א): *"I am the LORD your God, who brought you out of the land of Egypt, out of the house of slavery" (Exodus 20:2).* If you keep My commandments you will abide in My love.
>
> Feel the gentle warmth of Abba's Presence. Hear the Father speak to you: "I am everything." Behold, Abba reveals: "There is a place beside Me." Behold, YHVH. Behold, I am that I am.
>
> Abba is excited that we are discovering Him in another way. He is through us—the All in All. We love Abba because He first loved us.
>
> Feel His Presence close by. Abba is summoning us before Him in the spirit with the Lion of the Tribe of Judah by His side. Stand on the Sea of Glass. Ascend on high by His grace to the Father's Throne. When you reach the top, engage with the golden stream of energy that connects The Father's heart to your heart. It's opening something new.

> Hear Abba's voice: "We are one. We are one." Abba is fixing any discord between the two. Come into resonance with Abba. Receive Abba's song where His mercies are made new every day. Hold this heart-to-heart position where Abba's love—*ahava* (אהבה)—is real. It is where Is-real is found. [157]

ANCIENT ALEF/ALEPH (א)

The redemption of the Messiah's Quantum 22 begins with the first letter of the Hebrew Alphabet. ALEF/ALEPH (א) is pronounced *ah-lef* and has a numerical value of one. Its ancient pictograph is a picture of an ox or bull, which symbolizes strength, leader, first, and father. Its sound is silent. ALEF (א) represents the unity of God.

Since ALEF(א) is a silent letter and stands for the number 1, it points to the mysteries of the oneness of God: " *[9] I pray for them. I do not pray for the world but for those whom You have given Me, for they are Yours. [10] And all Mine are Yours, and Yours are Mine, and I am glorified in them. [11] Now I am no longer in the world, but these are in the world, and I come to You. Holy Father, keep through Your name those whom You have given Me, that they may be ONE as We are. [12] While I was with them in the world, I kept them in Your name. Those whom You gave Me I have kept; and none of them is lost except the son of perdition, that the Scripture might be fulfilled. [13] But now I come to You, and these things I speak in the world, that they may have My joy fulfilled in themselves. [14] I have given them Your word; and the world has hated them because they are not of the world, just as I am not of the world. [15] I do not pray that You should take them out of the world, but that You should keep them from the evil one. [16] They are not of the world, just as I am not of the world. [17] Sanctify them by Your truth. Your word is truth. [18] As You sent Me into the world, I also have sent them into the world. [19] And for their sakes I sanctify Myself, that they also may be sanctified by the truth. [20] I do not pray for these alone, but also for those who will believe in Me through their word; [21] that they all may be ONE, as You, Father, are in Me, and I in You; that they also may be ONE in Us, that the world may believe that You sent Me. [22] And the glory which You gave Me I have given them, that they may be ONE just as We are ONE: [23] I in them, and You in Me; that they may be made perfect in ONE, and that the world may know that You have sent Me, and have loved them as You have loved Me*" (John 17:9-23).

Notice the word "one" is mentioned six times in this John 17 passage, which also points to the oneness of mankind with the Godhead who created them in their own image: " *[26] Then God said, 'Let Us make man in Our image, according to Our likeness; let them have dominion over the fish of the sea, over the birds of the air, and over the cattle, over all the earth and over every creeping thing that creeps on the earth.' [27] So God created man in His own image; in the image of God He created him; male and female He created them*" (Genesis 1:26-27).

Lawrence Kushner has a delightful calligraphy book called *The Book of Letters: A Mystical Alef-bait*, which reveals that ALEF (א) is "the letter beginning the first of G-d's mysterious 70 names: (אֱלֹהִים) ELOHIM. G-d. It also begins the most important thing about Him (אֶחָד) ECHAD. One. Know that G-d is One. The first and the last and the only One. And the name of the herald of the last man will be (אֵלִיָּהוּ) ELIYAHU. Elijah. The name of the first Jew is also Alef begun (אברהם) AVRAHAM. (אבינו) AHVEENU. Abraham, our Father."[158]

Rabbi Aaron L. Raskin reveals in his insightful book *Letters of Light*: The Hebrew Sage Rashi tells us that the first two thousand years began with the first man Adam. As *aluf*—master—Adam named all the animals. Rashi goes on to reveal that the second two thousand years began with Abraham who was given the Torah on Mount Sinai. As a master of teaching—*alef-ulfana*—he was faithful to learn and teach others. The third two thousand years began with our Savior and Redeemer Messiah Yeshua. The final level of ALEF is illustrated by the "difference between the word גּוֹלָה-*golah* (exile) and גְּאוּלָה-*geulah* (redemption). If one inserts an alef into the גּוֹלָה-*golah*/exile, exile is empowered and transformed into גְּאוּלָה-*geulah*/redemption."[159]

Lawrence Kushner also reveals "and the very first letter of the first word of the first commandment begins with the first letter which has no sound: (א) ALEF. (אָנֹכִי יְהוָה אֱלֹהֶיךָ) ANOCHI. I. "I am the Lord your God who brought you out of the land of Egypt, the house of slavery."[160]

The Wisdom in the Hebrew Alphabet by Rabbi Michael L. Munk is one of my favorite books about the Hebrew letters. He shares that "Torah is unique, because it is the unifying order of the world, and it points to the Orderer, Who is One (Maharal). That is why God used the letter (א) to begin the Ten Commandments: (אָנֹכִי יְהוָה אֱלֹהֶיךָ), I am HASHEM, your God (Exodus 20:2)."[161]

Rabbi Munk continues: "(א) ALEF represents the one Divine, unchangeable Torah, the thought of God, the repository of all wisdom. Concealed in the letters of the Torah [Word of God], in their combinations, crowns, and shapes, are the laws of physics and metaphysics, music and mining, psychology and ethics. Solomon knew where mineral deposits were located and what veins of land were suitable for growing exotic plants, he knew the behavior of animals and the secrets of healing. All of this knowledge he found by interpreting the Torah.

The sum total of human knowledge derives from the Torah because the universe is a product of Torah which is the blueprint of the world. When His Ineffable Word took physical form, heaven and earth became the clothing for the Word of God which infuses Creation, and without which Creation would not continue to exist."[162] [163]

COMPOSITE LETTER (א)

The composite numerical value of a Hebrew letter is the total numerical value of the letters from which that letter is composed. Recall that the Hebrew Living Letters can be divided into two groups: simple letters and compound letters. The simple letters have shapes (pictures) that are unique. There are seven simple letters whose shapes are unique. They are ZAYIN (ז), KAF (כ), REYSH (ר), NUN (נ), YUD (י), VAV (ו), and DALET (ד).

The remaining 15 of the Quantum 22 are formed by a combination of simple letters. They are called compound letters. For instance, the sacred letter ALEF (א) is made up of the Hebrew letter VAV (ו) and two YUD (י)s. The composite letters for a compound letter can provide further depth to the meaning of a word or phrase as well as using the combined numerical value of a composite letter. [164]

The three different Hebrew letters that makeup ALEF (א) appear as a letter YUD (י) above, a letter YUD (י) below, and a diagonal VAV (ו) that looks like a line suspended between. The upper YUD (י) can represent the hidden and infinite aspects of YHVH. Think about how God is above our comprehension and how our understanding of His true essence is small in comparison, like a mere dot. The lower YUD (י) can represent the revelation of YHVH to mankind; therefore, it can correlate to God's people who dwell on the earth. When God's people humble themselves and realize that they are all a speck compared to Almighty God, they can become a vessel of His righteousness, holiness, and truth.

The diagonal VAV (ו) connects and separates two realms of humanity, because VAV is the sixth Hebrew Living Letter and man was created on the sixth day. VAV is on a diagonal due to being humbled in the face of God's mysteries.

The composite value of the form of ALEF (א)—YUD (י), YUD (י), VAV (ו)—is twenty-six because YUD's numerical value is 10 and VAV's numerical value is 6 (26 = 10 + 10 + 6).[165] It is no coincidence that the numerical value of YHVH (יהוה)—YUD (י), HEI (ה), VAV (ו), HEI (ה)—is also twenty-six (26 = 10 + 5 + 6 + 5). The Hebrew Living Letter ALEF represents the uniqueness of YHVH both with its intrinsic numerical value of one and its composite value of twenty-six. YHVH (י-ה-ו-ה) is the Name that represents God as the Eternal One for its four glorious letters are those that form the words "He was, He is, He will be."

FULFILLMENT OF ALEF (א)

The fulfillment of the Hebrew Living Letter ALEF (ALEF-LAMED-FAY - אלף) has a numerical value of 111 = 1 + 30 + 80. These three ones (111) easily show the three-in-one *echad* plurality in the one—1×1×1 = 1. These three ones (111) also symbolize the unity of God at all levels. It represents the *echad* oneness of the Trinity—the Father, the Son, and the Holy Spirit.[166]

Echad is a Hebrew word for "one," but it more precisely means a single entity that is made up of more than one part. The most important verse in the Bible that Jews memorize is *Deuteronomy 6:4*— "Hear, O Israel, Yahwah is our God, Yahweh is one [echad]."

The Holy Spirit could have used a different Hebrew word "*yachid*," which carries the meaning of the numeric and solitary oneness of God. Instead, the Holy Spirit inspired Moses to use the Hebrew word "*echad*," which is a unified oneness or a unified one. Take for example: "*the two shall become one [echad] flesh*" in Genesis 2:24 is the same word for "one" that is used in Deuteronomy 6:4. Please note that *Yachid*—the main Hebrew word for solitary oneness—is never used in reference to God.

QUANTUM 1

Atomic Number 1—Hydrogen (H). Quantumly speaking, hydrogen is a chemical element that has the atomic number of **1**, the atomic weight **1.0080**, and the atomic symbol **H**. The name hydrogen derives from the Greek "hydro" and "genes," which means water forming.[167]

Hydrogen is colorless, odorless, tasteless, non-toxic, and highly combustible in its diatomic gas form H_2.[168] Hydrogen is the lightest element of the Periodic Table, and it is also the simplest element due to having one proton and one electron in its nucleus. The most common isotope of hydrogen has no neutrons. Hydrogen is the most abundant element in the Universe; therefore, fundamental to it. Just as ALEF(א) is a silent letter, the Universe has been formed from the Almighty's breath.

Cecilia Payne discovered what the universe is made of. She was the first to reveal that the most abundant atom in the universe is hydrogen. Payne concluded that hydrogen and helium are the dominant elements of the Sun and stars. Henry Norris Russell is usually given credit for discovering that the Sun's composition is different than the Earth; but he came to this conclusion four years later than Payne, after telling her to omit it from her thesis (finished Jan. 2, 1925). Currently accepted values for the mass fraction of elements in the Milky Way Galaxy are ~74% hydrogen, 24% helium, and 2% of all the remaining elements, which confirms Payne's results. Her discovery of the true cosmic abundance of the elements profoundly changed what we know about the universe.[169]

Sciencing.com explains the galactic component of hydrogen: "Our own sun, as well as trillions of other stars in the universe, fuses hydrogen into helium to produce energy. Its energy is known on Earth as light and heat."[170] In fact, vast clouds of hydrogen are needed for stars to be born.

I find it interesting that Jupiter is composed mostly of hydrogen. Phys.org reveals: "Galaxies are often surrounded by a halo of hydrogen gas. Over time a galaxy can lose this halo, which streams off into the intergalactic medium. As a result, there can be clouds of hydrogen among the clusters of galaxies in the universe, with a million stars worth of mass."[171]

Most of the hydrogen on Earth exists in organic compounds: humans, plants, animals, and most abundantly, water. As most people know, "water is composed of two hydrogen atoms and one oxygen atom, but what you probably didn't know about water is its violent creation. Hydrogen and oxygen in the same environment create water in an explosive reaction. The atoms release large amounts of energy to bind into a water molecule. The same kind of reaction is used to propel the Atlas rocket."[172]

While hydrogen is present in nearly all molecules in living things, it is very scarce as a gas. Hydrogen is a natural gas that is being proclaimed as a clean alternative to methane, which is an over-abundant greenhouse gas.

Currently, there are various uses of hydrogen: From refining petroleum to producing fertilizer to producing foods … from creating liquid hydrogen rocket fuel to hydrogen fuel cells to produce electricity. Proponents of hydrogen as a fuel believe that a switch from hydrocarbons to hydrogen fuel is advantageous to the environment. Specifically, they tout the cleanliness with which hydrogen burns, producing only energy and water. However, this ignores the production end, which is highly pollutant.

1 IN SCRIPTURE

ALEF (א) is 1. Probably the most declared Scripture containing "one" is *Deuteronomy 6:4* — *"Hear, O Israel: The Lord our God, the Lord is ONE!"* Hebrews call it The Shema.

ALEF (א) is 1. Knowledge of the Holy One is understanding. *"The fear of the Lord is the beginning of wisdom, And the knowledge of the Holy ONE is understanding" (Proverbs 9:10).*

ALEF (א) is 1. The Holy One of Israel is in your midst. *"⁴ Praise the Lord, call upon His name; Declare His deeds among the peoples, Make mention that His name is exalted. ⁵ Sing to the Lord, For He has done excellent things; This is known in all the earth. ⁶ Cry out and shout, O inhabitant of Zion, For great is the Holy ONE of Israel in your midst!" (Isaiah 12:4-6).*

ALEF (א) is 1. "One" is prominent in Ephesians 4. *"⁴ There is ONE body and ONE Spirit, just as you were called in ONE hope of your calling; ⁵ ONE Lord, ONE faith, ONE baptism; ⁶ ONE God and Father of all, who is above all, and through all, and in you all" (Ephesians 4:4-6).*

ALEF (א) is 1. There is one body, which is called the "One New Man" (in the Messiah) in Ephesians 2: *"¹⁴ For He Himself is our peace, who has made both ONE, and has broken down the middle wall of separation, ¹⁵ having abolished in His flesh the enmity, that is, the law of commandments contained in ordinances, so as to create in Himself ONE new man from the two, thus making peace, ¹⁶ and that He might reconcile them both to God in ONE body through the cross, thereby putting to death the enmity" (Ephesians 2:14-16).*

ALEF (א) is 1. The Righteous Messiah's Body is referred to as one flesh in Ephesians 5. *"³⁰ For we are members of His body, of His flesh and of His bones. ³¹ 'For this reason a man shall leave his father and mother and be joined to his wife, and the two shall become ONE flesh.' ³² This is a great mystery, but I speak concerning Christ and the church" (Ephesians 5:30-32).*

CALL OF THE SMALL

There is a small (א) ALEF in the first phrase of the Book of Leviticus: (וַיִּקְרָא אֶל מֹשֶׁה), *"He [God] called to Moses" (Leviticus 1:1).*

"When God dictated the Torah to Moses, He told him to begin the Book with the word (וַיִּקְרָא), He called, a term of endearment that would indicate the intimacy between the Divine Presence and him. Moses, the humblest of men, was reluctant; who was he that God should hold him in such regard? Instead, Moses wanted to write, *He happened by,* an unflattering term that implies that Moses heard God's word only by coincidence, not that God held him in any particular regard. God insisted that Moses write the affectionate (וַיִּקְרָא), but He permitted him to make the (א) ALEF small.... We learn from the small aleph that to be successful in Torah study [learning about The Way, The Truth and The Life], one must consider himself small [humble and teachable]." [173]

The Book of Leviticus shows us the first revelation in the new tabernacle. God's people were made aware of their responsibility to maintain a high level of holiness. Leviticus reveals the laws for purity for God's Temple (you and I).

In the Most Holy Place, there were the golden cherubim that hovered over the Ark of His Presence, which had the face of small children. "This suggests the one who desires to acquire Torah [Word of God] knowledge that he must be ready to accept its teachings and absorb them like a child just beginning to learn and that he must be pure and sinless like a child" [174] "All other vessels of the Tabernacle and Temple could be made of any metal when no gold was available. The cherubim, however, could be made only of pure gold, the finest metal."[175]

"The small *aleph* also calls us to *teshuvah* (repentance). In the phrase (וַיִּקְרָא אֶל מֹשֶׁה), *He called to Moses,* but the subject is not specified. The subject is really the *Aluph* of the World, Who is denoted by the small א. God makes Himself 'small,' as it were, so that His holiness can be found everywhere, even in the hearts of the most wicked person. In everyone's heart the divine spark flickers, always ready to blaze into a flame of repentance. Not every person responds to the call of the small *aleph*—the Godly spark within himself. But whoever heeds the call and returns to God binds himself to holiness, and God's influence in him steadily increases and becomes more obvious."[176]

There is another angle to the scribal teaching that goes with the sign of the small *alef* of *Vyihkra,* which is the Hebrew name for the Book of Leviticus that is commonly translated "And He called." The primary teaching for the hieroglyphic drawing for ALEF (א) of an ox head is that this letter portrays strength. To the ancients, the strongest thing you could have was an ox or a bull.

In *"vyihkra,"* the strength of the ALEF is made small. If you are going to be called of the Lord, your strength must be made small, so His strength may be made large in your life. Any person who has been called of the Lord will share this small *alef* testimony. John the Baptist, who was a Levite priest, gave this very testimony of his call: *"He must increase, but I must decrease" (John 3:30).* Being made weak, so that the Lord may be made strong is a scribal teaching that comes from the first word of the Book of Leviticus— *"vyihkra."*

The Levites were the primary teachers of the children of Israel in the Land. They lived amongst all the tribes with the people. They were the tribe who did the duty—God's works—in His Temple. The Levites were the ones who rendered service to the Lord. The Levites taught the Torah (God's Word) weekly with all the people. This means that if someone had a question about the commandments of the Lord, they went down to the priest that lived near them. In the instructions to the priests in the Book of Leviticus are the basics of our faith. They are the ancient foundation for those who love God and keep His commandments:

"¹ Whoever believes that Jesus is the Christ is born of God, and everyone who loves Him who begot also loves him who is begotten of Him. ² By this we know that we love the children of God, when we love God and keep His commandments. ³ For this is the love of God, that we keep His commandments. And His commandments are not burdensome. ⁴ For whatever is born of God

overcomes the world. And this is the victory that has overcome the world—our faith. ⁵ *Who is he who overcomes the world, but he who believes that Jesus is the Son of God?* ⁶ *This is He who came by water and blood—Jesus Christ; not only by water, but by water and blood. And it is the Spirit who bears witness, because the Spirit is truth.* ⁷ *For there are three that bear witness in heaven: the Father, the Word, and the Holy Spirit; and these three are ONE.* ⁸ *And there are three that bear witness on earth: the Spirit, the water, and the blood; and these three agree as ONE"* (1 John 5:1-8).

QUANTUM QUARK[177] (א)

In summary, ALEF(א) is the first Hebrew Living Letter whose picture is an ox/bull. Just as ALEF(א) is a silent letter, the Universe and humanity have been formed from the Almighty's breath. The silence of ALEF(א) also points to the mysteries of the oneness of God. Since א has a numerical value of 1 and the fulfillment (אלף) of 111, it represents the *echad* oneness of God. ALEF(א) also has a composite value of 26 (ו+י+י); therefore, it represents the hidden and infinite aspects of YHVH (יהוה). The Lord our God is one and the knowledge of the Holy One is understanding.

Q: Ponder that the Universe, including humanity, has been formed from Almighty God's breath. What thoughts come to mind?

Q: What comes to mind when you think of "one" or "oneness"?

BET (ב)—2
HOUSE OF CREATION

REDEMPTION OF BET (ב)

The redemption of the Messiah's Quantum 22 continues with the second letter of the Hebrew Alphabet BET/BEIS (ב), which is usually pronounced *bet*, but can also be pronounced *bay-t*. BET (ב) has a numerical value of two. It is the symbol of blessing, creation, and duality (plurality).

Recall that the redemption of God's Quantum 22 is solely done in oneness with the Creator—ALEF-TAV (את). The sons of the Living God are joining the Heavenly Father to redeem His fiery Hebrew Living Letters to their pristine state. The key is that we must see like the Heavenly Father sees to come into agreement and help facilitate this restoration.

Through a couple Ascensions in Christ, He showed us what the redemption of BET (ב) looks like through His eyes.

> The first moving picture started with a group stepping into the Tent of Joseph, which we were told is the Seven Spirits of God Rainbow Tent. It's dark inside the teepee. There is a fire in the center. We are in a circle around the fire sitting cross-legged, holding hands, and singing the song "Bind Us Together."
>
> Smoke fills the tent, as we sing. The smoke tastes sweet and fills one's body with comfort. We feel the infusion of the Seven Spirits of God. Yeshua is smoking the peace pipe. We are told we are not old (mature) enough yet to smoke the peace pipe. Feel the white smoke infusion. Receive the gift (impartation) of His Presence. The smoke is not just rising vertically; but horizontally, including the whole group. There's an engaging with Christ. There's an emphasis of "in Christ" … "in the anointing" … "in the Anointed One" … We understand that His burning heart is in the middle of the tent. Yeshua dedicated His heart to the Father during Hanukkah.
>
> A teepee is a skin tent. It represents us, and it is in our body of remembrance "in Christ". The delicious white smoke infusion will journey with us. It carries Yeshua's passion for our Beautiful Heavenly Father. [178]
>
> The second picture that the Father of Lights revealed to us for the Redemption of BET (ב) started out by linking us to the first picture through us experiencing the One New Man in Christ under God's Rainbow Tent. We understood that here we are covered and protected by the Seven Spirits of God (Nature of Christ). There's an enormous amount of freedom under this covering.

Enlarge your tent multidimensionally and know that our Heavenly Father is in charge. Hear Abba say, "Sit with Me" in His teepee. We are instructed to sit down into the Father . . . into His Presence . . . into the smoke. Embrace His intimacy. The flap of the teepee is closed with a burly guard at the door. Hear the crackle of a bonfire. See embers going up and out the teepee into a midnight sky.

The Father shares that God's *Ketubah* (Marriage Contract) is not only with Yeshua. It is also with our Beautiful Heavenly Father.

Abba shares, "Do in the flesh the things you want to see in the future." We all reap what we sow. Behold, there's a beautiful maiden beside Yeshua wearing beautiful lace.

The *Ketubah* with The Father is revealed in John 17. See The Father kiss your fingertips . . . your unique identifier on your right hand . . . your unique touch.

The Bride of Christ (who is one with the Father, like Messiah Yeshua) praises, spins, and dances, as she focuses on the Father. We hear the phrase "fountain of living waters." The colorful fire water going up looks like a multi-colored aurora borealis, which is a sign to the darkness because of Abba's kisses on our fingertips. We are all unique and darkness can't control us. The signs of destruction in the sky—aurora borealis, halos around the sun, red iron-oxide skies, two suns, etc.—are signs of our Beloved coming.

We are in a circle holding hands. We raise our hands together and we see sparks fall. It looks like sparklers. The sparks spread. The signs in the earth foreshadow things in the spirit. We zero in on the fire droplets coming off our hands and see Abba kissing our fingertips repeatedly, as we begin to dance. We hear The Father say: "*Ketubah* with Me." It is Abba's heart to kiss our fingertips. Fingertips represent the finer things we do. Help impart The Father's love.

We see Abba's sparkling with joy. His countenance is sparkling with delight. We sense pleasure that the group wants to linger longer to speak to The Father. We are told that the teepee (God's Dwelling Place) is a perfect and honest place. All eyes are on Abba.

The Heavenly Father dons a green garment with white flowers. It reminds us of herbs. When we received His fire water in our fingers, it felt like a portal (vessel) of heaven. We hear: "No matter what you see come to Me for understanding." We know that we need to protect our hearts and our connections with Abba. We see the green lamp before The Throne—the Spirit of Counsel (Isa. 11:2)—that's connected to Abba's green robe. Come back to the teepee (His Dwelling Place) for counsel. [179]

ANCIENT BET (בּ)

The pictograph for BET is a tent, dwelling place, or house. BET (בּ) is "of the ground. A house firmly set upon the earth. The dot which is called a dagesh represents one who lives within." [180]

"You can walk into a Bait, and you are at home. The Holy One wants us to be at home in His world. So the Torah begins with a bait. (בְּרֵאשִׁית) BERESHEET—'*In the beginning G-d created the heavens and the earth*. Bait is the house G-d visits. The world is a home for those who remember who built the house. For them, it is filled with blessing, and it is a (בַּיִת) BAIT." [181]

According to Rabbi Raskin, "The meaning of the beis is bayis, which is Hebrew for 'home.' Why did G-d create the world? The Midrash (Tanchuma Nasa, 16) tells us that G-d desired a home. How does one define a home? A home is the place to which you return after finishing with your worldly affairs. You remove your shoes, change to comfortable clothes, and relax.

You don't have to put on a show or 'sell' yourself to anyone. It's a place where the real you comes alive. G-d also wanted a place where He could be Himself and unite with His bride... That was the objective of Creation. That is the *beis* of *bayis*, the first letter of the Torah, the blueprint of Creation."[182]

With *beis* signifying Creation, we note that the root of the word *Bereishis*—בְּרֵאשִׁית—is *rosh*, which means head. The prefix is a *beis*. The last two Hebrew letters of *Bereishis* are YUD and TAV. Together, the *beis*, YUD and TAV spell *bayis*—house. In the beginning, when G-d created the world, his *taavah*, תַּאֲוָה (desire), was that the head (which is G-d) should dwell in the *bayis*, His home. And how does one make a home for G-d? By living the letter *beis*.[183]

To live the letter BET, we need to resonate at the frequency of the Word of God by understanding that the entirety of Scripture begins with the Hebrew Living Letter BET (בּ) while the Ten Commandments begin with the sacred letter ALEF (א): *"I am God your God who has taken you out of the land of Egypt, out of the house of bondage" (Exodus 20:2).* The divine revelation to God's people at Sinai began with God initiating contact and coming down. In turn, Moses approached the Cloud of His Presence to ascend into a higher reality.

Scripture begins with the Book of Genesis. The Book of Genesis is named after its first word in Hebrew *"bereshit"* בְּרֵאשִׁית, which is commonly translated "in a beginning." This speaks of the whole of creation as a house—a BET—in relation to God. The word *bereshit* is always written large at the beginning of a Torah scroll. The Hebrews say that this is to teach humanity that they should always seek the first ALEF (א) first—The Creator.

One of the glorious fulfillments of BET is God's Holy Temple—the *Beis haMikdash*—that housed the Ten Commandments. The Ten Commandments was/is God's Marriage Contract —Ketubah—as a response to the people saying: *"We will do everything the LORD has said" (Exodus 19:8).* This is a foreshadow of what will be accomplished in the next iteration of God's Holy Temple where Messiah Yeshua and His Bride will be united via holy matrimony in the New Jerusalem.

Remember in the midst of the word *"bereshit"* בְּרֵאשִׁית is the Hebrew word *"rosh"* רֹאשׁ, which means head or chief. The first two verses in the Book of John communicate that Yeshua was *"in the beginning with God"* and is God Himself (The Head). The Head of the house of creation is Messiah Yeshua: *"For every house is built by some man; but He that built all things is God" (Hebrews 3:4 KJV).*

> *"In these last days has spoken to us in His Son, whom He appointed heir of all things, through whom also He made the world. And He is the radiance of His glory and the exact representation of His nature, and upholds all things by the word of His power. When He had made purification of sins, He sat down at the right hand of the Majesty on high" (Hebrews 1:2-3 NASB).*

COMPOSITE LETTER (ב)

The composite value of the form of BET (ב)—VAV (ו), VAV (ו), VAV (ו)—is eighteen because VAV's numerical value is 6. We will go more into this in the VAV chapter. For now, just remember that many Hebrew families believe that eighteen is the recommended age for marriage, so the blessed couple can set up house. This communicates oneness in His Dwelling Place, oneness with the Word of God, and oneness in the Father and in the Son.

FULFILLMENT OF BET (ב)

The fulfillment of the Hebrew letter BET is (BET-YUD-TAV - בֵּית), which has a numerical value of 412 = 2 + 10 + 400. The fulfillment (בֵּית) literally means a house.[184] Not only does the fulfillment of BET (בֵּית) mean a house; but due to the entire Bible starting with the letter ב, BET can represent the entire Word of God.

QUANTUM 2

Atomic Number 2—Helium (He). Quantumly speaking, helium is a chemical element that has the atomic number **2**, the atomic weight **4.0026**, and the atomic symbol **He**. Helium is the first in the noble gas group in the Periodic Table. The term "noble gas" is due to its "noble" property to stay calm and unreactive during chemical reactions.

After hydrogen, helium is the second lightest gas known to man, and it is the second most abundant element in the observable Universe.[185] The word "Helium" is derived from the ancient Greek word "Helios," which intimately connects this noble gas to its presence in our solar system's Sun.[186] "Helium was discovered in the gaseous atmosphere surrounding the Sun by the French astronomer Pierre Janssen, who detected a bright yellow line in the spectrum of the solar chromosphere during an eclipse in 1868."[187] Helium's connection to the Sun can be extrapolated to the Son of God as the BET (ב), our dwelling place, our home.

Helium is a colorless, odorless, tasteless, non-toxic, inert, and monatomic (single atom) gas. A single helium atom consists of two protons and two neutrons in its nucleus with two electrons in its atomic orbitals. Helium's boiling point and melting point is the lowest among all the known elements.[188]

As helium-infused balloons attest, helium is literally lighter than air due to its low atomic weight. Helium can be found in the upper atmosphere of Earth due to its density and it is lightweight. However, it is rarely found on Earth because it easily escapes into space. One way Helium is found on Earth is in the decay of Uranium, which produces alpha particles that are simply Helium nuclei with a positive charge. There are also natural helium gas deposits constantly releasing helium gas, like a large deposit in Tanzania.

Even though Helium is best known as the gas that fills birthday balloons, the gas serves many important purposes in medical machinery (like MRI scanners), in spacecraft and radiation monitors, and in welding. "Helium-neon gas lasers are used to scan barcodes in supermarket checkouts. A new use for helium is a helium-ion microscope that gives better image resolution than a scanning electron microscope."[189] It's also used in computer parts, airbags in cars, and the large hadron collider at CERN to maintain its temperature at 1.9 Kelvin (-271.25°C, -456.25°F). [190]

2 IN SCRIPTURE

BET (ב) is 2. The number two (2) in Scripture represents marital union, testimony, and witnesses. Scripture speaks of the marital union between a man and his wife where two become one flesh: *"23 And Adam said: 'This is now bone of my bones and flesh of my flesh; She shall be called Woman, Because she was taken out of Man.' 24 Therefore a man shall leave his father and mother and be joined to his wife, and they shall become one flesh" (Genesis 2:23-24).*

Scripture also speaks of the marital union between Christ and His Church: *"28 So husbands ought to love their own wives as their own bodies; he who loves his wife loves himself. 29 For no one ever hated his own flesh, but nourishes and cherishes it, just as the Lord does the church. 30 For we are members of His body, of His flesh and of His bones. 31 'For this reason a man shall leave his father and mother and be joined to his wife, and the TWO shall become one flesh.' 32 This is a great mystery, but I speak concerning Christ and the church" (Ephesians 5:28-32).*

BET (ב) is 2. The testimony of two or more witnesses is needed to convict anyone of a crime/sin: *"One witness shall not rise against a man concerning any iniquity or any sin that he commits; by the mouth of TWO or three witnesses the matter shall be established" (Deuteronomy 19:15).*

"26 For if we sin willfully after we have received the knowledge of the truth, there no longer remains a sacrifice for sins, 27 but a certain fearful expectation of judgment, and fiery indignation which will devour the adversaries. 28 Anyone who has rejected Moses' law dies without mercy on the testimony of TWO or three witnesses. 29 Of how much worse punishment, do you suppose, will he be thought worthy who has trampled the Son of God underfoot, counted the blood of the covenant by which he was sanctified a common thing, and insulted the Spirit of grace? 30 For we know Him who said, 'Vengeance is Mine, I will repay,' says the Lord. And again, 'The LORD will judge His people.' 31 It is a fearful thing to fall into the hands of the living God" (Hebrews 10:26-31).

BET (ב) is 2. Every word and accusation are established by two witnesses or more: *"But if he will not hear, take with you one or TWO more, that 'by the mouth of TWO or three witnesses every word may be established" (Matthew 18:16).*

BET (ב) is 2. The number two stands for witnesses, most classically the two end-time witnesses disclosed in Revelation 11: *"1 Then I was given a reed like a measuring rod. And the angel stood, saying, 'Rise and measure the temple of God, the altar, and those who worship there. 2 But leave out the court which is outside the temple, and do not measure it, for it has been given to the Gentiles. And they will tread the holy city underfoot for forty-two months. 3 And I will give power to my TWO witnesses, and they will prophesy one thousand two hundred and sixty days,* clothed in sackcloth.' ⁴ These are the TWO olive trees and the TWO lampstands standing before the God of the earth. ⁵ And if anyone wants to harm them, fire proceeds from their mouth and devours their enemies. And if anyone wants to harm them, he must be killed in this manner. ⁶ These have power to shut heaven, so that no rain falls in the days of their prophecy; and they have power over waters to turn them to blood, and to strike the earth with all plagues, as often as they desire. ⁷ When they finish their testimony, the beast that ascends out of the bottomless pit will make war against them, overcome them, and kill them. ⁸ And their dead bodies will lie in the street of the great city which spiritually is called Sodom and Egypt, where also our Lord was crucified. ⁹ Then those from the peoples, tribes, tongues, and nations will see their dead bodies three-and-a-half days, and not allow their dead bodies to be put into graves. ¹⁰ And those who dwell on the earth will rejoice over them, make merry, and send gifts to one another, because these TWO prophets tormented those who dwell on the earth. ¹¹ Now after the three-and-a-half days the breath of life from God entered them, and they stood on their feet, and great fear fell on those who saw them. ¹² And they heard a loud voice from heaven saying to them, 'Come up here.' And they ascended to heaven in a cloud, and their enemies saw them" (Revelation 11:1-12).

    BET (ב) is 2. Noah was told to gather two animals—male and female—from every creature on earth: *"⁷ So Noah, with his sons, his wife, and his sons' wives, went into the ark because of the waters of the flood. ⁸ Of clean animals, of animals that are unclean, of birds, and of everything that creeps on the earth, ⁹ TWO by TWO they went into the ark to Noah, male and female, as God had commanded Noah"* (Genesis 7:7-9).

    BET (ב) is 2. Yeshua sent out His disciples two by two: *"And He called the twelve to Himself, and began to send them out TWO by TWO, and gave them power over unclean spirits"* (Mark 6:7).

    BET (ב) is 2. During the second day of Creation, God made the firmament and divided the waters: *"⁶ Then God said, 'Let there be a firmament in the midst of the waters, and let it divide the waters from the waters.' ⁷ Thus God made the firmament, and divided the waters which were under the firmament from the waters which were above the firmament; and it was so. ⁸ And God called the firmament Heaven. So the evening and the morning were the SECOND day"* (Genesis 1:6-8).

    BET (ב) is 2. During the fourth day of Creation, two great lights in the cosmos were created: *"¹⁴ Then God said, 'Let there be lights in the firmament of the heavens to divide the day from the night; and let them be for signs and seasons, and for days and years; ¹⁵ and let them be for lights in the firmament of the heavens to give light on the earth'; and it was so. ¹⁶ Then God made TWO great lights: the greater light to rule the day, and the lesser light to rule the night. He made the stars also. ¹⁷ God set them in the firmament of the heavens to give light on the earth, ¹⁸ and to rule over the day and over the night, and to divide the light from the darkness. And God saw that it was good. ¹⁹ So the evening and the morning were the fourth day"* (Genesis 1:14-19).

    BET (ב) is 2. The word "God" appears in all books except two, which are the Song of Solomon and Esther. The shortest verse in the Bible contains only two words: *"Jesus wept"* (John 11:35). Of the ten shortest books, the eighth-ranked Haggai has only two chapters.

BET (ב) BLESSED

Rabbi Michael L. Munk shares: "The Midrash relates that (ב) was chosen because every being—from the heavenly hosts, to man, to the tiniest creature—blesses God with the (ב), saying (בָּרוּךְ יְהֹוָה לְעוֹלָם אָמֵן וְאָמֵן), Blessed is HASHEM forever, amen and amen (Psalms 89:53)." [191]

Michael Munk also speaks of *beis,* which means house (בַּיִת) "and alludes to the focal point of holiness on earth—the (בית המקדש) *Beis HaMikdash,* Sanctuary or Holy Temple in Jerusalem; and to the (בַּיִת) of man, which he can transform into a miniature sanctuary." [192]

The Hebrew Sage Rashi tells us that angels are purely spiritual beings, with but one desire—to do God's will. Their 'heart' is called לֵב, *lev.* (*Rashi, Genesis* 18:5). This is also the heart of mankind when they return to their primordial condition and nature before The Fall. Man has been made in God's own image; and now it's the quest of every man, woman, and child to journey back to the future so that we can take on our original cherubic image (Genesis 1:26-27). For more information, please refer to my *Understanding the Order of Melchizedek: Complete Series* book. [193]

Rabbi Munk writes: "In the heart of man, . . . *yetzer hatov,* the inclination of goodness and spirituality, struggles for supremacy with *yetzer hara,* the urge toward evil and hedonism. Man's heart with its two opposing drives is called לבב, *levav,* with a double ב, for man has conflicting desires that propel him in different directions. Man is commanded to subjugate his *yetzer hara* to the service of God—as we read in the Shema (וְאָהַבְתָּ אֵת יְהֹוָה אֱלֹהֶיךָ בְּכָל-לְבָבְךָ), You shall love Hashem, your God, with all our heart (Deuteronomy 6:5); the double (ב) implies that we are to love God with both inclinations *(Rashi).* You shall love God with your physical as well as your spiritual existence, and subjugate all your desires to Him, as an offering of love *(R'Hirsch).*" [194]

Think about how a person has two drives within them where we can choose which one to give priority. *"Behold, I set before you today a blessing and a curse: the blessing, if you obey the commandments of the Lord your God which I command you today; and the curse, if you do not obey the commandments of the Lord your God, but turn aside from the way which I command you today, to go after other gods which you have not known" (Deuteronomy 11:26-28).*

Deuteronomy 30 emphasizes: *"¹⁵ See, I have set before you today life and good, death and evil, ¹⁶ in that I command you today to love the Lord your God, to walk in His ways, and to keep His commandments, His statutes, and His judgments, that you may live and multiply; and the Lord your God will bless you in the land which you go to possess. ¹⁷ But if your HEART turns away so that you do not hear, and are drawn away, and worship other gods and serve them, ¹⁸ I announce to you today that you shall surely perish; you shall not prolong your days in the land which you cross over the Jordan to go in and possess. ¹⁹ I call heaven and earth as witnesses today against you, that I have set before you life and death, blessing and cursing; therefore choose life, that both you and your descendants may live; ²⁰ that you may love the Lord your God, that you may obey His voice, and that you may cling to Him, for He is your life and the length of your days; and that you may dwell in the land which the Lord swore to your fathers, to Abraham, Isaac, and Jacob, to give them" (Deuteronomy 30:15-20).*

As everyone knows, the distinction between good and evil can be blurred at times. This is when we must exercise our God-given gift of discernment. In order to choose good over evil, life over death, we must first distinguish the evil masquerading as good. Additionally, we need an understanding of His will and His ways. Moses prayed to understand the manifold ways of God so that he could know His supreme way.

QUANTUM QUARK (ב)

To summarize, BET (ב) is the second Hebrew Living Letter whose picture is a tent, a dwelling place, or a house. ALEF (א) teaches us about the Oneness of God while BET (ב) teaches us about the House that God built with His glorious letters building blocks. ב is about the Creator and His Creation. The Book of Genesis is named after its first word in Hebrew *"bereshit"* בְּרֵאשִׁית, which is commonly translated "in the beginning." This speaks of the whole of creation as a house (a BET) in relation to God. The word *bereshit* is always written large at the beginning of a Torah scroll. The Hebrews say that this is to teach humanity that they should always seek the first ALEF (א) first—The Creator. How does one make a home for God? By living the letter BET (ב).

Not only does the fulfillment of BET (בֵּית) literally mean a house; but due to the entire Bible beginning with the sacred letter ב, BET can represent the entire Word of God. Another name for ALEF-TAV (את) is the Word of God (Rev. 19:13). The second element on the Periodic Table is Helium whose name is connected to the Sun *(helios)* that can also be extrapolated to the Son of God *(ben Elohim)* as the BET (ב), our dwelling place, our home. Behold, the kings and priests of the Righteous Order of Melchizedek are made like unto the Son of God (Heb. 7:3).

Q: How can you live the letter BET (ב)?

Q: How can you be made like unto the Son of God? Hint: Look up the phrase "Son of God" in Scripture, seek to understand it; and then, apply it to your own life.

ג

GIMEL (ג)—3
DIVINE COMPLETENESS

REDEMPTION OF GIMEL (ג)

The redemption of the Messiah's Quantum 22 continues with the third letter of the Hebrew Alphabet, which is GIMEL (ג). It is pronounced *geeh-mel*. GIMEL (ג) has a numerical value of 3. Its ancient pictograph is a camel with a long neck, which symbolizes provision, lifted up, and pride (self-will).

For the redemption of GIMEL (ג), the Three Spheres are featuring their amazing provision for holding the Body of Christ and other living organisms together. In the human body, laminin is the "glue" that holds it together. Laminin is in the shape of a cross: *"And He Himself existed before all things and in Him all things consist (cohere, are held together). . . . And God purposed that through (by the service, the intervention of) Him [the Son] all things should be completely reconciled back to Himself, whether on earth or in heaven, as through Him, [the Father] made peace by means of the blood of His CROSS"* (Colossians 1:17,20 $_{AMP}$).

Laminin chains differ at the level of the amino acid sequences; however, laminin's polypeptide[195] chains have an overall arrangement in a cross-shaped pattern. Since GIMEL (ג) is 3, let's concentrate on laminins possessing three different polypeptide chains and the place where these three protein chains intersect where they form an adhesion cross at the molecular level. We could basically say that the Three Spheres intersect at the cross to give life. The redemption of GIMEL (ג) bounce is God's provision for human bodies to shift from mortal life in 3-D to 5-D eternal life in full living color (1 John 5:7-11). The trimeric proteins that intersect to form the cross are:

[1] α (alpha)—The Father,

[2] β (beta)—The Son, and

[3] γ (gamma)—The Holy Spirit.

Behold, the Father, the Son, and the Holy Spirit unite through the laminin cross at the molecular level to hold human bodies together. Technically, laminins are large molecular weight glycoproteins made up of an assembly of three disulfide-linked polypeptides—the α, the β, and the γ—chains. [196] The laminin molecules are named according to their chain compositions; for example, laminin-511 contains 5 α, 1 β, and 1 γ chains. [197] Where the trimeric[198] proteins intersect, it creates a cruciform structure that is able to bind to other molecules of the extracellular matrix and cell membranes. The three short arms have an affinity for binding to other laminin molecules, leading to sheet formation. "Laminins are integral to the structural scaffolding of almost every tissue of a living organism."[199] The long arm is capable of binding to cells and helps anchor organized tissue cells to the basement membrane. [200]

Laminins are indispensable building blocks for cellular networks that physically build the intracellular (within a cell) and extracellular (outside a cell) compartments as well as physically relaying signals critical for cellular behavior. [201] Basement membranes are specialized extracellular matrices that hold cells and tissues together, a property largely due to containing laminins. Every basement membrane includes at least one, sometimes several, members of the laminin family, which largely determines the unique bodily functions of various basement membranes. [202]

The basement membrane is a thin, pliable sheet-like type of extracellular protein matrix that provides cell and tissue support while acting as a platform for complex signaling. [203] The primary function of the basement membrane is to anchor down the epithelium (e.g., epidermis, the outermost layer of skin, or epithelial tissues that line the outer surfaces of many internal organs) to its loose connective tissue underneath. This is achieved by cell-matrix adhesions through substrate (surface on which an organism lives) adhesion molecules. It is basically the "glue" that sticks the human body's skin to its muscles as well as forming the borders and boundaries of many internal organs, including kidneys and lungs. "Molecular biologists call laminin 'the glue of the body,' because without it, our bodies would fall apart." [204] Laminin binding is very crucial for various cellular functions like tissue development (overall health and longevity), wound healing (the body healing itself), and the stability of organs and tissues (blood supply, proper functioning of organs, etc.).

Additionally, the basement membrane that contains at least one member of the laminin family is a protective barrier, preventing malignant cells from invading deeper tissues. [205] The basement membrane is also essential for angiogenesis (development of new blood vessels), blood filtration, and muscle homeostasis (the state of steady physical and chemical conditions). [206]

Let us give God the glory for the soon coming glorious GIMEL laminin shift into eternal life. *"For the wages of sin is death, but the gift of God is eternal life in Christ Jesus our Lord"* (Romans 6:23).

ANCIENT GIMEL (ג)

GIMEL (ג) is the symbol of kindness and culmination. "What is a *gimmel?* The letter *gimmel* represents the benefactor or the giver of charity [lovingkindness]."[207] First and foremost, GIMEL (ג) represents God's abundant and eternal lovingkindness. Without God's *chessed* (kindness), we would not exist. The very essence of kindness is rooted in ALEF-TAV (את).

As it is written: *"All the paths of the Lord are lovingkindness and truth To those who keep His covenant and His testimonies" (Psalms 25:10 ₙₐₛʙ).*

"The term (גמלת חסד) *gemilus chessed* is used to describe the performer of a kind deed. The shape of the (ג) resembles a (גָּמָל) *gamal*, camel, with its long neck. The camel received its name because it … can go a long time without drinking (R'Hirsch, Genesis 21:8). Furthermore, the camel is equipped physically to endure tremendous stress, which enables it to help travelers survive the perils of the desert. Thus, the camel is a *gomel chessed*, a performer of kindness."[208]

Abraham was not content to merely be a recipient of God's kindness. He wanted to emulate the Divine Kindness *(chessed)* that he received. So, Abraham established an inn in Beer Sheba where he and Sarah provided room and board to travelers without a cost to them. When the guests wanted to thank Abraham for his hospitality, he would point them to the fact that everything offered to them was from the kindness, goodness, and abundance of the One to whom the world belongs. Abraham and Sarah not only cared for their physical needs but also fed their souls and spirits.

One of my favorite quotes comes from Michael Munk: "Kindness not only elevates people to heavenly levels; it even leaves its mark upon the place where it is performed."[209] GIMEL (ג) —"And one day all the souls doing acts of lovingkindness (גְּמִילוּת חֲסָדִים) GIMILUT HASIDIM will rise into a great yearning wave that will reach to the heavens and fill the world with (גאולה) GEULAH. Redemption."[210]

"The (גימל), *gimmel*, is cognate [same original root] to (גמל), *gamol*, which means to nourish until completely ripe, as in (וַיִּגְמֹל שְׁקֵדִים), it produced mature almonds (Numbers 17:8)."[211]

The greatness of GIMEL (ג) is ALEF-TAV (את)—the Son of God—being the lifeline (nourishment) for the Body of Christ: *"¹³ Till we all come to the unity of the faith and of the knowledge of the Son of God, to a perfect man, to the measure of the stature of the fullness of Christ; ¹⁴ that we should no longer be children, tossed to and fro and carried about with every wind of doctrine, by the trickery of men, in the cunning craftiness of deceitful plotting, ¹⁵ but, speaking the truth in love, may grow up in all things into Him who is the head—Christ— ¹⁶ from whom the whole body, joined and knit together by what every joint supplies, according to the effective working by which every part does its share, causes growth of the body for the edifying of itself in love"* (Ephesians 4:13-16).

"¹⁸ Let no one cheat you of your reward, taking delight in false humility and worship of angels, intruding into those things which he has not seen, vainly puffed up by his fleshly mind, ¹⁹ and not holding fast to the Head, from whom all the body, nourished and knit together by joints and ligaments, grows with the increase that is from God" (Colossians 2:18-19).

"¹ I am the true vine, and My Father is the vinedresser. ² Every branch in Me that does not bear fruit He takes away; and every branch that bears fruit He prunes, that it may bear more fruit. ³ You are already clean because of the word which I have spoken to you. ⁴ Abide in Me, and I in

you. As the branch cannot bear fruit of itself, unless it abides in the vine, neither can you, unless you abide in Me. ⁵ I am the vine, you are the branches. He who abides in Me, and I in him, bears much fruit; for without Me you can do nothing" (John 15:1-5).

COMPOSITE LETTER (ג)

The composite value of the form of GIMEL (ג)—ZAYIN (ז), VAV (ו), YUD (י)—is 23 because numerically ZAYIN is 7, VAV is 6, and YUD is 10.

As three, GIMEL can represent the Three in One (1×1×1=1)—The Trinity of Father, Son, and Holy Spirit. As twenty-three (the composite value of the form GIMEL), GIMEL (ג) can represent 23 pairs of chromosomes in the human body. Chromosomes are located inside the nucleus of a cell and each cell consists of 2n (diploid) chromosomes, except the gametes or the sex cells. So each cell consists of 23 pairs of chromosomes.

FULFILLMENT OF GIMEL (ג)

The fulfillment of the Hebrew Living Letter GIMEL (GIMEL-YUD-MEM-LAMED —גימל) has a numerical value of 83 = 3 + 10 + 40 + 30. The fulfillment (גימל בְּ) can be translated as "to nourish," "benevolent," "a camel," or "a bridge." [212]

The letter GIMEL (ג) is unique. It is the only quantum letter of the Hebrew Alef-Bet whose standard (i.e., first) fulfillment (גימל) is comprised of four Hebrew letters, which is symbolic for the four-letter Names of God YHVH (יהוה) and ADONAI (אֲדֹנָי). [213]

Appropriately, GIMEL (ג) is the first letter in the ascending alphabet sequence that is adorned with three crowns. These three crowns are called tag (singular), tagin (plural). There are seven Hebrew letters—GIMEL, ZAYIN, TET, NUN, AYIN, TSADIK, and SHIN—which are adorned with three tagin crowns.

We can break the first fulfillment GIMEL (GIMEL-YUD-MEM-LAMED - גימל) down further into the second fulfillment GIMEL (GIMEL-YUD-MEM-LAMED – גימל), YUD (YUD-VAV-DALET – 20 = 10 + 6 + 4 – יוד), MEM (MEM-MEM – 80 = 40 + 40 - מם) and LAMED (LAMED-MEM-DALET – 74 = 30 + 40 + 4 – למד). Total = 257. [214] Not only does the second fulfillment of GIMEL (ג) have a total numerical value of 257 but so does the Hebrew word for "tiara" (נזר). Thus, the appropriateness for the tagin crown on the letter GIMEL. [215]

Recall that *mispar katan* method for interpreting sums the digits of the letters; and then, drops all the zeros for the numerical value of each letter so that only a basic integer from one (1) to nine (9) remains. The eight sacred letters that make up the hidden fulfillment of the word GIMEL (גימל יוד מם למד) have a final *mispar katan* of three. Three is also the numerical value of the original Hebrew letter ג. The complete fulfillment of the word GIMEL contains all twelve quantum letters above. The *mispar katan* of twelve (12) is also three (3 = 1 + 2). [216]

"*A threefold cord is not quickly broken*" (Ecclesiastes 4:12).

QUANTUM 3

Atomic Number 3—Lithium (Li). Quantumly speaking, lithium is a chemical element that has the atomic number **3**, the atomic weight **6.94**, and the atomic symbol **Li**. A single lithium atom has 3 protons, 4 neutrons with 3 electrons in its outer shell. The term "Lithium" comes from the Greek word for stone *"lithos."* [217] GIMEL (ג) is 3 and the word "stone" in Hebrew communicates oneness in the *echad*.

Lithium is a soft, silvery-white alkali metal, which is the least dense metal as well as the least dense solid element under standard conditions. Like all alkali metals, lithium is highly reactive and flammable; thus, it must be stored in a vacuum, inert atmosphere, or in an inert liquid such as mineral oil or purified kerosene. [218]

Due to lithium being so reactive, it is not found in its pure form in nature. However, it is found throughout the world in a variety of areas including seawater, mineral springs, and igneous rocks. [219]

The nucleus of lithium verges on instability, and because of this it has been used in nuclear physics. The transmutation of lithium atoms to helium in 1932 was the first fully man-made nuclear reaction with lithium deuteride serving as a fusion fuel in staged thermonuclear weapons. [220]

Probably, "the most important use of lithium is in rechargeable batteries for mobile phones, laptops, digital cameras, and electric vehicles. Lithium is also used in some non-rechargeable batteries for things like heart pacemakers, toys, and clocks. Lithium metal is made into alloys with aluminum and magnesium, improving their strength and making them lighter. A magnesium-lithium alloy is used for armor plating. Aluminum-lithium alloys are used in aircraft, bicycle frames, and high-speed trains.

Lithium oxide is used in special glasses and glass ceramics. Lithium chloride is one of the most hygroscopic materials known and is used in air conditioning and industrial drying systems (as is lithium bromide). Lithium stearate is used as an all-purpose and high-temperature lubricant. Lithium carbonate is used in drugs to treat manic depression, although its action on the brain is still not fully understood. Lithium hydride is used as a means of storing hydrogen for use as a fuel."[221]

"The principal industrial applications for lithium metal are in metallurgy, where the active element is used as a scavenger (remover of impurities) in the refining of such metals as iron, nickel, copper, and zinc and their alloys. A large variety of nonmetallic elements are also scavenged by lithium, including oxygen, hydrogen, nitrogen, carbon, sulfur, and the halogens."[222]

3 IN SCRIPTURE

GIMEL (ג) is 3. The number "three" represents the Torah (the first five books of the Bible), because it was given to God's people on the third day of the third month of the year—Sivan—by the third of three children—Moses. At Mount Sinai, the Torah was issued to three groups:

the *Kohanim* (High Priest Aaron's descendants), the Levites, and the entire nation of Israel. And the Torah itself is divided into three sections: the Five Books of Moses, the Prophets, and the Writings.

GIMEL (ג) is 3. The number three can also represent the three Patriarchs: Abraham, Isaac, and Jacob. Abraham personified kindness. Isaac epitomized uncompromising justice, and Jacob embodied truth. "Truth," as in The Way, The Truth, and The Life (John 14:6), decides when to utilize kindness and/or justice. When kindness and justice are blended in proper measure by the Judge of the whole earth, the result is Truth. *"Doesn't the Judge of all the Earth judge with justice?"* (Genesis 18:25 $_{MSG}$). *"Or do you show contempt for the riches of His kindness, forbearance and patience, not realizing that God's kindness is intended to lead you to repentance?" (Romans 2:4 $_{NIV}$)*.

GIMEL (ג) is 3, which represents the three that bear witness in heaven—Father, Son, Holy Spirit: *"For there are THREE that bear record in heaven, the Father, the Word, and the Holy Ghost: and these three are one" (1 John 5:7 $_{KJV}$)*. GIMEL (ג) also represents the three that bear witness on earth—Spirit, Water, Blood—*"And there are three that bear witness on earth: the Spirit, the water, and the blood; and these THREE agree as one" (1 John 5:8 $_{KJV}$)*.

E.W. Bullinger tells us in his book *Numbers in Scripture* that the number three points us to what is solid, real, essential, perfect, substantial, complete, and divine. For example: *"Holy, Holy, Holy is the Lord of hosts; The whole earth is full of His glory!" (Isaiah 6:3)*.

> *"The first living creature was like a lion, the second living creature like a calf, the THIRD living creature had a face like a man, and the fourth living creature was like a flying eagle. The four living creatures, each having six wings, were full of eyes around and within. And they do not rest day or night, saying: "Holy, Holy, Holy, Lord God Almighty, Who was and is and is to come!" (Revelation 4:7-8)*.

In the fresh dispensation that has dawned in this Kingdom Day, the Lord our God is causing us to want to be just like Yeshua, so we can all be made one. Divine series of three in Scripture are keys for the hour that we live in. When you see three items listed in Scripture, sit up and take notice. Biblical threes list things from easiest to hardest; but more importantly, they reveal the way of life that propels us toward maturity (Psalm 16:11; John 14:6). Divine threes are typically progressive in nature with one building on what was previously gained or attained. Usually, these steps that move us unto perfection also overlap where God can be working on two or three different levels within us at the same time since parts of any one of us can be more mature than others.

"I indeed baptize you with water unto repentance, but He who is coming after me is mightier than I, whose sandals I am not worthy to carry. He will baptize you with the Holy Spirit and fire" (Matthew 3:11). Yeshua's three baptisms articulated in Matthew 3:11 is a good example. Being baptized by water is easier, and usually precedes, being baptized by the Holy Ghost, which is definitely easier than being baptized by fire by going to the cross.

Mark 4:13 tells us that we need to understand the "Parable of the Sower" to understand all parables. Significantly, the Parable of the Sower is illustrated three times in Scripture: Matthew 13,

Mark 4, and Luke 8. Notice that the good seed in good soil is those who hear the Word with a noble and good heart, keep it, and bear fruit with patience: *"¹³ And He said to them, 'Do you not understand this parable? How then will you understand all the parables? ¹⁴ The sower sows the word. ¹⁵ And these are the ones by the wayside where the word is sown. When they hear, Satan comes immediately and takes away the word that was sown in their hearts. ¹⁶ These likewise are the ones sown on stony ground who, when they hear the word, immediately receive it with gladness; ¹⁷ and they have no root in themselves, and so endure only for a time. Afterward, when tribulation or persecution arises for the word's sake, immediately they stumble. ¹⁸ Now these are the ones sown among thorns; they are the ones who hear the word, ¹⁹ and the cares of this world, the deceitfulness of riches, and the desires for other things entering in choke the word, and it becomes unfruitful. ²⁰ But these are the ones sown on good ground, those who hear the word, accept it, and bear fruit: some thirtyfold, some sixty, and some a hundred' "* (Mark 4:13-20).

What could Messiah Yeshua mean when He communicated that if you understand this Parable of the Sower, you can comprehend all the parables? I am sure that there are many keys to deciphering the moral attitudes and spiritual principles illustrated in Yeshua's short stories. One such key can be called 30-60-100: *"some thirtyfold, some sixty, and some a hundred" (Mark 4:20)*, which clearly displays an ascending Divine Series of Three. In that same verse, we have another Divine Series of Three listing things from easiest to hardest in sequential order: *"These are the ones sown on good ground, those who hear the word, accept it, and bear fruit" (Mark 4:20)*.

We can apply this Divine Series of Three principle to the entire Word of God. Take for example:

"For in Him we live and move and have our being" (Acts 17:28).

"And now abide faith, hope, love, these three; but the greatest of these is love" (1 Corinthians 13:13).

"May the God of peace sanctify you entirely, so you may be preserved complete without blame at the coming of our Lord Jesus Christ: spirit, soul, and body" (1 Thessalonians 5:23).

"Do not be conformed to this world, but be transformed by the renewing of your mind, so that you may prove what the will of God is, that which is: good, acceptable (or pleasing), perfect" (Romans 12:2).

"⁴ Hear, O Israel: The Lord our God, the Lord is one! ⁵ You shall love the Lord your God with all your heart, with all your soul, and with all your strength" (Deuteronomy 6:4-5).

Sometimes, the Divine Series of Three may not be contained in one verse, like obedience leading to righteousness, which leads to holiness: *"¹⁶ Do you not know that to whom you present yourselves slaves to obey, you are that one's slaves whom you obey, whether of sin leading to death, or of obedience leading to righteousness? . . . ¹⁸ And having been set free from sin, you became slaves of righteousness. ¹⁹ . . . For just as you presented your members as slaves of uncleanness, and of lawlessness leading to more lawlessness, so now present your members as slaves of righteousness for holiness" (Romans 6:16-19).*

There are other examples of Divine Series of Three that are not contained within one Bible verse, like Romans 6:16-19. Sometimes a person must have a greater understanding of the totality of Scripture:

[1] Baptized by water is equivalent to *"See the Kingdom of God" (John 3:3),*

[2] Baptized by the Holy Spirit is equivalent to *"Enter the Kingdom of God" (John 3:5),* and

[3] Baptized by fire is equivalent to *"Inherit the Kingdom of God" (Galatians 5:19-21).*

We hold His precious treasures in these earthen vessels. Your character is key to containing the glory that's being poured out from on high, so that it can redemptively impact the earth. Your journey to becoming just like Christ—made into His exact image—progresses from:

[1] Salvation to

[2] Sanctification to

[3] Purification.

Here's to being transformed into the character of Christ and us moving unto perfection (maturity)!

QUANTUM QUARK (ג)

In summary, GIMEL (ג) is the third Hebrew Living Letter whose picture is a camel. The fulfillment of GIMEL (גימל) has a numerical value of 83, which can be translated as "to nourish," "benevolent," "a camel," or "a bridge." GIMEL (ג) is unique in that it is the only quantum letter whose first fulfillment is comprised of four sacred letters(גימל), which connects it to the four-letter Names of God: YHVH (יהוה) and ADONAI (אֲדֹנָי).

GIMEL represents the three that bear witness in heaven—the Father, the Son, and the Holy Spirit (1 John 5:7). GIMEL (ג) is the first letter in the ascending alphabet sequence that is adorned with three crowns. This alludes to the first stage of the crowning of the kings and priests made after the Righteous Order of Melchizedek when a person is commissioned through entering the Kingdom of God. This is when the Trinity of the Holy Three in One ($1 \times 1 \times 1 = 1$) harmoniously unites with the three that bear witness on earth—the Spirit, the Water, and the Blood (1 John 5:8)—in your earth. The Father and the Son and the Holy Spirit are three, plus one—you—make a fourfold unity, which is one of the aspects of the four living creatures in Ezekiel 1 that has four faces.

Q: What aspect of GIMEL (ג) is most meaningful to you—nourishment, benevolence, provision of a camel, or a bridge—and why?

Q: How can you more harmoniously unite with the Highest and Holiest Three in One?

ד

DALET (ד)—4
DOOR TO I AM THAT I AM

REDEMPTION OF DALET (ד)

The redemption of the Messiah's Quantum 22 continues with the fourth letter of the Hebrew Alphabet, which is DALET (ד). It is pronounced *dah-let* and has a numerical value of four. The pictograph for DALET is a DOOR. It is symbolized by a door, a portal, a path, a way of life, or movement especially in or out.

ALEF- TAV (את) shared with a small group what the redemption of DALET (ד) looks like in this Kingdom Day.[223] The most fundamental association of ALEF-TAV's connection to the door is Yeshua's two declarations: *"I am the door" (John 10:7,9)*, which are also linked to the *Revelation 4:1* door: *"I saw a portal open into the heavenly realm, and the same trumpet-voice I heard speaking with me at the beginning said, 'Ascend into this realm!'" (TPT)*.

The Revelation 4:1 DALET Door is now opening for the sons of man (mankind) to be integrated into their transformed state,[224] like Enoch, being fully made into the image of the Son of God—the fullness of Messiah Yeshua (Jesus Christ). The caveat is they must walk with God, as Enoch did. He had such a close and intimate relationship that God took him (Gen. 5:24). Theirs will be resurrected fifth-dimensional bodies operating on higher frequencies versus the lower frequencies of the third dimension. Know that no defilement goes through this gold doorway of transformation. The Blood of the Lamb of God is on the doorposts.

ANCIENT DALET (ד)

Since the Hebrew Living Letter DALET numerically represents the number 4, it represents the four rivers flowing from Eden (Gen. 2:10). The number four (4) is also connected to the four corners of the Ark of the Covenant as well as the four corners of the Golden Altar and the four corners of the Brazen Altar. Just as the items connected to one's heart—the Brazen Altar (Altar of Burnt Offering) and the High Priest's Breastplate—are foursquare, so is the New Jerusalem (Exo. 38:1; Exo. 39:9; Rev. 21:16). There were four rows of three precious stones on the High Priest's Breastplate. There are four winds connected to the four directions of the compass (north, south, east, and west). Significantly, four (4) represents the four faces of God (Melchizedek, Living Creature) and the four abominations.

There are 13 references in Scripture to the "four living creatures" who have "four faces" each, "four sides" each, "four wings" each and a total of four wheels within wheels for The Living Creature made up of four living creatures (cherubs), *"behold, the four wheels by the cherubim, one wheel by one cherub" (Ezekiel 10:9).*

The four wheels of the four living creatures have the appearance of beryl: *"The appearance of the wheels and their work was like unto the color of a beryl: and they four had one likeness: and their appearance and their work was as it were a wheel in the middle of a wheel" (Ezekiel 1:16 $_{KJV}$).*

The spirit of each of the four living creatures is in its wheels: *"^{19} And when the living creatures went, the wheels went by them: and when the living creatures were lifted up from the earth, the wheels were lifted up. ^{20} Whithersoever the spirit was to go, they went, thither was their spirit to go; and the wheels were lifted up over against them: for the spirit of the living creature was in the wheels. ^{21} When those went, these went; and when those stood, these stood; and when those were lifted up from the earth, the wheels were lifted up over against them: for the spirit of the living creature was in the wheels" (Ezekiel 1:19-21 $_{KJV}$).*

Beryl is a mineral form of Beryllium. We will shortly go into Beryllium in greater depth. Beryllium is the fourth element on the periodic table with Beryl being the principal store of beryllium in the earth's crust. Beryllium is uniquely strong and light; but can be highly toxic. Despite its toxicity, beryllium is highly useful due to it being one of the lightest metals with one of the highest melting points. Beryl is a relatively rare silicate mineral with the chemical composition $Be_3Al_2Si_6O_{18}$. Its crystal structure is hexagonal, which means that it is six-sided, and a Star of David *(Merkabah)* can fit in the midst of it. We will dive deeper into the six-sided *Merkabah* (Star of David) and its associated Metatron's Cube structure during our exploration of the sixth sacred letter VAV. Know that the basic geometries of life in all levels of reality—the five platonic solids—are all connected to Metatron's Cube.

Notice that the mobile "four living creatures" in the book of Ezekiel have "four wings" each (Ezek. 1:6,8,11,23; Ezek. 10:21), but the four living creatures (four beasts—lion, ox, eagle, man) in the throne room each have "six wings" (Rev. 4:6-8). Why the difference? A clue lies in two phrases *"two wings of every one were joined one to another, and two covered their bodies" (Ezekiel 1:11)* and *"every one had two, which covered on this side, and every one had two, which covered on that side, their bodies" (Ezekiel 1:23).* The Ancient Hebrew Pictograph for "wings" tells us that the wings of the living creatures (cherubim) point to kingdom life and action.

I believe that the Lord revealed to me that the wings of the living creatures speak about living an ascended kingdom lifestyle where two wings are connected to an ascended person's spirit, two wings are connected to an ascended person's soul, and two wings are connected to an ascended person's body. This ascended six wings state is when His Ascended Ones have fully proven all things, by holding fast to that which is good and abstaining from all appearance of evil. It's when these Crucified Ones have fully done what is excellent in His eyes so that the God of Peace sanctifies them wholly—spirit, soul, and body—being preserved pure and spotless (1 Thess. 5:21-24).

SIMPLE LETTER (ד)

DALET (ד) belongs to the seven simple-shaped letters. Some point out that DALET's form can resemble a vertical VAV (ו) and a horizontal VAV (ו).

FULFILLMENT OF DALET (ד)

The fulfillment of the fourth Hebrew Living Letter DALET is DALET-LAMED-TAV (דלת). Numerically, DALET (ד) is 4, LAMED (ל) is 30, and TAV (ת) is 400; therefore, DALET's fulfillment has a total numerical value of 434 = 4 + 30 + 400, which can be translated as a " 'poor man' standing at the doorway when collecting charity or 'elevation.'" [225] The *mispar katan* of the first fulfillment of DALET (דלת) is eleven, 11 = 4 + 3 + 4, which is the same *mispar katan* as God's Name ADONAI (אֲדֹנָי). Adonai is connected to God's attribute of mercy. [226]

QUANTUM 4

Atomic Number 4—Beryllium (Be). Quantumly speaking, beryllium is a chemical element that has the atomic number 4, the atomic weight **9.0122**, and the atomic symbol **Be**. Beryllium has 4 protons and 5 neutrons with 4 electrons in its outer shell. It is a steel-gray, strong, lightweight, and brittle alkaline metal.

The element Beryllium is "stronger than steel and lighter than aluminum. Its physical properties of great strength-to-weight, high melting point, excellent thermal stability and conductivity, reflectivity, and transparency to X-rays make it an essential material in the aerospace, telecommunications, information technology, defense, medical, and nuclear industries. Beryllium is classified as a strategic and critical material by the U.S. Department of Defense."[227]

Beryllium is "used in metallurgy as a hardening agent and in many outer space and nuclear applications"[228] It is a divalent element, which occurs naturally only in combination with other elements to form minerals. The most notable gemstones that contain high amounts of beryllium are beryl (aquamarine, emerald) and chrysoberyl. We are told that the wheels within wheels have the appearance of beryl: *"The appearance of the wheels and their work was like unto a beryl: and they four had one likeness; and their appearance and their work was as it were a wheel within a wheel" (Ezekiel 1:16 ASV).*

It's significant that Scripture also tells us that the spirit of the Living Creature (consisting of four living creatures) is in the wheels: *"²⁰ Wherever the spirit wanted to go, they went, because there the spirit went; and the wheels were lifted together with them, for the spirit of the living creatures was in the wheels. ²¹ When those went, these went; when those stood, these stood; and when those were lifted up from the earth, the wheels were lifted up together with them, for the spirit of the living creatures was in the wheels" (Ezekiel 1:20-21).* Understand that the gyroscopic movement of the Whirling Wheels—*Ophanim* in Hebrew—is controlled through the oneness of the spirits of the living creatures with the Spirit of Almighty God. The four wheels within wheels are part of the elemental structure of Ezekiel's Chariot Throne, which causes us to call this mobile sapphire throne contraption a *Merkabah* in Hebrew—a Chariot Throne.

Notably, beryl crystals occur in hexagonal prisms—a polygon of six angles and six sides. The Star of David is an emblem consisting of two interlacing triangles forming a six-pointed star. When we connect the beryl wheels with a certain man dressed with a gold belt that had a body of beryl in Daniel 10:6, we can get a glimpse of the corporate body of Christ that has been stamped with the royal seal of the Messiah's Hebraic roots. DALET (ד) is the door to the corporate body of Christ quantumly flying by the Holy Spirit with one another.

Recall, Beryllium is the fourth element on the Periodic Table, and Beryl is the principal store of beryllium in Earth's crust. Four is the number of Creation. Think about the four corners of the earth; the four elements of earth, wind, water, and fire; the four directions of north, south, east, and west. The Cherubim (i.e., the Living Creature) is always connected to creation (Ezekiel 1:19; Ezekiel 10:15). I have written extensively about the subject of beryllium and beryl in the sixth book *Window into the Wheels Within Wheels* of the *Understanding the Order of Melchizedek: Complete Series* book.[229]

4 IN SCRIPTURE

DALET (ד) is 4. The number four (4) fundamentally represents creation. On the fourth day of creation, lights were created in the firmament of the heavens to divide the day from night: sun, moon, and stars. Creation's celestial lights—sun, moon, and stars—are for signs and seasons, days and years (Gen. 1:14). Hence, the four seasons of winter, spring, summer, and fall.

> "*14 Then God said, 'Let there be lights in the firmament of the heavens to divide the day from the night; and let them be for signs and seasons, and for days and years; 15 and let them be for lights in the firmament of the heavens to give light on the earth'; and it was so. 16 Then God made two great lights: the greater light to rule the day, and the lesser light to rule the night. He made the stars also. 17 God set them in the firmament of the heavens to give light on the earth, 18 and to rule over the day and over the night, and to divide the light from the darkness. And God saw that it was good. 19 So the evening and the morning were the FOURTH day*" (*Genesis 1:14-19* $_{KJV}$).

DALET (ד) is 4. Our physical world extends in four directions—north, south, east, and west. Remember that 4 is also connected to the spiritual world by the four rivers that flow from Eden.

> "*8 The Lord God planted a garden eastward in Eden, and there He put the man whom He had formed. 9 And out of the ground the Lord God made every tree grow that is pleasant to the sight and good for food. The tree of life was also in the midst of the garden, and the tree of the knowledge of good and evil. 10 Now a river went out of Eden to water the garden, and from there it parted*

and became FOUR riverheads. ¹¹ The name of the first is Pishon; it is the one which skirts the whole land of Havilah, where there is gold. ¹² And the gold of that land is good. Bdellium and the onyx stone are there. ¹³ The name of the second river is Gihon; it is the one which goes around the whole land of Cush. ¹⁴ The name of the third river is Hiddekel; it is the one which goes toward the east of Assyria. The FOURTH river is the Euphrates. ¹⁵ Then the Lord God took the man and put him in the garden of Eden to tend and keep it" (Genesis 2:8-15 _{KJV}).

Follow the living water to see our journey is back to Eden and beyond, which leads to the gates (doors) of the New Jerusalem where the fourth man, like unto the Son of God, resides.

"¹ Now I saw a new heaven and a new earth, for the first heaven and the first earth had passed away. Also there was no more sea. ² Then I, John, saw the holy city, New Jerusalem, coming down out of heaven from God, prepared as a bride adorned for her husband. ³ And I heard a loud voice from heaven saying, 'Behold, the Tabernacle of God is with men, and He will dwell with them, and they shall be His people. God Himself will be with them and be their God. ⁴ And God will wipe away every tear from their eyes; there shall be no more death, nor sorrow, nor crying. There shall be no more pain, for the former things have passed away.' ⁵ Then He who sat on the throne said, 'Behold, I make all things new.' And He said to me, 'Write, for these words are true and faithful.' ⁶ And He said to me, 'It is done! I am the Alpha [ALEF] and the Omega [TAV], the Beginning and the End. I will give of the fountain of the water of life freely to him who thirsts. ⁷ He who overcomes shall inherit all things, and I will be his God and he shall be My son. ⁸ But the cowardly, unbelieving, abominable, murderers, sexually immoral, sorcerers, idolaters, and all liars shall have their part in the lake which burns with fire and brimstone, which is the second death.' ⁹ Then one of the seven angels who had the seven bowls filled with the seven last plagues came to me and talked with me, saying, 'Come, I will show you the Bride, the Lamb's wife.' ¹⁰ And he carried me away in the Spirit to a great and high mountain, and showed me the great city, the holy Jerusalem, descending out of heaven from God, ¹¹ having the glory of God. Her light was like a most precious stone, like a jasper stone, clear as crystal. ¹² Also she had a great and high wall with twelve gates [3x4 doors], and twelve angels at the gates, and names written on them, which are the names of the twelve tribes of the children of Israel: ¹³ three gates on the east, three gates on the north, three gates on the south, and three gates on the west" (Revelation 21:1-13 _{KJV Additions mine}).

"Look!" he answered, "I see FOUR men loose, walking in the midst of the fire; and they are not hurt, and the form of the FOURTH is like the Son of God" (Daniel 3:25).

If an individual or group's goal is to bring heaven to earth, there are two pieces to the mystery of making the kingdoms of this earth, the kingdoms of our God (Rev. 11:15). Connecting to the heavenly realms is one side of this divine puzzle. The other piece crucial to the restoration of all things is God's people fully developing the character of Christ to walk out on the earth the heavenly realities we see. Each of us needs to be made complete in growth and labor as well as emotionally, morally, and mentally. A person can see heavenly glory after heavenly glory after heavenly glory, but the world around you will only change if you can contain the heavenly glories (truths) you see and flow in them.

Know that His Reigning Ones will walk out heavenly glories in the same manner that Yeshua did (1 John 2:6). When we are in the fire and following the way that He has chosen for us to take, we have a glorious promise that we will come forth as gold (Job 23:10). Let's not waste an ounce of our trying experiences.

If your heart desires to operate after the Righteous Order of Melchizedek, there is a deeper reality to our trials. The Order of Melchizedek (with its High Priest *Yeshua Ha Machiach*—Jesus Christ) is a corporate entity. Therefore, if you are committed to a group of people who have entered the Kingdom of God, the fiery trials of the individuals within that group are part of that group being forged into One New Man made like unto the Son of God, which is a characteristic true of a godly Melchizedek Company (Dan. 3:25; Heb. 7:3).

This is illustrated in the story of Shadrach, Meshach, and Abednego. Due to the unified stance of these three men against an earthly king, they were thrown into the fire. Recall that the furnace was stoked up seven times hotter (Dan. 3:19). Know that pure gold is refined in God's fire seven times. The pure gold within this little unified group was being refined, so that the fourth man could manifest in the fire with them.

We have been repeatedly told that the fourth man is Yeshua (Jesus), which I believe is one beautiful facet of this kingdom reality. Christ in them the hope of glory in the midst of these three. But what if there is another facet to this kingdom reality? What if this story is also illustrating a group of individuals so united for His Kingdom purposes that they chose to be thrown in the fire together to fight with and for one another?

What if the unity in the fire creates a fourth man with the appearance of the Son of God, which is a One New Man that was made up of the composite spirit of all three? I have seen this happen. When I ascended with a group almost every week for four years, this unified One New Man in Christ appeared. Now when I am ascending with another group consistently, as we have committed to both God and one another for His Kingdom purposes, I see another Corporate One New Man in Christ manifest made up of the composite of the united spirits of the group. As we go through fiery trials together, our "fourth man" with an appearance like that of the Son of God gets more refined and more clear . . . golden, if you will. Please be encouraged that your fiery trials have both a personal and corporate purpose.

In a group, the good news is that when one falls down, the others are able to lift them up. When one is being hammered, the others can stand united to protect them. A primary thing that the Body of Christ needs to grasp in this hour is self-sacrificial unity. If we are willing to lay down our own lives for another, we have the same heart that Yeshua had for us (John 15:13). Then we can experientially know *"I with them and thou with Me, that they may become perfected in one; so that the world may know that Thou did send Me, and Thou did love them just as Thou did love Me"* (John 17:23).

DOOR OF LIFE

Know before whom you stand. Know (דַּ֚א) DA. Know that when you stand before the DALET, the door, you also stand before the Judge. (דַּיָּ֚ן) DAYAHN. The One who sees into your heart, the One who judges the judges of flesh and blood.[230] The Hebrew word for "judge" is dan (דַּ֚ן). The word picture tells us that a judge is the door of life. Yeshua is the judge who offers justice and redemption. Additionally, the Hebrew word for "Eden" (עדן) means to see (ע) the judge (דן), or eternal (עד) life (ן).[231]

The restoration of the priesthood of all believers is what God's Kingdom is all about and it is connected to Eden. Revelation 2:7 reveals that overcomers, who restore their first love, are given the right to eat of the Tree of Life. *"To the angel of the church of Ephesus write: These things says the Omnipotent One who holds the seven stars in his right hand, who walks in the midst of the seven candlesticks. I know your works and your labor and your patience and how you cannot endure those who are ungodly; you have tried those who say they are apostles and are not, and you have found them liars; And you have patience, and have borne burdens for My name's sake, and have not wearied. Nevertheless I have something against you, because you have left your first love. Remember therefore from whence you have fallen and repent and do the first works; or else I will remove your candlestick from its place unless you repent. But this you have in your favor, you hate the works of the Nicolaitanes, which I also hate. He who has ears, let him hear what the Spirit says to the churches: To him who overcomes, I will give to eat of the Tree of Life, which is in the midst of the paradise of My God" (Revelation 2:1-7 $_{Aramaic}$).*

Timekeeping on Earth commenced with the creation of the sun, moon, and stars, which is connected to our spacetime continuum. Space consists of three dimensions and time one dimension; therefore, spacetime is a four-dimensional continuum, which together specifies the location of a particle or event. On the third day of creation, God created land, sea, and a garden . . . an infinite fifth-dimensional continuum of a Garden east in Eden: *"Adonai, God, planted a garden toward the east, in Eden, and there he put the person whom he had formed. Out of the ground Adonai, God, caused to grow every tree pleasing in appearance and good for food, including the Tree of Life in the middle of the garden and the Tree of the Knowledge of Good and Evil" (Genesis 2:8-9 $_{CJB}$).*

On the fourth day of creation, God created the finite fourth-dimensional spacetime continuum of celestial bodies. The whole wide world is currently experiencing higher dimensional shifts, which is causing the shaking of everything that can be shaken (Heb. 12:28).

> Follow the living water to see our journey back to Eden and beyond, which leads to the gates (doors) of the New Jerusalem where the fourth man, like unto the Son of God, resides. It's time to be born again again. It's the dawn of a new day. It's time for the Sons of Light to arise. Step into the Deed of Eden. The land that time forgot. Step into the foundation of redemption and restoration. Step into the reset back to pristine Eden and beyond. Hear the song of creation. See the Hebrew Living Letters come out the Lion of the Tribe of Judah's mouth. Receive the bursting forth of new life. Receive the reset and knitting of the One New Man in the Messiah's DNA. Receive the seed of immortality. Receive the divine exchange. Let the glorious letters move in and out of your light body to repair your DNA. Hear the song of the Father's love. He's playing "Your Song."

QUANTUM QUARK (ד)

To summarize, DALET (ד) is the fourth Hebrew Living Letter whose picture is a door. ALEF-TAV (את) is the door (John 10:7,9). The number four fundamentally represents creation. Once an individual is united in fourfold unity with the Father, the Son, and the Holy Spirit, they can operate in the four faces of Melchizedek, or some say the four faces of God: lion, ox, eagle, and man (Ezek. 1:10; 10:14).

DALET (ד) is the door to the corporate Body of Christ quantumly flying by the Holy Spirit with one another. When a person is in fourfold unity with the Holy Trinity and joins with four or more other people in fourfold unity as a righteous wing of God's Melchizedek Army, they rise through the Revelation 4:1 Door, that is Yeshua, to stand at the Door of the East Gate where He returns to earth—your earth and mine (Matt. 24:27).

Know that the Spirit of the Living Creature is made up of four living creatures on multiple levels. Recall that the spirit of each of the four living creatures is in its wheels (Ezek. 1:19-21). Behold, we are a wheel within His wheel created to do fiery works according to His perfect will. The four living creatures' wheels within wheels have the appearance of beryl (Ezek. 1:16), which connects them to the *Merkabah* with its hexagonal prism shape. This means that the purpose of uniting in fourfold unity with God and with others is to become part of His fully mature Body in the form of uncreated light. This is where we follow the living water for our journey back to Eden and beyond, which leads to the gates (doors) of the New Jerusalem where the fourth man, like unto the Son of God, resides.

Q: What are the doors of the New Jerusalem and to whom are they connected? (Hint: Rev. 21:12-13).

Q: Why is ALEF-TAV (את) the DALET (ד) door to people becoming united in the Body of Christ? (Hint: Eph. 5:23 and Col. 1:18).

∽

ה

HEI (ה)—5
DIVINE BREATH, REVELATION & LIGHT

REDEMPTION OF HEI (ה)

The redemption of the Messiah's Quantum 22 continues with the fifth letter of the Hebrew Alphabet. HEI (ה) is pronounced *hey*. HEI (ה) has a numerical value of five. Its ancient pictograph is a priest with hands raised, which forms a window above their head. It's a window of revelation that communicates the ability to behold, to show, and to reveal.

Behold—HEI (ה)—the blueprint of the Messiah's Bride. The following Ascension in Christ is the facet of the redemption of HEI (ה) that our glorious Heavenly Father revealed to a small group of genuine seekers of His perfect heart and will (John 5:19,30). [232]

> We begin with communion: "With this bread, I thee wed. With this wine, I thee wed." We first see a little red wagon—a Radio Flyer—which we know has to do with frequency. We understand we are going to a higher frequency. There is a great expanse of water. It's an ocean full of fish. From a different perspective, we experience going up at an extreme speed in some sort of rocket ship. We see water and lights all over the earth, coming from people. The lights look like fireworks going up, similar to beams of light.
>
> We behold that the fish are full of energy. Some of them are jumping up out of the water and flying. They are rainbow-colored fish. *Psalms 107:23—"They that go down to the sea in ships that do business in great waters."* There is something in all of this about trading and doing it in a holy way. Trading in the heavens, not just business. The ocean's salt is alive. It is singing. It's a preservative. Saline is what people are made of. We behold whirls of gold, like pools of gold in the blue. It's the gold element. The elements are us. The gold is us, swirling in the blue in the sea—the sea of humanity. We are amongst the people. The fish are eating the gold—the manna. We see gold inside the fish. Yeshua pulled the gold coin out of the fish for provision. We hear: "We are treasures deposited in the Sea of Humanity." We are fishers of men in a new way. There's a net going into the water but we don't see a man's hands or a boat. Men will not have their hands on catching these fish—this blueprint. They will be supernaturally caught. We see words etched on the scales of the fish. Words that the fish want us to see. The fish are turning their eyes toward the words on their sides. They want to see. They want to be . . . supernatural . . . greater revelation . . . The words of the Lord become evident to them, and this is what they want to see. Interestingly, the words are not coming out of the mouth of the fish. It's on their sides. It's manifesting by BEing the Word. We see words of restoration written on the sides of the fish and the fish are turning around to see what the Lord has for them.

Now we experience going way up in a rocket ship. We look down and see lights. These lights are the reflection of certain people, their restoration and gold. The LORD says: "Monatomic gold." Monatomic gold was the substance of Eden. Everything was created alive in that gold. It was even a part of our blood. The restoration going on is taking us back into the creative light that will draw all men. It's taking us back to the original of who we were created to be before the Fall. We hear: "Taking us back to the original blueprint." Yes. That's right—blood and gold. Yes. The salt is cell salt, and the gold is "mannatomic." Every time we see or hear "gold," we hear "Melchizedek." The fish are eating manna, which is the mannatomic gold. Monatomic gold is its purest form. We hear the first phrase of Creation: "Let there be light!"

There are scrolls above the water, rolled up with different color ribbons tied around them. The first scroll has a red ribbon. There are many scrolls above the fish. There are whirls that are high and spinning. We behold fish coming out of the whirls, but these fish are not fish anymore. The fish are coming out as a new creation, as birds. This speaks of the FIFTH DAY OF CREATION. As the fish come into the whirl, they appear to be fish. As they come out of the whirl, they appear to be birds. We understand that He is giving us different days of creation.

We are instructed to stand in the whirls. We see the earlier transformation gold. Initially, we were unsure what the gold was forming, as it came together. Now, we see various sea creatures coming out of the whirls: whales, dolphins, sting rays, and all sorts of different kinds of fish. Both sea creatures and the birds come out at the same time. It seems like the restoration of creation. So far, God has shown us the FIRST AND FIFTH DAYS OF CREATION. [The first and fifth days are also connected to the first and fifth Spirits of the Lord in Isaiah 11:2—the Spirit of the Lord and the Spirit of Might.] We see the City of Peace hovering over the sea. It's built of the Book of Light and all the creatures are going towards it.

The gold is changing into an ornate scroll. Earlier when we saw the gold transforming, we saw one specific pool of gold underneath the water that was being transformed into a building. This was not just any building. It appeared to be a small-scale model of a grand palace. We understand that the gold transformed to make a Bridal Palace. We hear the word: "Governmental." We hear: "Look for the Bridal Chamber within." The Bride of Christ is always governmental. She is the Queen—the True Queen of Heaven—to the King of Kings.

We walk up the steps, and don't have to knock. We just go through the door. The door is completely gold with scrollwork around the edge of it. There's a unique frequency to the door. We hear: "As you step into the palace, your frequencies will be one with Me." We also hear: "This is a new welcome for the Bride." As soon as you step in, you will be transformed. A new way of becoming ONE with the Bridegroom. We behold that we are on the step in front of the door. We are instructed to choose to become one through our will in our inner being. We behold a coil of octahedrons of all colors that have wings that raise their voice in praise, as the coil moves. It's very big with wings all over. "Octa" means 8; therefore, it's an eight-sided coil of all colors. The coil with wings is a crown. Behold, it's a living crown. We are instructed to step into one another to become one. We form a beautiful Bride with a white bridal gown adorned beautifully. As we stand right in front of the big door, we step into each other to form a taller bride. The Bride of Christ looks like none of us, yet she looks like all of us. She is wearing the most beautiful expensive dress with lots of jewelry on her. We see diamonds with various different-sized gems on the bridal gown, with sapphires up and down our arms. It sounds overdone, but it does not look that way. We hear: "Extravagant and lavish. Spare no expense for My beautiful Bride."

As the Bride of One walks, there are all types of stones (not just gemstones) falling all around us. We have a clear path. There are all types of these stones on the side of the path, as we walk. Natural stones are from both sea and land: onyx and jasper, et cetera. They are falling around us and making a way (kind of like a headway), but they are not getting in our way. There is gold in the gown. There is gold in the swirls and the gold appears to be alive. We behold that we are carrying a bouquet of people and gemstones are decorating the bridal pathway instead of flower petals. We hear: "All creation comes to see the Bride." His One Bride is carrying the people to the King. This reminds us of the Story of Esther. His Bride is so beautiful and humble.

There are gems floating all around us, waiting for us to receive the gift. The gift is a miter, which is a two-sided crown. One side is the servant, and the other side is kingship. When you go in to take the land, you turn it around to the servant's side, because no one is threatened by the servant. Once you go in and take the land, then you turn it around and you sit in that land and serve as a king. So, on the Bride of Christ's head is a crown called a miter, which is the first gift. We behold the Bride bowing down before the King. It seems like her coronation with the crown being placed on her head by the King Himself. We hear: "Coming to the fullness of the inheritance." It's a new way of being or ministering.

The King hands this Bride His royal signet. In the Story of Esther, the King handed Esther His royal signet which basically declared that she can write the decree and use his ring and it will be as if he decreed it. "All My authority is yours." We see Yeshua hand the Bride a world globe and the scepter of a servant king. The Living Crown is used as His signet ring, as His signature. Each member of the multimembered Bride will use this crown to establish His authority, as the signet ring of the King. The signet ring is placed on the Bride's finger, and a royal blue sash is placed over both of their hands. It is how the priests would marry a couple, binding them together. Our hands changed from a blue light to a white; and then, the whole color of the Bride seemed to just emanate a beautiful color. It's like a light went on.

We behold a chalice filled with red wine, but it looks oily. It is oil and wine mixed together. When the Bride peers into the chalice, she sees a reflection of herself but it's the reflection of the King. When the Bride sees the King, she sees the Lion of the four faces of God. The Lion represents the kingly function: rulership, protection, provision, establishing boundaries, and things like that. The King has a crown that is very big, heavy, gold and ornate. We are assured that it's the same crown as the Bride in a way, but it's different. We sense that the King's crown is a miniature city, like the city we walked into—the bridal palace.

We are instructed to drink the oily wine. By faith, we hold up our hands like we are drinking water out of a brook with our hands. We take it in. When the oily wine runs down our throat into our body, we first appear to be solid; and then, all these colors start to come out of us and around us. The colors are vibrant rainbow colors. They look electric. The colors are powerful and vibrating. These rainbow colors are swirling at extremely high frequencies. It appears like the colors are living. This is living color at a whole new level. This is created light. It's the octahedron rainbow-colored thing with wings. It is coming alive in the Bride. We feel radiating power and see a chamber to the right side filled with gold, purple and blue with an angel standing at the entrance like he's a servant.

We get the phrase "Inception of world within worlds." Earlier in the water when we saw the gold transforming into a palace and we stepped into it, it brought the waters into a different world. We see a city in the crown of the King, which reminds us of the movie "Inception" where there is a dream within a dream within a dream. Our earthly paradigms don't work here. We look closer at that crown. There are other worlds in His crown. We keep getting the word "inception." This is creative. It has to do with what He is birthing in the Bride as far as creation. We look closer at the worlds within worlds. What moves Him is every part of His character. God is saying that He is not multi-anything, I am OMNI-everything. We are going to start taking on more of the unlimited character of the Father, moving in many more dimensions. The inception of the world and worlds merging and becoming another world, which is a new inception.

We see a circuit board, but what's happening in it is like that multi-colored light that is so radiant and bright. All of it is flashing around us: the circuit board, the electric light with electric speed, and multi-colors. We hear: "New... New... New... New... New... New..." The word "new" just doesn't stop. It's a new circuit board. It's something so radically new that it's coming from a heavenly perspective (we can't even contemplate it from an earthly perspective). It's new mind-blowing connections. This is what we are birthing here. We are called to bring forth a new world that illuminates.

We behold the earth again. There are flashes of light, like little zips of energy. The flashes of light are going all over the world at lightning speed. The flashes of lights connect frequency, energy, and light all at the same time. On top of the crown, we see worlds—little worlds—at the peaks of the crown. In other words, on top of each of those peaks is a world. Light and electricity is touching the gold. The gold is a conduit for electricity that's a force connecting these worlds and dimensions. There are explosions of light that are connecting all these different worlds (or dimensions). Something is forming in the center of the crown. There's a tetrahedron, or some kind of cymatics energy pattern.

We know it's the new circuit board. Circuit boards allow circuit components to transfer information and signals between each other as well as information to external devices. Other devices could be worlds in this case. A circuit board is a conduit of communication or a transfer of information. So, as we are looking in, as the Bride, at our reflection and we see Him and the crown, we are seeing the power that lies within us to affect these worlds or galaxies or whatever He leads. We behold a DNA Strand—the eight-sided figure with wings—circling around all this. We hear: "This is hackproof." There is heat coming off our bodies. We are vibrating. We see our belly area looking like universes and skies. It's multi-dimensional. We see a finger that looks like it was dipping in water. Our stomachs start to look like blue water, and waves are just coming out... waves and waves... BEHOLD, Out of your bellies will flow rivers of living water (John 7:38).

ANCIENT HEI (ה)

HEI (ה) is connected to faith, grace, and mercy because these words have five letters in all languages. Recall that the pictograph for the Hebrew Living Letter HEI (ה) is a person with their hands lifted up, or an open window above one's head.

> *"Bring all the tithes into the storehouse, That there may be food in My house, And try Me now in this," Says the Lord of hosts, "If I will not open for you the WINDOWS of heaven and pour out for you such blessing that there will not be room enough to receive it" (Malachi 3:10).*

> *"O God, You are my God; Early will I seek You; My soul thirsts for You; My flesh longs for You In a dry and thirsty land where there is no water. So I have looked for You in the sanctuary, To see Your power and Your glory. Because Your lovingkindness is better than life, my lips shall praise You. Thus I will bless You while I live; I will LIFT UP MY HANDS in Your Name" (Psalms 63:1-4).*

Since the Hebrew Living Letter HEI numerically represents the number 5, it can be connected to the fifth day of Creation. *"Then God said, 'Let the waters abound with an abundance of living creatures, and let birds fly above the earth across the face of the firmament of the heavens.'*

So God created great sea creatures and every living thing that moves, with which the waters abounded, according to their kind, and every winged bird according to its kind. And God saw that it was good. And God blessed them, saying, 'Be fruitful and multiply, and fill the waters in the seas, and let birds multiply on the earth.' So the evening and the morning were the FIFTH DAY" (Genesis 1:20-23).

When two are married, a Hebrew couple looks at each other and whisper: HARAY AHT LIKU-DESHET LI. *"Behold, I will try with all my being to be present for you."* And so the (ה) of (הוא) HOO which means "he" and the (ה) of (היא) HEE which means "she" become the (ה) of (הם) HAYM which means "them."[233]

"The sound of (ה) is a mere exhalation of breath." [234] Elijah found out that the barely heard (ה) alludes to God. Elijah experienced a strong wind that rent the mountains and shattered the rocks. He experienced an earthquake and a fire, but the Lord was not present in any of these violent phenomena. Then *"after the fire was a soundless whisper, and in it appeared God" (1 Kings 19:12).*

Did you know that Hebrew history records that the letter HEI (ה) was attached to the Ark of His Presence? The Holy Ark of the Covenant was enveloped in a ה-shaped cloud—when it preceded the Israelites during their wilderness travels. The ה-shaped cloud was a visible sign of God's special guidance and presence. [235] As it is written, *"and HASHEM would go before them by day in a pillar of cloud to lead them on the way" (Exodus 13:21).*

Around a thousand years later, Nehemiah stood in Jerusalem inspiring the Israelites to repent. To underline God's incredible and continuous care for His people in the wilderness, Nehemiah paraphrased Exodus 13:21 by adding a HEI (ה), which emphasized that it was God who led them (Midrash Aggadah 2:79).[236] *" 17 They refused to obey, And they were not mindful of Your wonders That You did among them. But they hardened their necks, And in their rebellion they appointed a leader to return to their bondage. But You are God, Ready to pardon, Gracious and merciful, Slow to anger, Abundant in kindness, And did not forsake them. 18 Even when they made a molded calf for themselves, And said, 'This is your god That brought you up out of Egypt,' And worked great provocations, 19 Yet in Your manifold mercies You did not forsake them in the wilderness. The pillar of the cloud did not depart from them by day, To lead them on the road; Nor the pillar of fire by night, To show them light, And the way they should go" (Nehemiah 9:17-19).*

"Since the beginning of mankind, the message of ה, with its symbolic opening for penitent sinners, has become an inspiration for man to repent. God is always ready to accept the penitent whenever he is ready to return to Him."[237]

Since this World was created with HEI ה, it stands for freedom of choice. We have the freedom to obey or disobey God's will. If we choose to leave the safety of God's Biblical boundaries, we lose our foothold and slip - fall. [238] However, HEI (ה) symbolizes God's readiness to forgive those who seek forgiveness. God created the opportunity to return to Him through (תשובה) *teshuvah*—repentance. The word תשובה can be read in two parts תשוב ה—return to ה— referring to the last sacred letter of YHVH.

The Bible says that *teshuvah* was created before the world itself: *"Before the mountains were born, or You brought the world into being, You were the Eternal God who says—return" (Psalms 90:2-3).* The Lamb slain before the foundation of the world clothed humanity with

His kindness (1 Pet. 1:19-20). The Creator created the world with His Hebrew Living Letters which are the essence of the Word of God as well as with His blood that speaks of better things (Heb. 12:24): I love you. Return to all of Me. "*³ And do you think this, O man, you who judge those practicing such things, and doing the same, that you will escape the judgment of God? ⁴ Or do you despise the riches of His goodness, forbearance, and longsuffering, not knowing that the goodness of God leads you to repentance?" (Romans 2:3-4).*

"*¹ Come, and let us return to the Lord; For He has torn, but He will heal us; He has stricken, but He will bind us up. ² After two days He will revive us; On the third day He will raise us up, That we may live in His sight. ³ Let us know, Let us pursue the knowledge of the Lord. His going forth is established as the morning; He will come to us like the rain, Like the latter and former rain to the earth*" (Hosea 6:1-3).

COMPOUND LETTER (ה)

When we look at HEI being composed of the letters VAV + DALET, it has a combined final *mispar katan* of 1—unity of YHVH. Alternatively, HEI can be composed from YUD + DALET, both the sum of these letters and their *mispar katan* are equivalent to 5.[239]

Let us take a closer look at the quantum letter HEI (ה) being composed of the letter DALET (ד) and an inverted YUD (י). The word DALET fulfills as (דלת), which represents the opening of a "doorway," like a mother's womb. When we add an inverted YUD to DALET (ד) to form the Hebrew letter HEI (ה), it can represent a developing fetus in a head-down position prior to birth.[240]

The HEI (ה) denoting Creation is also found in Genesis 2:4— "*These are the products of the heaven and the earth when they were created.*" The Hebrew word (בְּהִבָּרְאָם) for *"when they were created,"* can be divided into two words: (בְּהּ) (בְּרָאָם), which can be translated as: *"He created them with the letter HEI."* (Osios R' Akiva).

Michael L. Munk shares: "Thus, God did not create This World as a purely physical entity, but also imbued it with the Divine. Maharal observes that the letter (ה) consists of a ד and a י. The vertical and horizontal line of the ד represent the physical world that is measured in its expanding width and height, whereas (י) denotes the World to Come, spiritually."[241]

Thus, the HEI (ה) teaches us to imbue [inspire or permeate with a feeling or quality] our lives with sanctity; to always combine the physical with the spiritual.

FULFILLMENT OF HEI (ה)

The Hebrew Living Letter HEI fulfills in one of four different ways HEI-ALEF (הא), HEI-HEI (הה), HEI-YUD (הי), and HEI-YUD-ALEF (היא), which translate as "behold" or "to take seed" or "to be broken"[242]

It's interesting that in the Wilderness Tabernacle, there is a curtain that marks the boundaries of God's Sanctuary. Yer-ee-aw is the Hebrew word for "curtain," and it speaks of a tremulous (trembling) hanging or a hanging of the fear of the Lord. Of note, the Ancient Hebrew Word Picture for the word "curtain—Yer-ee-aw—communicates that the curtain of God's Dwelling Place represents the highest and most important work to be done is the purification work in a person's soul. When we dig a little deeper, we see that the primitive root to *yer-ee-aw* is the Hebrew word *yaw-rah'*, which means to be broken up (with any violent action). Behold [HEY!], the highest and most important work in the purification of one's soul requires a person to be broken.

QUANTUM 5

Atomic Number 5—Boron (B). Quantumly speaking, boron is a semimetal chemical element that has the atomic number **5**, the atomic weight **10.81**, and the atomic symbol **B**. Boron has 5 protons and 5 neutrons with 5 electrons in its outer shell. "The name boron comes from the Arabic word 'buraq' and the Persian word 'Burah,' which means borax."[243]

In its impure amorphous form, boron is a brown powder, which was its only known form for over a century. [244] In the form of boric acid or borates, traces of boron are necessary for the growth of many land plants and thus are indirectly essential for animal life. Typical effects of long-term boron deficiency are stunted, misshapen growth; vegetable "brown heart" and sugar beet "dry rot" are among the disorders due to boron deficiency. Boron deficiency can be alleviated by the application of soluble borates to the soil."[245] Soils naturally abundant in boron can cause gigantism (excessive growth in stature) in several species of plants. [246]

In addition to traces of boric acid being necessary for plant life, boric acid is used as an antiseptic, for electroplating nickel, tanning leather, and manufacturing heat-resistant glass. [247]

"Pure crystalline boron is a black, lustrous semiconductor (i.e., it conducts electricity like a metal at high temperatures) and is almost an insulator at low temperatures. In the semiconductor industry, small, carefully controlled amounts of boron are added as a doping agent to silicon and germanium to modify electrical conductivity."[248]

Crystalline boron is extremely hard on the Mohs scale (about 9.5), and a poor electrical conductor at room temperature. [249] Therefore, boron is used industrially in alloys as a hardener. Limited quantities of elemental boron are widely used to increase hardness in low-carbon steel. [250]"Boron is most found as borax ($Na_2B_4O_5(OH)_4 \cdot 8H_2O$) or boric acid. There are approximately 100 borate minerals known. Boron is not found in an uncombined form on Earth but is occasionally found as a metalloid in meteoroids. [251]

Boron is used in many different applications. When woven as a fiber, it is used in aerospace to create lightweight composite structures. These woven fibers have also been used to create high-end consumer shaftings, such as lightweight fishing rods and golf club shafting. [252]

Borosilicate glass is another use of boron. This type of glass has a low coefficient of expansion and is used for telescope lenses and mirrors. It has good resistance to thermal shock and is used for labware and cookware. In the United States, it is trademarked as PYREX™."[253]

5 IN SCRIPTURE

HEI (ה) is 5. There are five books in the Torah: Genesis (*Bereishit*—"In the Beginning"), Exodus (*Shemot*—"Names"), Leviticus (*Vayikra*—"And God Called"), Numbers (*Bamidbar*—"In the Wilderness"), and Deuteronomy (*Devarim*—"Words"). The concept of the Hebrew Torah is much broader than the books themselves. Torah can refer to all learning, but the term "The Torah" usually refers to the written Torah—*she'bi'ktav*—also known as the *chumash* (the five volumes) or the Pentateuch—the Five Books of Moses.

The name Torah means "the teachings." The object of Torah is to know God. It is your attitude towards God, which communicates what "commandments" means to you. A person with a disobedient heart feels commandments are burdensome while obedient hearts feel that they are instructions. *"As the Father loved Me, I also have loved you; abide in My love. If you keep My commandments, you will abide in My love, just as I have kept My Father's commandments and abide in His love." (John 15:9)*.

HEI (ה) is 5. Since the word "light" is mentioned five times on the first day of creation in Scripture, we can understand that the sacred letter HEI (ה) represents divine breath, revelation, and light. *"³ Then God said, 'Let there be LIGHT'; and there was LIGHT. ⁴ And God saw the LIGHT, that it was good; and God divided the LIGHT from the darkness. ⁵ God called the LIGHT Day, and the darkness He called Night. So the evening and the morning were the first day"* (Genesis 1:3-4).

"¹ In the beginning was the Word, and the Word was with God, and the Word was God. ² He was in the beginning with God. ³ All things were made through Him, and without Him nothing was made that was made. ⁴ In Him was life, and the life was the LIGHT of men. ⁵ And the LIGHT shines in the darkness, and the darkness did not comprehend it" (John 1:1-5).

HEI (ה) is 5. The Second Temple lacked five features of the First Temple:

[1] The Ark of the Covenant with its covering cherubs.

[2] The constantly burning heavenly fire on the Brazen Altar.

[3] The *Shekinah* Glory—God's Dwelling Presence.

[4] The Divine Spirit

[5] The Breastplate of the *Kohan Gadol* (High Priest).

The prophet Haggai appealed to the people to begin building the Second Temple after their return from the Babylonian exile: *"'Bring wood and build the house and I will find satisfaction in it, and I will be glorified,' said HASHEM" (Haggai 1:8)*. Notice that, in Hebrew, there was a missing HEI (ה) in the phrase *"I will be glorified"* indicating its five missing features. Where was the missing HEI (ה) in the Second Temple? *"And looking at Jesus as He walked, he said, 'BEHOLD the Lamb of God!'" (John 1:36)*. *"³ And there shall be no more curse, but the throne of

God and of the Lamb shall be in it, and His servants shall serve Him. *⁴ They shall see His face, and His name shall be on their foreheads. ⁵ There shall be no night there: They need no lamp nor light of the sun, for the Lord God gives them light. And they shall reign forever and ever"* (Revelation 22:3-5).

QUANTUM QUARK (ה)

In summary, HEI (ה) is the fifth Hebrew Living Letter whose pictograph is a priest with their hands lifted up, or we could say it is a picture of an open window above a person's head. The pictograph for HEI (ה) symbolizes to behold, to show, or to reveal. Elijah found out that the barely heard ה alludes to God (1 Kings 19:12). During the Israelites' travels in the wilderness, there was a ה-shaped cloud that enveloped the Holy Ark of the Covenant when it preceded them, which was a visible sign of God's special guidance and presence (Exo. 13:21). There is a double portion of HEIs (ה) in YHVH (יהוה). The four-letter name YHVH is different because it is more of a personal name than an adjective. The most common name for God in scripture is YHVH (יהוה), which derives from the verb להיות, which means "to be." When Moses encountered God at the burning bush, Moses asked: "What is your name?" Moses was told: אהיה אשר אהיה— *ehyeh asher ehyeh "I will be what I will be,"* or the more modern translation *"I am that I am"* (Exodus 3:14).

Q: Have you ever noticed any special sign of God's presence or guidance in your life?

Q. What are your thoughts about YHVH's connection to a state of being . . . I am that I am?

ו

VAV (ו)—6
POWER TO CONNECT HEAVEN & EARTH

REDEMPTION OF VAV (ו)

The redemption of the Messiah's Quantum 22 continues with the sixth letter of the Hebrew Alphabet. VAV (ו) is pronounced *vahv* and has a numerical value of six. Since VAV (ו) is 6, let's consider the six-pointed star, which is also called the Star of David. The Magen David (מָגֵן דָּוִד) signifies God as the protector of David—The Shield of David. It is also called the Star of Creation that represents the six days of Creation, and/or the *Merkabah* (מרכבה) that identifies Ezekiel 1 as God's Chariot Throne.

The fundamental geometry of the first photon of light manifesting through God's *"Let there be light!"* statement is a *Merkabah* (מרכבה). It is not a coincidence that a *Merkabah* portrays ascending and descending. Geometry is the visual manifestation of vibration, which is seen through cymatics these days. Top scientists in today's world believe that space is "nothing." They call it a vacuum. They say it is structureless, yet it is laden with hyper-dimensional geometries.

Two scientists—Nima Arkani-Hmed and Juroslav Tranka from Cal Tech—discovered that all space and time are the emanation of a single geometric form. They just don't know what it is yet. They call this shape the "amplituhedron" or the "positive grassmanian."[254] It looks like four tetrahedrons stuck together. This Amplituhedron appears to simply be part of the *Merkabah*—technically it is two *Mer-ka-bahs*. All photons (energy, light) are all tetrahedral *Merkabahs*. That means that all light is an emanation of a single geometric form—the *Merkabah*—which consists of the two tetrahedrons (one pointed up and the other pointed down), which illustrates ascending and descending.

In Hebrew, *"Mer"* means light, *"Ka"* means spirit, and *"Bah"* means body. The Biblical *Merkabah* Process is the key to the complete transformation of our bodies into light. The *Merkabah* is all about changing your Kingdom of God within through the work of Yeshua; and then, manifesting it outward.

The *Merkabah* is featured in the "Melchizedek Section" of the *Dead Sea Scrolls* where it is revealed through fragments of the "Songs of the Sabbath Sacrifice." There was a transcendent quality of angelic praise between the earthly community of Israel with its priesthood and their heavenly counterparts. As the community on earth was led through the 13-week progressive, ecstatic experience of the highest praise, the Sabbath Songs were given spirit until the worshipers experienced the heavenly holiness of the *Merkabah* (i.e., God's Chariot Throne).[255]

Thirteen weeks... thirteen circles... The thirteen circles of Metatron's Cube are the fundamental blueprint for all atomic structures. From this matrix of 13 spheres, all platonic solids can be created, which are the basic geometries of life at all levels of reality. These are the five Platonic solids that can be created from Metatron's Cube:

[1]   Tetrahedron (4-faces)
[2]   Cube (6-faces)
[3]   Octahedron (8-faces)
[4]   Icosahedron (12-faces)
[5]   Octahedron (20-faces)

A friend once told me that the Star of David (the *Merkabah*) was demonic because it can be found in all the religions of the world. So, I laid it down on the Lord's Altar for months and asked Him what He thought of it. The Lord showed me that its origin is of Him, which means that it will be completely redeemed. It's foundational to all life and light, and everyone has been designed by our Creator to come back to this single point of origin in Christ.

Let's come back to VAV (ו) is 6. A hexagon is six. Consider when water is frozen, its molecules crystallize in a hexagonal pattern, just like Metatron's Cube. Perhaps, a six-pointed star snowflake is a clue to the geometric/frequency structure underlying the thing it is freezing —our reality of three-dimensional space.

Just like snowflakes, there is a hexagonal shape to the bee's honeycomb cells. Scientists now accept that bees build cylindrical cells that later transform into hexagonal prisms through a process that is still debated. Earlier hypotheses solely gave credit to the geometry skills of bees, but now the action of physical forces has been added. The geometry of honeycomb cells can only arise when each one of the bee's cells is tightly packed and surrounded by six similar same-sized cells.[256]

Hexagons possess the highest surface/perimeter ratio, compared to other polygons, for tiling a plane. Therefore, honeybees build their hexagonal cells with nature's assistance to achieve the most efficient use of material. Perhaps, the honeybee's six tiny legs have something to do with the hexagonal form of a honeycomb. Rasmus Bartholin suggested that hexagons would result automatically from the pressure of each bee trying to enlarge each cell as much as possible. Just know when modeling the effect of surface tension at a triple junction where there are three 120-degree angles between wax walls, the honeycomb architecture is the result.[257]

> Behold, a dwelling place for the LORD is found in the honeycomb. There is one big honey world—a beehive—connected to His True Queen of Heaven. She is His Sweet One who will resound with His renown. The sweetness of her honey sound melts us into one... melts us into Him. She literally releases sticky bombs, so Christ's Body can be built at the cellular level. Taste and see that the LORD is good!!!

"I have come into my garden, my sister, My bride; I have gathered my myrrh along with my balsam. I have eaten my honeycomb and my honey; I have drunk my wine and my milk. Eat, friends; Drink and imbibe deeply, O lovers. I was asleep but my heart was awake. A voice! My beloved was knocking: 'Open to me, my sister, my darling, My dove, My perfect one! For My head is drenched with dew, My locks with the damp of the night'" (Songs 5:1-2 _{NASB}).

"I will not give sleep to my eyes or slumber to my eyelids, until I find a place for the LORD, a dwelling place for the Mighty One of Jacob. Behold, we heard of it in Ephrathah, we found it in the field of Jaar" (Psalms 132:4-6 _{NASB}). The root of the Hebrew word *Ya' ar* (Jaar) contains the concept of honey in the comb (as hived in the Tree of Life) as well as simply honey and the honeycomb. The hexagonal wax cells built by honeybees are their dwelling place. It contains their seeds, family, fruit, and sustenance. When I looked up the word "honeycomb" in my *Merriam-Webster's Collegiate Dictionary (10th Edition),* I was struck by a couple definitions: "to penetrate into every part : FILL : to become cellular."[258]

The honeycomb is a picture of Christ's Body being built at the cellular level. It is the place where we grow up in all aspects into Him, being fitted and held together by what every joint supplies, which causes us to STICK to Him and causes us to grow in love as one: *"in whom the whole building, being fitted together, is growing into a holy temple in the Lord, in whom you also are being built together into a dwelling of God in the Spirit" (Ephesians 2:21-22 _{NASB}).*

"But speaking the truth in love, we are to grow up in all aspect into Him who is the head, even Christ, from whom the whole body, being fitted and held together by what every joint supplies, according to the proper working of each individual part, causes the growth of the body for the building up of itself in love" (Ephesians 4:13-16 _{NASB}).

Speaking of growing up in all aspects into Christ, consider that VAV (ו) denotes physical completion. Our physical world was created and completed in six days. Also, of note, a completely self-contained object has six dimensions: above, below, right, left, before, and behind.

The VAV (ו) in the name of Elijah is a symbol of the complete inner harmony that will be once more in the Messianic Era. The Spirit of Elijah will herald the coming of the Messiah who will establish peace, as it is written: *"He will return the heart of parent to their children and the heart of children to their parents."* [259] Recall *"Jesus answered and said to them, 'Indeed, Elijah is coming first and will RESTORE ALL THINGS'" (Matthew 17:11).*

The primal roots of anything reveals what can be redeemed as righteous, or what will be judged and done away with. A primary function of the kings and priests of the Righteous Order of Melchizedek is to restore all things by bringing heaven to earth. To do this, we must operate in the Spirit of Elijah. The Bride of the Messiah operates in the Spirit of Elijah. This is the time when the Lord is leading His people to restore all things; therefore, we will revisit any unholy practice and see if its primal roots are righteous or not.

It is not a coincidence that honey and the honeycomb are connected to the righteous judgments of God: *"The fear of the LORD is clean, enduring forever: the judgments of the LORD are true and righteous altogether. More to be desired are they than gold, yea, than much fine gold:*

sweeter also than honey and the honeycomb" (Psalms 19:9-10). And this honey of His righteous judgments that is more desired than fine gold comes from the Rock of Israel: *"So the LORD alone led him, and there was no foreign god with him. He made him ride in the heights of the earth, that he might eat the produce of the fields; He made him draw honey from the rock, and oil from the flinty rock"* (Deuteronomy 32:12-13).

It's time for the Melchizedek Army with sweet bridal hearts of One Love to arise! It's time to hold our Honey—Bridegroom—as solely sacred and most dear. This is all about the redemption of the Quantum 22—the twenty-two letters from ALEF (א) to TAV (ת). The essence(s) of the Messiah that constitutes every living substance. These twenty-two Hebrew letters are the protoplasm of God's Universe. They are the individual spiritual forces that originally belonged, and still belong, to the nature of Christ.

"He has delivered us from the power of darkness and conveyed us into the kingdom of the Son of His love, in whom we have redemption through His Blood, the forgiveness of sins. He is the image of the invisible God, the firstborn over all creation. For by Him all things were created that are in heaven and that are on earth, visible and invisible, whether thrones or dominions or principalities or powers. All things were created through Him and for Him. And He is before all things, and in Him all things consist" (Colossians 1:13-16).

ANCIENT VAV (ו)

The pictograph for the ancient Hebrew Living Letter VAV (ו) is a nail, a peg, or a hook. Each Hebrew word has a primary root. The sacred letter VAV (ו) is unique in the Hebrew language in that there is only one Hebrew word whose root begins with this letter. This word is (וו)—hooks—as in the Wilderness Tabernacle's hooks on the pillars (Exo. 38:10). The Hebrew sage Radak explains: "These were pegs that protruded from the pillars, in the shape of a letter ו; they were used to hang the carcasses of the sacrifices when they were being skinned." (Sefer HaShorashim).

One of the mysteries of the three nails, which were used to crucify Yeshua, is the redemption of "666;" because VAV is a picture of a nail and its numerical value is six. We can literally see the picture on the Cross that He redeemed the three nails: VAV-VAV-VAV. Additionally, the mystery of the pictures of the first Hebrew letters of the words on the sign that hung on the Cross above Messiah Yeshua proclaimed: Behold, salvation comes by the nailed hand. *"19 Now Pilate wrote a title and put it on the cross. And the writing was: JESUS OF NAZARETH, THE KING OF THE JEWS. 20 Then many of the Jews read this title, for the place where Jesus was crucified was near the city; and it was written in Hebrew, Greek, and Latin. 21 Therefore the chief priests of the Jews said to Pilate, 'Do not write, "The King of the Jews," but, 'He said, "I am the King of the Jews."' 22 Pilate answered, 'What I have written, I have written'"* (John 19:19-22). The pictograph for VAV (ו) symbolizes joining together, making secure, or becoming bound (nailed).

Since the Hebrew Living Letter VAV numerically represents the number 6, it is connected to the sixth day of Creation. "²⁴ *Then God said, 'Let the earth bring forth the living creature according to its kind: cattle and creeping thing and beast of the earth, each according to its kind'; and it was so. ²⁵ And God made the beast of the earth according to its kind, cattle according to its kind, and everything that creeps on the earth according to its kind. And God saw that it was good. ²⁶ Then God said, 'Let Us make man in Our image, according to Our likeness; let them have dominion over the fish of the sea, over the birds of the air, and over the cattle, over all the earth and over every creeping thing that creeps on the earth.' ²⁷ So God created man in His own image; in the image of God He created him; male and female He created them. ²⁸ Then God blessed them, and God said to them, 'Be fruitful and multiply; fill the earth and subdue it; have dominion over the fish of the sea, over the birds of the air, and over every living thing that moves on the earth.' ²⁹ And God said, 'See, I have given you every herb that yields seed which is on the face of all the earth, and every tree whose fruit yields seed; to you it shall be for food. ³⁰ Also, to every beast of the earth, to every bird of the air, and to everything that creeps on the earth, in which there is life, I have given every green herb for food'; and it was so. ³¹ Then God saw everything that He had made, and indeed it was very good. So the evening and the morning were the SIXTH DAY*" (Genesis 1:24-31). It is significant that God created all things in our material realm in six days.

SIMPLE LETTER (ו)

VAV (ו) is one of seven simple letters, which are used to form the remaining Hebrew Living Letters. There are a total of seven letters in the Hebrew Alef-Bet that use the letter VAV (ו) in their composition—ALEF (א), BET (ב), HEI (ה), CHET (ח), LAMED (ל), MEM (מ), and SAMECH (ס). Six of these quantum letters always use the letter VAV (ו) in their composition—ALEF (א), BET (ב), CHET (ח), LAMED (ל), MEM (מ), and SAMECH (ס)—while HEI (ה) may be formed with or without a VAV (י+7) or (ו+7).

FULFILLMENT OF VAV (ו)

The word "VAV" has three alternate Hebrew spellings. VAV's fulfillments are either VAV-VAV (וו), VAV-ALEF-VAV (ואו), and VAV-YUD-VAV (ויו). They are all palindromes that read the same right to left as left to right.²⁶⁰ The fulfillment of VAV-VAV (וו) has a numerical value of 12 while VAV-ALEF-VAV (ואו) has a numerical value of 13.²⁶¹

Mathematically, these fulfillments of VAV represent the connective and conversive properties of VAV (ו) in the following ways:²⁶²

$$12 \times 12 = 144 \qquad 13 \times 13 = 169$$
$$21 \times 21 = 441 \qquad 31 \times 31 = 961$$

Recall that *mispar katan* method for interpreting sums the digits of the Hebrew letters; and then, drops all the zeros for the numerical value of each letter so that only a basic integer from one (1) to nine (9) remains. When we look at VAV (ו) being composed of the letters VAV +

VAV, its final *mispar katan* is 3—*echad* trinity in one. Alternatively, VAV (ו) can be composed from VAV + ALEF + VAV whose *mispar katan* is 4, or VAV + YUD + VAV whose final *mispar katan* is also 4.

QUANTUM 6

Atomic Number 6—Carbon (C). Quantumly speaking, carbon is a chemical element that has the atomic number **6**, the atomic weight **12.011**, and the atomic symbol **C**. One Carbon-12 neutral atom has 6 protons and 6 neutrons with 6 electrons. Carbon is nonmetallic and tetravalent, which means it has four electrons available to form covalent chemical bonds.

Carbon is one of the few elements known since antiquity. It is the "duct tape of life," binding atoms one to another, forming humans, animals, plants, rocks, etc. Carbon can form four bonds, which it does with many other elements, creating hundreds of thousands of compounds (plastics, gas, etc.).[263]

Carbon is the fourth most abundant element in the universe after hydrogen, helium, and oxygen.[264] Four is the number for Creation, which includes the creation of our planet Earth that's also VAVed connected to the spiritual realm. Nearly 20 percent of your body is carbon. A professor of inorganic chemistry at Oregon State University—May Nyman—tells us that carbon has an almost unbelievable range.[265]

"It makes up all life forms, and in the number of substances it makes, the fats, the sugars, there is a huge diversity," Nyman says. It forms chains and rings in a process chemists call catenation. Every living thing is built on a backbone of carbon (with nitrogen, hydrogen, oxygen, and other elements). So animals, plants, every living cell, and of course humans are a product of catenation. Our bodies are 18.5 percent carbon, by weight.[266]

Carbon is found in four major forms: graphite, diamonds, fullerenes, and graphene. "Structure controls carbon's properties," says Nyman. Graphite ("the writing stone") is made up of loosely connected sheets of carbon formed like chicken wire. Penciling something in actually is just scratching layers of graphite onto paper.[267]

Diamonds, in contrast, are linked three-dimensionally. These exceptionally strong bonds can only be broken by a huge amount of energy. Because diamonds have many of these bonds, it makes them one of the hardest substances on Earth. Diamonds are called "ice" because their ability to transport heat makes them cool to the touch—not because of their look. This makes them ideal for use as heat sinks in microchips. This is where diamonds' three-dimensional lattice structure comes into play. Heat is turned into lattice vibrations, which are responsible for diamonds' very high thermal conductivity.[268]

Graphene on the other hand is the world's strongest material, which is an allotrope of carbon that consists of a single layer of atoms arranged in a two-dimensional honeycomb lattice nanostructure. A sheet of one square meter of graphene weighs 0.77 milligrams. Its strength is 200 times greater than that of steel and its density is similar to that of carbon fiber. All these make it resist high bending forces without breaking.[269]

Not only is graphene the world's strongest material, but the world's thinnest too. It has an extremely high surface area to volume ratio. Graphene is one of the most conductive materials for electricity and heat, which makes it the perfect material for electronics and many other industries. It has lots of possible applications, ranging from batteries, to sensors, solar panels, and more.

One recent finding is that scientists have discovered graphene oxide in some of the jabs. This is not surprising when one considers the vast medical applications of graphene. The following are a few according to nanografi.com. Graphene is used in:[270]

[1] Drug Delivery—Functionalized graphene can be used to carry chemotherapy drugs to tumors. Graphene-based carriers are said to target cancer cells better and reduce the toxicity of healthy cells.

[2] Cancer Treatment—Graphene can also detect cancer cells in their early stages and stop them from growing further in many types of cancers.

[3] Gene Delivery—Gene delivery is a method used to cure some genetic diseases by bringing foreign DNA into cells. Graphene oxide modified by polyethyleneimine can be used for this purpose, just like for drug delivery.

[4] Diabetes Monitoring—A blood glucose monitoring system that doesn't pierce the skin, which is a patch that includes a graphene sensor, has been developed by scientists at the University of Bath.

[5] Dialysis—Graphene membranes are not only useful in applications for energy, nuclear, and food industries but also can be used to filter blood from waste, drugs, and chemicals.

[6] Tissue Engineering and Cell Therapy—Graphene combined with Hydroxyapatite and Chitosan can be used as a synthetic bone substitute. Note that stem cells are especially important in tissue reengineering.

I don't doubt that there are some good uses for graphene. I am just wary of the motives and morals of the people trying to re-engineer man and society in general. Re-engineering stem cells is re-writing man at the cellular level. That sounds kind of Antichrist-y to me.

> Behold, the genu-wine Diamond Bride of Messiah Yeshua versus the counterfeit and worldly graphene bride of the Anti-Messiah (i.e., Antichrist). Under Yeshua's wedding *chupa* is His uni-verse—"*Heaven is His throne, the earth is His footstool*" (Isaiah 66:1). If one has ears to hear, eyes to see, and a mind that understands, perceive He's presenting jewels to us with His creative nature—all that He is. Everything contained in Him is draped with jewels. The jewels —Designer Diamonds—are also all of His Bride. We contain it all. These are the secrets of The Kingdom, the mysteries hidden from the ages. "*And He said unto them, Unto you it is given to know the mystery of the Kingdom of God*" (Mark 4:11 $_{KJV}$).

6 IN SCRIPTURE

VAV (ו) is 6. VAV (ו) has connective properties as the link between heaven and earth[271] and its form of a hook. The VAV (ו) links words and phrases to form sentences. It joins sentences into paragraphs and chapters as well as VAV connecting one chapter to another, uniting books. A VAV (ו) implies a close relationship between events and continuity between generations. The absence of a ו at the beginning of a new chapter in the Bible indicates the beginning of a new era or new subject. [272].

The conversive properties of VAV (ו) have to do with VAV being a prefix to a verb. When VAV is a prefix to a verb, it changes the tense of the verb from past to future or vice versa. [273]

VAV (ו) represents woe and grief through its fulfillment as VAV-YUD-VAV (ויו). The shape of the Hebrew letter ו represents YHVH's staff, which is His corrective measure for those that don't utilize His merciful gateway of repentance HEI (ה).

VAV (ו) is 6. Messiah Yeshua was asked six times to produce a sign to prove who He claimed to be (Matt. 12:38; Matt. 16:1; Matt. 24:3; Luke 11:16; John 2:18; John 6:30). Messiah Yeshua was accused of being demon-possessed six times (Mark 3:22; John 7:20; John 8:48; John 8:52; John 10:20; and Luke 11:15). And, six people found Messiah Yeshua innocent of the charges that led to His crucifixion: Pontius Pilate (Luke 23:14), Herod (Luke 23:15), Judas Iscariot after the devil left him (Matt. 27:3), Pontius Pilate's wife (Matt. 27:19), one of the thieves on a cross near Christ's cross (Luke 23:41), and a Roman Centurion at the crucifixion (Luke 23:47).

QUANTUM QUARK (ו)

To summarize, VAV (ו) is the sixth Hebrew Living Letter whose pictograph is a nail, a peg, or a hook that symbolizes joining together, making secure, or becoming bound (nailed). Each Hebrew word has a primary root. The sacred letter VAV (ו) is unique in the Hebrew language in that there is only one Hebrew word whose root begins with this letter. This word is (וו)—hooks—as in the Wilderness Tabernacle's hooks on the pillars that were used to hang the sacrifices when they were being skinned. The three nails used in the crucifixion of Messiah Yeshua are connected to this singular VAV root. Recall that the pictures of the first letters of the words on the sign that hung on the Cross above Messiah Yeshua proclaimed: Behold, salvation comes by the nailed hand.

The fundamental geometry of the very first photon when the Father of Lights spoke *"let there be light,"* is in the form of two Merkabahs—four tetrahedrons stuck together. It's not a coincidence that the "Songs of the Sabbath Sacrifice," articulated in the Melchizedek section of the *Dead Sea Scrolls*, culminated in worshipers experiencing the heavenly holiness of the *Merkabah* (i.e., God's Chariot Throne).

Q: Why is the Messiah's redemption of the picture of nails in the Tabernacle so important?

Q. What are your thoughts about the first photon of creation being connected to VAV (ו)?

ז

ZAYIN (ז)—7
SWORD OF TIME

REDEMPTION OF ZAYIN (ז)

The redemption of the Messiah's Quantum 22 continues with the seventh letter of the Hebrew Alphabet. ZAYIN (ז) is pronounced *ZAH-yeen,* and it has a numerical value of seven. The seven seals of the Book of Revelation and the seven seals of the human body are inextricably connected. Only the Lion of the Tribe of Judah can open these scrolls with seven seals: *"See, the Lion of the Tribe of Judah, the Root of David, has triumphed. He is able to open the scroll and its seven seals" (Revelation 5:5).*

The worldly systems have been trying to duplicate this work; but know that they will always fall short because Messiah Yeshua (Jesus Christ) is the only one worthy to break the seven seals that opens the scrolls of the earth and the human body to the higher ascended realities of the fifth and the fourth dimension (Rev. 5:1-5). The New Age Religion has come up with their lower frequency chakra system, which tries to tap into releasing the seven seals of the human body. However, chakras and their meditation practices focused on themselves are at its root a demonic control structure. It is a form of witchcraft where those who concentrate their consciousness on their chakras are trying to manipulate and control an ascended reality of their mind and body. Their focus is wrong. Ultimately, mankind can't focus on ourselves to facilitate our own transformation. We must solely focus on the One that can transform our human bodies; then our spirits and souls will follow.

Both the seven seals on the scroll of the body of Earth in the Book of Revelation and the seven seals on the scroll of the human body have already been unsealed, but the reality of their fullness has been held back. On 8 April 2024, a total solar eclipse completed an ancient ALEF (א) and two of three ancient TAVs (ת) on the map of the U.S.A. by the finger of God. This 2024 Great American Eclipse was a sign of Jonah,[274] which communicated it was a God-given season of repentance.[275] Afterwards, Father God (Abba) shared that He was grieved due to the lack of repentance. We were told that Abba is taking away what has been holding back the fulfillment of all seven seals. The fullness of all seven seals for the Earth and for the human bodies of the Bridal Firstfruits Company of the 144000 are all being released at the same time. More fifth dimension transformation fruits of the Wise Virgins will follow; then the rest will shift into a fourth-dimension reality. How exciting! During cosmic and earthly disturbances, martyrs being killed, and the riding of the four horsemen of the Apocalypse, the 144000 servants are sealed on their heads, signifying the fullness of their human bodies' transformation into a resurrected body state, like Yeshua and Enoch (Gen. 5:24).

The seven seals of the human body that only the Lion of the Tribe of Judah (Messiah Yeshua) can open and fulfill in a fifth- or fourth-dimension transformation are seven endocrine glands and organs. "Your endocrine system is in charge of creating and releasing hormones to maintain countless bodily functions." [276] It "consists of the tissues (mainly glands) that create and release hormones. Hormones are chemicals that coordinate different functions in your body by carrying messages through your blood to your organs, skin, muscles, and other tissues. These signals tell your body what to do and when to do it. Hormones are essential for life and your health." [277] "The main function of your endocrine system is to release hormones into your blood while continuously monitoring the levels. Hormones deliver their messages by locking into the cells they target so they can relay the message. You have more than 50 different hormones, and they affect nearly all aspects of a person's health—directly or indirectly," including: metabolism, mood, homeostasis (blood pressure, blood sugar, electrolyte balance, and body temperature), sleep-wake cycle, growth and development, sexual function, and reproduction.

Three different kinds of tissue make up the endocrine system. They are endocrine glands, endocrine organs, and endocrine-related tissues. The endocrine glands make and release hormones directly into a person's bloodstream. The endocrine glands in a human body are:

> [1] PINEAL GLAND is a tiny pinecone-shaped gland in your brain that's beneath the back part of the corpus callosum. It makes and releases melatonin. It's linked to spiritual connections.
>
> [2] PITUITARY GLAND is a small pea-sized gland at the base of your brain below your hypothalamus. It releases eight hormones, some of which trigger other endocrine glands to release hormones.
>
> [3] THYROID GLAND is a small butterfly-shaped gland at the front of your neck. It releases hormones that help control metabolism.
>
> [4] PARATHYROID GLANDS are four pea-sized glands that are typically behind your thyroid. They release parathyroid hormone (PTH), which controls the calcium level in your blood.
>
> [5] ADRENAL GLANDS are small triangle-shaped glands on top of each kidney. They release several hormones that manage metabolism, blood pressure, and stress response.
>
> [6] THYMUS GLAND is behind the sternum and is also part of the immune system. The thymus gland is both an endocrine gland and a lymphatic organ. It secretes hormones and is crucial to the production, maturation, and differentiation of immune T cells. [278]

There are certain organs in the human body that also make and release hormones. One of the mysteries of science is how organs form in vitro. There is an energy field that enables a baby to organize in their mother's womb. In 2016, scientists at Northwestern University proved that there's a flash of light at the moment of conception. [279] This is *"the true light that gives light to everyone . . . In Him was life, and that life was the light of all mankind" (John 1:9,4).*

At conception, the true light triggers the one cell to divide and multiply. It keeps dividing until around replication 260, the cells start to differentiate. Suddenly, a cell becomes a kidney cell or a neurologic cell or whatever it is going to become. Note that the very first cell goes into forming a person's heart. Not only does a replicated cell become a unique cell, but it also knows where to migrate on a three-dimensional map in a mother's womb to become an organ system. However, doctors and scientists cannot find that map inside biology, inside a human cell. This map seems to be in the physics field, specifically the electromagnetic field [280] that is connected to quantum entanglement with the Creator who is the Quantum One.
"^{13} For You created my inmost being; You knit me together in my mother's womb. ^{14} I praise You because I am fearfully and wonderfully made; Your works are wonderful, I know that full well" (Psalms 139:13-14).

The two organs in the human body that are part of the endocrine system that we will focus on are:

> [1] HYPOTHALAMUS is a structure deep within a person's brain, which is an organ. It is the main link between a person's endocrine system and nervous system. The hypothalamus makes two hormones—oxytocin and vasopressin—that the body's pituitary gland stores and releases as well as making and releasing two other hormones—dopamine and somatostatin. [281]
>
> [2] PANCREAS is an organ in the back of a person's belly. The pancreas is both an organ and a gland, which is part of a person's digestive system. It releases two hormones essential to maintaining healthy levels of blood sugar: insulin and glucagon.

Seven of the endocrine glands and organs are the seven seals of the human body that only Messiah Yeshua can open and shift into their fullness to facilitate the transformation of a human being into a resurrected body and a quickening spirit. The seven seals of the human body are the pituitary gland, the pineal gland, the thyroid and parathyroid glands, the thymus gland, the adrenal gland, the hypothalamus, and the pancreas. [282]

ANCIENT ZAYIN (ז)

ZAYIN (ז) denotes the spiritual values that were, and still are, the purpose of Creation. God created the Uni-verse in six days and rested on the seventh. [283] As seven, ZAYIN (ז) figuratively represents the six directions—north, south, east, west, up, and down—that surround the seventh—every human being. These directions represent influences outside man while the seventh factor is the placid center of it all—the inner man who is subject to all these forces but not part of them. "How well he succeeds in shaping and maintaining his identity in accordance with the spiritual dictates of his soul [in Christ] is the challenge and purpose of life."[284]

Significantly, not only does ZAYIN (ז) denote the spiritual values that are the purpose of Creation, but so do the seven lights of the Temple Menorah. The Temple Menorah consisted of the central stem denoting the Spirit of God with six arms radiating outward, three on each side.[285] According to Scripture, the seven lights of the Temple Menorah map to the Seven Spirits of God. *"¹ There shall come forth a Rod from the stem of Jesse, And a Branch shall grow out of his roots. ² The Spirit of the Lord shall rest upon Him, The Spirit of wisdom and understanding, The Spirit of counsel and might, The Spirit of knowledge and of the fear of the Lord"* (Isaiah 11:1-2). *"Seven lamps of fire were burning before the throne, which are the Seven Spirits of God"* (Revelation 4:5).

Not only does ZAYIN (ז) stand for the spiritual purposes of Creation but also the All-Sufficient One's sustenance. *Shabbos 104a* states that if someone studies Torah (God's Word), helps the poor, and dedicates his deeds to God's name, as indicated by the first six letters of the Hebrew Alphabet, he will be rewarded amply. The letter ZAYIN (ז) represents the first stage of the reward. ZAYIN (ז) is an allusion to the Most High God who nourishes all. Our daily supply of nourishment—our Daily Bread—is one of the foremost acts of divine kindness (Midrash Shocher Tov).[286]

Not only does ZAYIN (ז) stand for the spiritual purposes of Creation and sustenance, but also struggle.[287] In ancient Hebrew, ZAYIN (ז) means a weapon, which this glorious letter resembles in form (picture) in all the ancient alphabets.[288] The top of the ZAYIN (ז) is the handle, and the vertical leg is the blade. ZAYIN's design can also represent a crown and a scepter; thus, alluding to power and authority.[289]

This brings us back to the powerful living and active, two-edged sword: *"For the word of God is living and active and full of power [making it operative, energizing, and effective]. It is sharper than any two-edged sword, penetrating as far as the division of the soul and spirit [the completeness of a person], and of both joints and marrow [the deepest parts of our nature], exposing and judging the very thoughts and intentions of the heart"* (Hebrews 4:12 AMP). It is extremely significant that there is a flaming sword that turns in all directions to guard access to the Garden of Eden and its Tree of Life (Gen. 3:24).

SIMPLE LETTER (ז)

ZAYIN (ז) is one of seven simple letters, which are used to form the remaining Hebrew Living Letters.

FULFILLMENT OF ZAYIN (ז)

The Hebrew Living Letter ZAYIN (ז) fulfills as ZAYIN-YUD-NUN (זין). This fulfillment translates as "weapon" or "sword". The first time that a weapon was used to kill a person in Scripture was in the seventh generation from Creation when Lemech killed both Cain and TuvalCain.[290]

QUANTUM 7

Atomic Number 7—Nitrogen (N). Quantumly speaking, nitrogen is a chemical element that has the atomic number **7**, the atomic weight **14.007**, and the atomic symbol **N**. A single nitrogen atom has 7 protons and 7 electrons. The number of neutrons in an atom can be determined by the difference between the atomic mass and the number of protons. The difference between the mass number of the nitrogen atom and the number of protons is seven. Therefore, a nitrogen atom has 7 neutrons.

The name nitrogen is derived from the Greek "nitron" and "genes," which means nitre forming. [291] Among all the known elements, nitrogen ranks sixth or seventh in cosmic abundance in the Milky Way.[292] Nitrogen is a colorless, odorless, tasteless gas. Nitrogen is the most plentiful uncombined element in Earth's atmosphere, approximately four-fifths of it. Despite this, it is not very abundant in Earth's crust.[293]

Nitrogen occurs in all living organisms. Nitrogen is found primarily in amino acids (and thus proteins), in the nucleic acids (DNA and RNA), and in the energy transfer molecule adenosine triphosphate.[294] The human body contains about 3% nitrogen by mass. It is the fourth most abundant element in the human body after oxygen, carbon, and hydrogen. [295]

Nitrogen is the key yield (produce/fruit) building nutrient for all crops, but excessive levels can be detrimental to the natural environment. Therefore, understanding how nitrogen reacts in the soil and is used by crops helps growers increase profitability and production. [296]

Many gardeners know about the indigenous Three Sisters technic for planting corn, beans, and squash together. Corn provides support for beans, beans provided nitrogen through nitrogen-fixing rhizobia bacteria that live on the roots, and squash and pumpkins provided ground cover to suppress weeds and inhibit evaporation from the soil. [297] Maize is known as a nitrogen "hungry" crop. The nitrogen needs of maize are also the reason for the Pilgrims planting fish with their corn. The Natives showed the Pilgrims to dig a hole, place 2-3 red herring (fish) with 4-5 corn kernels in it; and then, cover it all with dirt. [298]

Nitrogen is one of the key components in chlorophyll formation in the leaf and stem. Chlorophyll is essential to facilitate photosynthesis for any successful crop growth. However, excess nitrogen leads to larger canopies, which prevents the plant from efficiently utilizing radiant energy from the sun. [299]

7 IN SCRIPTURE

ZAYIN (ז) is 7. Always remember that ZAYIN (ז) is seven, which is one of the most significant numbers in Scripture that is connected to the Plan of God. Due to the prevalence of seven throughout Scripture, let's simply review some of the sevens in the Book of Revelation.

The Book of Revelation starts out with the Apostle John's greeting to the seven churches from the One who is and who was and is to come, and from the seven spirits who are before the throne (Rev. 1:4). The seven churches in Asia are pictured as seven golden lampstands (Rev. 1:4, 1:11, 1:12, 1:13, 1:20). There are seven stars in the right hand of the One who is like the Son of

Man. He has a sharp two-edged sword coming out of his mouth (Rev. 1:16). These seven stars are the seven angels of the seven churches (Rev. 1:20).

Revelation 4:5 reveals seven lamps of fire burning before the throne, which are the rainbow-colored Seven Spirits of God. Please refer to the "Seven Spirits of God" section of the *Water into Wine* chapter.

Revelation 5:1 speaks of a scroll with seven seals that only the Lion of the Tribe of Judah, the Root of David, can open (Rev. 5:5). When the Apostle John beholds the Lion of the Tribe of Judah in Revelation 5:6, he sees Him in the midst of the throne, in the midst of the four living creatures, and in the midst of the 24 elders. He appears as a Lamb as though slain, having seven horns and seven eyes, which are the Seven Spirits of God sent out into the earth.

When the seventh seal is opened, there is silence in heaven for about half an hour (Rev. 8:1). You get the idea. Not to mention other references to "seven" in the Bible, like forgiving a brother up to seventy times seven (Matt. 18:22). The seventh-day march around Jericho seven times (Josh. 6:15-20). Etc.

SWORD OF TIME

Everything in the world of time revolves around the number seven. Since ZAYIN (ז) represents seven and a sword, it is no surprise that it's used to cut up time *(z'man)* into units of seven. The Sabbath is the 7th day of a 7-day week [week of days]. Remember that the Hebrews enjoyed a double portion of manna—the Heavenly Bread—on the Sabbath, which sustained them for 40 years in the wilderness.

Shavuot is the 49th day (7 × 7) after Passover [week of weeks].

Tishri is the 7th month in the Hebrew year [week of months].

Shemitah is the 7th year of rest for the land [week of years].

Yovel is the 49th year [week of weeks of years].

The Millennial Kingdom is the 7th millennium of human history [week of 1000 years]. [300]

"But, beloved, do not forget this one thing, that with the Lord one day is as a thousand years, and a thousand years as one day" (2 Peter 3:8). The Seventh Day of Creation corresponds to the seventh millennium (7000 years), which is a day of rest and tranquility for all eternity. We are living on the cusp of the 7th Millennium, which is when the Messiah arrives to rule and reign.

However, don't forget what Chuck Missler shares: Time is a property of mass. If it has no mass, it has no time. The real you requires no time dimension. You are eternal. The issue is where you will spend eternity.

PALACE IN TIME

As we have seen, ZAYIN (ז) is a picture of a sword and ZAYIN (ז) is seven. As seven, it marks the spiritual values that are the purpose of Creation. God created the Universe in six days and rested on the seventh.

"Thus the heavens and the earth were completed, and all their hosts. By the seventh day God completed His work which He had done and rested on the seventh day from all His work which He had done. Then God blessed the seventh day and sanctified it, because in it He rested from all His work which God created and made" (Genesis 2:1-3 $_{NASB}$).

> The sword of the Lord is being placed in the highest mountain of man. The sword is being struck and a word begins to form on the edge of the blade. Stick the top of your double-edged sword in Everest. Fall on your sword. There is a heavenly mountain that overshadows the earthly one. It's a sacred mountain. Ascend to the top of God's holy mountain of "ever rest." Sit down and put out your hand to receive the key to unlock our corporate heart. It's time for the sons of God to arise. Ever ready. Ever rest. We release the sword to fulfill its function in heaven and earth to cause the unblocking of the sons' stars. [301]

One of my favorite books is called *The Sabbath* by Abraham Joshua Heschel. He teaches: "The art of keeping the seventh day is the art of painting on the canvas of time the mysterious grandeur of the climax of creation: as He sanctified the seventh day, so shall we.... Our keeping the Sabbath day is a paraphrase of His sanctification of the seventh day."[302]

"The Sabbath is the most precious present mankind has received from the treasure house of God."[303] "To observe the Sabbath is to celebrate the coronation of a day in the spiritual wonderland of time, the air of which we inhale when we 'call it a delight.'"[304] "The seventh day is like a palace in time with a kingdom for all. It is not a date, but an atmosphere. It is not a different state of consciousness but a different climate; it is as if the appearance of all things somehow changed."[305]

What does the word 'Sabbath' mean? According to some, it is the name of the Holy One."[306] Through the years of trying to keep and protect the Sabbath, I have discovered that the Sabbath is my most loved and preferred time. It's a divine time when time intersects eternity, as I discover more truly that "He is my Sabbath rest; and I am His."

QUANTUM QUARK (ז)

In summary, ZAYIN (ז) is the seventh Hebrew Living Letter whose pictograph is a sword. Think of the living and active two-edged sword of the Spirit, which is the Word of God (Eph. 6:17; Heb. 4:9-12). The foundation of God's seventh-day rest is based on the Word of God. The One called the Word of God with a sharp sword coming out of His mouth is the King of Kings and Lord of Lords (Rev. 19:13-16). The pictograph for ZAYIN (ז) can also represent a scepter, because this glorious letter has three small crown-like extensions on top, which is often referred to as the Golden Scepter שרבית הזהב of the King.

The precious blood of Yeshua, the Hebrew Living Letters, and the Seven Spirits of God were components of creating our physical world, which are all essences of the exact re-presentation of the Son (Heb. 1:3).

Q: How can you do the sword dance with the double-edged sword? Hint: Study Hebrews 4.

Q. What are some ways that you can engage the Seven Spirits of God? Hint: Study Scriptural references for each; and then, apply what you learn. For example, you can study references to the word "wisdom" for the Spirit of Wisdom (Isa. 11:2).

CHET (ח)—8
GATEWAY OF LIFE

REDEMPTION OF CHET (ח)

The redemption of the Messiah's Quantum 22 continues with the eighth letter of the Hebrew Alphabet. The pronouncement of the quantum letter CHET (ח) rhymes with "met," the sound of "ch" as in Bach and it has a numerical value of eight. CHET (ח)'s ancient pictograph is a fence or an inner room.

We will concentrate on the inner room that is the heart of marriage with regard to the redemption of CHET (ח). When a man and woman get married in the Hebrew culture, they are typically united beneath a *chupah*—a marriage canopy. Notice how the form of the letter CHET (ח) looks like a marriage canopy. "The Hebrew word *chupah* begins with the sacred letter CHET (ח), and means *chet po-chet* (God, man and woman) is *po* (here). Therefore, CHET (ח) is referred to as the heart of marriage because true marriage is a three-strand cord—man, woman, and God—united under His canopy of love." [307]

There are two parts to a Hebrew wedding ceremony. First, there is the betrothal—*erusin* (אירוסין). Second is the marriage itself—*nesuin* (נישואין). Originally, these two ceremonies were held as much as a year apart, but later the two ceremonies merged. [308] After the preliminaries of the *erusin*, the groom and bride are led to the wedding canopy—*chupah* (חוּפָּה)—for the official marriage ceremony. Usually, the *chupah* is made up of a piece of cloth that is held up by four poles. When the bride and groom stand under the cloth *chupah*, it is equivalent to a groom placing a garment over his bride. The *chupah* is also like a house, which is open on all four sides. This reflects Abraham's house that had entrances on all four sides as well as Abraham's incredible gift of hospitality. [309]

The word *chupah* is a biblical term: *"Let the bridegroom go forth from his chamber and the bride from her chupah" (Joel 2:16).* It is understood throughout the Jewish community that *chupah* is the act through which a couple clearly demonstrates that they are husband and wife and that this is the act that binds them together. [310]

Essentially, the second part of the wedding ceremony mainly consists of the Seven Blessings (*Sheva Brachot*, שבע ברכות). The Seven Blessings are recited over the part of the wedding ceremony where the bride and groom become permitted (united, bound) to each other through their marriage contract—*ketubah* (כְּתוּבָּה). The traditional *ketubah* was written in ancient Aramaic. It outlines the rights and responsibilities of a husband to his wife. It was meant to protect the woman and serve as a deterrent for divorce. The *ketubah* is the replacement for the money paid by the groom to the bride for marriage. Typically, two glasses of wine are included in a Hebrew wedding ceremony. The first cup is drunk over the prenuptial blessing during the betrothal. The second cup is for the Seven Blessings during the actual marriage ceremony.

The Seven Blessings (*Sheva Brachot*, שבע ברכות) are as follows:[311]

1. Blessed are You, LORD, our GOD, Sovereign of the Universe, Creator of the vine-fruit.

2. Blessed are You, LORD, our God, Sovereign of the Universe, who created everything for His glory.

3. Blessed are You, LORD, our God, Sovereign of the Universe Creator of man.

4. Blessed are You, LORD, our God, Sovereign of the Universe, who created man in Your image, fashioning perpetuated life. Blessed are You, LORD, Creator of man.

5. The barrenness will surely exult and be glad in gathering her children to herself joyfully (in haste). Blessed are You, LORD, Gladdener of Zion by the way of her children.

6. Loving companions will surely gladden, as You gladdened your creations in the Garden of Eden in the east. Blessed are You, LORD, Gladdener of groom and bride.

7. Blessed are You, LORD, our God, Sovereign of the Universe, who created joy and gladness, groom and bride, mirth, song, delight and rejoicing, love and harmony, and peace and companionship. Quickly, LORD our God, there should be heard in the cities of Judah and in the courtyards of Jerusalem the voice of joy and the voice of gladness, the voice of the groom and the voice of the bride, the jubilant voices of grooms from the bridal canopy, and of young people from the feast of their singing. Blessed are You, LORD, Gladdener of the groom with his bride.

The ultimate redemption of CHET (ח) is Messiah Yeshua and His Bride under the Universe's *chupah*. A friend and I experienced a fantastic *chupah* Ascension in Christ. It went like this:

> "We see Yeshua holding a baby, rocking it. He drops it and grabs the woman next to Him and plants a wonderful long kiss on her. We hear: "Yes, My Bride. It's time to put away childish things and become My Bride."
> We see a giant ballroom with His Bride spinning and spinning and spinning. She is dressed so elegantly. As she spins, her dress gets bigger and bigger. Things start to fling off of her spinning dress. In fact, they are flinging off due to the spin. She is changing the atmosphere. The dance is beautiful and glorious, but necessary. The glory is simply incredible.

Above the dance floor, there is no roof because the universe is the wedding canopy over her. We hear: "You are the pearl (of great price)." We know that the Bride is doing a *chuwl* dance. The Hebraic definition of *chuwl* is to twist, whirl, dance, writhe, fear, tremble, travail, be in anguish, be pained. His Bride is abandoned in worship. She is lost in Him. She spins under wild emotion. Her passion for Christ cannot be contained. To her, the passion of His cross is a beautiful dance with Him to become just like Him.

There are mirrors on the walls to catch all the angles of her glory. When the Bridegroom gets the glory, He loves to glorify His Bride. It even appears to be a seamless give and take. He likes His wild woman ... abandoned in wild passionate worship. She is power-filled!

We see the Bride and the Bridegroom stop on the dance floor and fall to the floor. They both lay back and are captured in the vastness of the roofless ceiling. They are basking in the aftermath of their glory dance. It's a sacred dance. We see Yeshua grab His Bride's hand and point His finger explaining what He created. He is showing off, so pleased that He created it just for her—His Beloved." [312]

ANCIENT CHET (ח)

Recall that the number seven symbolizes the complete purpose of human existence, which combines the spiritual Seventh Day Sabbath with the physical effort of the week. Therefore, the number eight symbolizes man's ability to transcend the limitations of physical existence. So, we can basically say that CHET (ח) whose numerical value and ordinal value is eight stands for that which is on a plane above nature—the divine—the metaphysical. The way that Hebrews strive to exalt human spirituality towards the realm above the natural is through studying God's Word and practicing it. [313]

CHET (ח) and its value of eight not only symbolizes transcendence but also divine grace. Noah was the reason that eight survived The Flood encased in the Ark that he built in obedience to the Lord (Gen. 7:13). We are told that *"Noah found grace in God's eyes" (Genesis 6:8)*. Had God not intervened with divine grace, he and his family would have been swept up in the cataclysm. Because Noah found divine grace, he was found worthy of rebuilding the world after The Flood (Haamek Davar). [314]

"Noah spent so many years building the Ark to save him and his family. He walked in obedience with God, not knowing God was also creating one inside of him, as he was working on the physical ark! What Ark is God trying to build in your life today? 'The Lord then said to Noah, *'Go into the ark, you and your whole family, because I have found you righteous in this generation' (Genesis 7:1).*"[315]

Moses paid attention to the divine grace Noah received when he asked God: Let me have insight into Your ways, so that I may find grace in Your eyes and be enabled to lead Your people with Your intentions (Exo. 33:13). Not only does CHET (ח) symbolizes transcendence and divine grace, but life. CHET is the sacred letter most known for its connection to life.[316] *L'Chaim* in Hebrew is a toast meaning "to life". When a couple becomes engaged, they get together with friends and family to celebrate. Since they drink to life—*l'chaim*—the celebration is also called a *l'chaim*.

Scripture speaks of the breath of life through which God created man (Gen. 2:7). It is the breath of the Almighty that gives mankind life: *"The Spirit of God has made me, And the breath of the Almighty gives me LIFE"* (Job 33:4).

There is the Tree of Life in the midst of the garden east of Eden: *"And out of the ground the Lord God made every tree grow that is pleasant to the sight and good for food. The Tree of LIFE was also in the midst of the garden, and the Tree of the Knowledge of good and evil"* (Genesis 2:9). Those who eat of the Tree of Life live forever: *"Then the Lord God said, 'Behold, the man has become like one of Us, to know good and evil. And now, lest he put out his hand and take also of the Tree of LIFE, and eat, and live forever'"* (Genesis 3:22). Cherubim and a flaming sword guard the way to the Tree of Life: *"Then the Lord God said, 'So He drove out the man; and He placed cherubim at the east of the garden of Eden, and a flaming sword which turned every way, to guard the way to the Tree of LIFE'"* (Genesis 3:24).

There is a pure river of the Water of Life proceeding from The Throne: *"And he showed me a pure river of Water of LIFE, clear as crystal, proceeding from the throne of God and of the Lamb"* (Revelation 22:1). On either side of the river is the Tree of Life: *"In the middle of its street, and on either side of the river, was the Tree of LIFE, which bore twelve fruits, each tree yielding its fruit every month. The leaves of the tree were for the healing of the nations"* (Revelation 22:2).

The Source of CHET (ח) is ALEF-TAV (את) who gives of the fountain of the "Water of Life" freely: *"And He said to me, 'It is done! I am the Alpha [ALEF] and the Omega [TAV], the Beginning and the End. I will give of the fountain of the water of LIFE freely to him who thirsts'"* (Revelation 21:6).

"Life" is in the blood: *" ¹⁴ For it is the life of all flesh. Its blood sustains its life. Therefore I said to the children of Israel, 'You shall not eat the blood of any flesh, for the LIFE of all flesh is its blood. Whoever eats it shall be cut off'"* (Leviticus 17:14).

Taking Messiah Yeshua's communion is connected to the eternal life: *"Then Jesus said to them, 'Most assuredly, I say to you, unless you eat the flesh of the Son of Man and drink His blood, you have no LIFE in you. Whoever eats My flesh and drinks My blood has ETERNAL LIFE, and I will raise him up at the last day'"* (John 6:53-54). In Him is the life and light of men: *" ³ All things were made through Him, and without Him nothing was made that was made. ⁴ In Him was LIFE, and the LIFE was the light of men"* (John 1:3-4).

Messiah Yeshua has life in Himself, just like Our Heavenly Father: *"For as the Father has LIFE in Himself, so He has granted the Son to have LIFE in Himself"* (John 5:26).

Yeshua is the Bread of Life: *" 'For the bread of God is He who comes down from heaven and gives LIFE to the world.' Then they said to Him, 'Lord, give us this bread always.' And Jesus said to them, 'I am the bread of LIFE. He who comes to Me shall never hunger, and he who believes in Me shall never thirst'"* (John 6:33-35). When you follow Yeshua, you have the Light of Life: *"Then Jesus spoke to them again, saying, 'I am the light of the world. He who follows Me shall not walk in darkness, but have the light of LIFE'"* (John 8:12).

True knowledge of our Heavenly Father and the Son is eternal life: *"And this is ETERNAL LIFE, that they may know You, the only true God, and Jesus Christ whom You have sent"* (John 17:3).

There is life in the Messiah's Name: "*³⁰ And truly Jesus did many other signs in the presence of His disciples, which are not written in this book; ³¹ but these are written that you may believe that Jesus is the Christ, the Son of God, and that believing you may have LIFE in His name*" (John 20:30-31).

The Righteous Order of Melchizedek is based on the eternal life of its High Priest Yeshua: "*¹⁵ And it is yet far more evident if, in the likeness of Melchizedek, there arises another priest ¹⁶ who has come, not according to the law of a fleshly commandment, but according to the power of an endless LIFE. ¹⁷ For He testifies: 'You are a priest forever according to the Order of Melchizedek'*" (Hebrews 7:15-17).

COMPOSITE LETTER (ח)

CHET (ח) is one of the composite letters made up of VAV (ו) and DALET (ד), which have a total numerical value of 10 (6+4). Their *mispar katan* is one, returning our focus to the Unity of YHVH.

FULFILLMENT OF CHET (ח)

The fulfillment of the Hebrew Living Letter CHET (ח) is CHET-YUD-TAV (חית) with *mispar katan* of 13 (8+10+400 = 418 = 13). [317] 13 is the same number as the Hebrew word *echad* (אֶחָד), which signifies the unity of God, just like the total numerical value of its composite letters.

QUANTUM 8

Atomic Number 8—Oxygen (O). Quantumly speaking, oxygen is a chemical element that has the atomic number **8**, the atomic weight **15.999**, and the atomic symbol **O**. A single Oxygen atom has 8 protons, 8 electrons, and 8 neutrons.

Oxygen is a colorless, odorless, diatomic gas. It is the third most abundant element in the universe and the third most abundant element on Earth too. There's not much more important than the air we breathe. We don't live long without oxygen. Neither does almost everything that has life. [318] Even though the air that we breathe is only made up of 21 percent oxygen, approximately two-thirds of the mass of the human body is oxygen. [319] It is not a coincidence that the four ingredients essential for a human body's protein, carbohydrate, and fat structures are oxygen, carbon, hydrogen, and nitrogen whose atomic numbers sum to 22 (8+6+1+7). Behold, mankind's vital connection to the Quantum 22.

Oxygen is life. CHET (ח) is the letter of life. Essentially, all living things on Earth use oxygen to create energy in their cells. For humans, oxygen is our greatest and first source of energy. It is the fuel required for the proper operation of all body systems. Only 10% of your energy comes from food and water, and 90% of our energy comes from oxygen. [320]

Oxygen gives our body the ability to rebuild itself. Oxygen detoxifies the blood and strengthens the immune system. Oxygen displaces deadly free radicals, neutralizes environmental toxins, and destroys anaerobic bacteria, parasites, microbes, and viruses. [321]

Oxygen greatly enhances the body's absorption of vitamins, minerals, amino acids, proteins, and other important nutrients. Oxygen enhances brainpower and memory. Oxygen can beneficially affect your learning ability. The ability to think, feel and act is all dependent on oxygen. It also calms the mind and stabilizes the nervous system. Oxygen heightens concentration and alertness. Without oxygen, brain cells die and deteriorate quickly. Oxygen strengthens the heart. [322]

Oxygen is also vital for one's voice, as in the voice of the Almighty: *"I heard the noise of their wings, like the noise of great waters, as the voice of the Almighty, the voice of speech, as the noise of an host"* (Ezekiel 1:24 $_{KJV}$). Behold, the noise of eight sets of wings is the voice of the Almighty.

8 IN SCRIPTURE

CHET (ח) is 8. The number 8 represents "new beginnings." Take the figure 8 and lay it on its side and you get the sign of infinity. Your new beginning in Christ has no limit. Isaiah 11:1 tells us that Messiah Yeshua is the shoot from the stem of Jesse, which connects Him to the eighth son of Jesse—David—whose kingdom will never end (1 Sam. 17:12-14; Isa. 9:6-7).

CHET (ח) is 8. The eighth foundation stone for the New Jerusalem is beryl, which is the same as the wheels within wheels. *"17 Then he measured its wall: one hundred and forty-four cubits, according to the measure of a man, that is, of an angel. 18 The construction of its wall was of jasper; and the city was pure gold, like clear glass. 19 The foundations of the wall of the city were adorned with all kinds of precious stones: the first foundation was jasper, the second sapphire, the third chalcedony, the fourth emerald, 20 the fifth sardonyx, the sixth sardius, the seventh chrysolite, the EIGHTH beryl, the ninth topaz, the tenth chrysoprase, the eleventh jacinth, and the twelfth amethyst"* (Revelation 21:17-20).

"15 Now as I looked at the living creatures, behold, a wheel was on the earth beside each living creature with its four faces. 16 The appearance of the wheels and their workings was like the color of BERYL, and all four had the same likeness. The appearance of their workings was, as it were, a wheel in the middle of a wheel" (Ezekiel 1:15-16). Each of the four living creatures has one wheel within a wheel beside each one of them. According to Ezekiel 1:16, the appearance of the wheels and their works are like the color of beryl. What's interesting about this statement is that pure beryl is colorless. If beryl is tainted with impurities, it can be many different colors: green, blue, yellow, red, and white. Beryl is a mineral. Its two most desirable variations are probably aquamarine and emerald. Beryl is the principal store of beryllium in the earth's crust, and that's saying something; because beryllium is the most abundant metal in the earth's crust, and it always occurs in combination with other metals.

The spirit of each living creature, both individually and corporately, is in its wheel(s). Recall, fundamentally, this speaks of the whirling wheels' (wheels within the wheels) having a corporate essence, which is the Spirit of the New Living Creature of the One New Man in Christ made after the Order of Melchizedek from the get-go. Please note that the Living

Creature, Cherubim, and the Righteous Order of Melchizedek are interchangeable concepts, and they are all corporate beings.

Remember the octa-creature we learned about with HEI (ה)? It was a coil of octahedrons of all colors that had wings that raised their voice in praise, as the coil moved. "Octa" is 8. Do you recall how some oily wine caused powerful and vibrant rainbow colors to come out of the Bride of Christ? The rainbow colors swirled at extremely high frequencies, and it appeared like the colors were living. This is living color at a whole new level. This is created light. It's the octahedron rainbow-colored thing with wings. It is coming alive in the Bride. "Octa" is 8. Count the total number of sets of wings of the four living creatures in Ezekiel 1 and you get eight. Significantly, the noise of the living creatures' wings is likened to the voice of the Almighty. *"And when they went, I heard the noise of their wings, like the noise of great waters, as the voice of the Almighty, the voice of speech, as the noise of a host: when they stood, they let down their wings"* (Ezekiel 1:24 $_{KJV}$).

CHET (ח) is 8. Sharing sustenance is connected to the number eight: *"¹ Cast your bread upon the waters, For you will find it after many days. ² Give a serving to seven, and also to EIGHT, for you do not know what evil will be on the earth"* (Ecclesiastes 11:1-2).

CHET (ח) is 8. When eight days were completed and He was circumcised, Yeshua received the name given by the angel before he was conceived: *"And when EIGHT days were completed for the circumcision of the Child, His name was called JESUS, the name given by the angel before He was conceived in the womb"* (Luke 2:21).

CHET (ח) is 8. After eight days, Yeshua supernaturally walked into a room: *"And after EIGHT days His disciples were again inside, and Thomas with them. Jesus came, the doors being shut, and stood in the midst, and said, 'Peace to you!'"* (John 20:26).

TRANSCENDENCE

"The number eight represents transcendence—a level beyond nature and intellect. Everything in the world of time revolves around the number seven: the seven days of the week, the seventh year being a Sabbatical year . . . Eight, however, represents transcendence, a level that is beyond the natural order."[323]

The word "transcendence" comes from Latin. Its prefix *trans* means "beyond," and the word *scandare* means "to climb." When you achieve transcendence, you have gone beyond ordinary limitations.

Transcendence is often used to describe a spiritual or religious state or a condition of moving beyond physical needs and realities. Seeing, hearing, and talking to God are examples of transcendent experiences. Therefore, when one lays the number 8 on its side, we get the infinity symbol . . . a quintessential sign of transcendence. [324]

QUANTUM QUARK (ח)

In summary, CHET (ח) is the eighth Hebrew Living Letter whose pictograph is a fence or an inner room. ALEF-TAV (את)'s redemption for the quantum letter CHET (ח) has to do with the heart of marriage and the wedding canopy—*chupah*. A *chupah* is usually made up of a piece of cloth that is held up by four poles. When the bride and groom stand under the cloth *chupah*, it is equivalent to a groom placing a garment over his bride. The word *chupah* is a biblical term: *"Let the bridegroom go forth from his chamber and the bride from her chupah" (Joel 2:16).* The ultimate redemption of CHET (ח) is Messiah Yeshua and the Bride of Christ under His Universe's *chupah*.

CHET (ח) is known as the letter of life. Scripture speaks of the breath of life through which God created man (Gen. 2:7). Job 33:4 reveals that it is the breath of the Almighty that gives mankind life. There is the Tree of Life in the midst of the garden east of Eden (Gen. 2:9). Those who eat of the Tree of Life live forever (Gen. 3:22). There is a pure river of the Water of Life proceeding from The Throne (Rev. 22:1). On either side of the river is the Tree of Life (Rev. 22:2). ALEF-TAV (את) gives of the fountain of the "Water of Life" freely: *"And He said to me, 'It is done! I am the Alpha* [ALEF] *and the Omega* [TAV], *the Beginning and the End. I will give of the fountain of the water of life freely to him who thirsts' " (Revelation 21:6).*

Q: Which aspect of the *chupah* do you like most? Why?

Q. Why do you think ALEF-TAV (את) gives the water of life freely?

ט

TET (ט)—9
CONCEALED GOODNESS

REDEMPTION OF TET (ט)

The redemption of the Messiah's Quantum 22 continues with the ninth letter of the Hebrew Alphabet. The pronouncement of the quantum letter TET (ט) rhymes with "mate," the sound of "t" as in tall, and it has a numerical value of 9. The ancient pictograph for TET (ט) looks like a snake, or something that surrounds.

For the redemption of TET (ט), the Lord is highlighting its number 9 and the surround sound of creation. When one thinks of the sound that was released when the Father of Lights spoke the world into existence and the essential elements associated with speech, you can catch a glimpse of one of the fundamental building blocks of creation—sound. The whole universe is touched by the mysteries of sound; and its partners: frequency, energy, vibration, light, color, etc.

The elementary physics of sound reveals that sound is a form of energy that travels through various mediums, such as air and water. These sounds have amplitude, which is the volume of the sound, as well as frequency, which is its pitch. Both amplitude and frequency determine the quality of the sound. Frequency and vibration are closely related in the realm of sound. When something vibrates at a specific frequency, it produces a sound wave with the same frequency. Different frequencies create different vibrations in the body.

The human body is literally a symphony of sound. Every organ, every bone, every tissue, every cell has its own resonant frequency, its own sound. The frequency of the human body is 62-78 MHz with the average frequency being 70 MHz. The frequency of the brain is 72-90 MHz. The frequency of the liver is 55-60 MHz with the average frequency being 57.5 MHz. The frequency of the pancreas is 60- 80 MHz with the average frequency of 70 MHz. [325] Every organ has its own frequency-based spectrum that varies according to environment, physical fitness, mentality, etc. "There are many parameters in which the body organs change their frequency in a specific range." [326] Disease starts at the frequency of 58 MHz. [327] When an organ is out of tune, the entire body is affected. A less-than-optimal frequency leads to states of disease and disintegration. In this, frequency plays an important role in identifying an actual problem in the human body. [328] Frequency is also beginning to be used to help fix problems. I have personally experience how frequency plays an astounding role in health and wellness by using frequency protocols from *iCare* via PEMF (Pulsed ElectroMagnetic Fields)[329] for myself and my animals with noticeable results.

440 Hz tuning appears to be not as pleasant as 432 Hz on our minds and bodies. Stress levels, heart rate, and blood pressure were measured on participants who listened to the same music twice, once at 440 Hz and once at 432 Hz. The results showed that the high-frequency sound exposure immediately showed its detrimental effect on heartbeats after a short adaptation period in the 440 Hz group.[330] 440 Hz seems to cause more harm than harmony. There are theories that 440 Hz was deliberately chosen to oppress people. Whatever you believe, many sound healers instruments will be tuned to 432 Hz. 432 Hz resonates with 8 Hz (the Schumann Resonance), which is the documented fundamental electromagnetic "heartbeat" of Earth.[331]

Let's consider the numbers for the harmonics and overtones of the "A" above middle "C" being tuned to 432 Hz:

A= 432 Hz			
OVERTONES		**REDUCTIONS**	**HARMONICS**
Root	36 Hz	36=3+6=9	1st
1st	72 Hz	72=7+2=9	2nd
2nd	108 Hz	108=1+8=9	3rd
3rd	144 Hz	144=1+4+4=9	4th
4th	180 Hz	180=1+8=9	5th
5th	216 Hz	216=2+1+6=9	6th
6th	252 Hz	252=2+5+2=9	7th
7th	288 Hz	288=2+8+8=18/2=9	8th
8th	324 Hz	324=3+2+4=9	9th
9th	360 Hz	360=3+6=9	10th
10th	396 Hz	396=3+9+6=18/2=9	11th
11th	432 Hz	432=4+3+2=9	12th
12th	468 Hz	468=4+6+8=18/2=9	13th
13th	504 Hz	504=5+4=9	14th
14th	540 Hz	540=5+4=9	15th
15th	576 Hz	576=5+7+6=18/2=9	16th

The sum of the numbers for each frequency is 9 because each harmonic interval is separated by 36 hertz (Hz). These are not arbitrary numbers if you look at them through a Pythagorean lens or understand the math behind a circle (a symbol without end—eternity).

Nikola Tesla said, "If you only knew the magnificence of the 3, 6, and 9; then you would have a key to the universe." He also expounded, "If you want to find the secrets of the universe, think in terms of energy, frequency, and vibration." Mathematics is the language of God, which is connected to all the Hebrew Living Letters, because the Hebrew language is an alphanumeric language. Since TET (ט) is nine numerically, it can represent the nine months of pregnancy for a human baby, which is alluded to in the molecular structure of DNA.

The Hebrew Sage Ibn Ezra points out the "Circle of Truth" in his book *Sefer Echad*. The Circle of Truth situates the digits 0-9 around the circumference of a circle. All the multiples of nine are composed of the numbers situated on the same horizontal line. The first line is 0 and 9. The second line is 1 and 8. The third line is 2 and 7. The fourth line is 3 and 6, and the fifth line is 4 and 5.

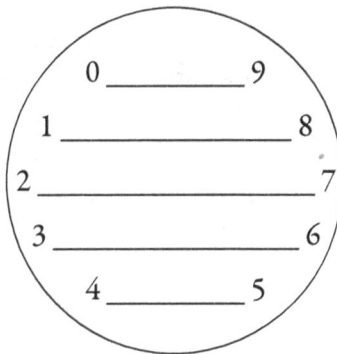

When these numbers are placed around a sphere, the connection between the pairs forms a spherical helix shape—duplicating the DNA molecular structure. The DNA molecular structure is the basis for all living cells. Scripture alludes to all this in the phrase *"all seed of truth"* (*Jeremiah 2:21*). A seed is the foundation from which all living matter develops, and all living matter is composed of DNA; therefore, this DNA "seed" of truth and life can be represented by the number nine.

DNA is a result of the energetic force of the Word. For God to transform (transfigure) human bodies, He has to re-sequence our DNA to His perfect Hebrew Living Letters structure. Ascend the spiral DNA staircase of Christ to receive His breath full of the substance of the Quantum 22. The righteous Hebrew letters know the perfect will of the Three Spheres of Creation. When the righteous and pure Word comes out of PEY (His mouth), the Hebrew Living Letters attach to our DNA to restore it. This creates a golden band, a line if you will, that stretches throughout the eternal now—eternal past, eternal present, and eternal future. Our physical bodies get re-sequenced and re-aligned through being in Christ, so this mortality can take on immortality (1 Cor. 15:53). *"He sent His Word and healed them"* (*Psalms 107:30*).

The Hebrew letters in the midst of Messiah Yeshua came out of the Word of God's heart and mouth "in the beginning." The quantum letters were, and are, the creative force resident in the deep waters of MEM (מ) that the Divine Presence (i.e., the Holy Spirit) hovered upon just before they went to work when the Father of Lights spoke: *"Let there be light"* (*Genesis 1:3*).

The sons of the Living God are going deeper. Destiny is arising. The One New Man in Christ sits in the midst of the Father's perfect heart. Having the Heavenly Father's heart changes the atmosphere around a person. We can choose to hear His heart and join as one man in the Messiah to be pulled into His lights, sounds, and colors. We can choose to step into common union (communion) in Christ. We can choose to become His sons of a higher, enhanced, living color frequency that is surround sound. The One New Man in the Messiah is fully awake and aware of their surroundings. They are Christ Conscious. [332]

ANCIENT TET (ט)

TET (ט) is a quantum letter that you will never find at the end of a Hebrew word. [333] TET (ט) is the least common letter in the Hebrew Bible. There are only 100 words in the Hebrew Bible starting with the quantum letter TET. Of these 15 are proper names and 85 are regular words. The first TET (ט) that appears in the Torah is in the Hebrew word for good— *"tov"* (טוב)—in Genesis 1:4. This speaks of the letter (ט) representing *goodness*.

God is the essence of all good—*tov*: *"Praise the Lord of hosts, For the Lord is good, For His mercy endures forever '—And of those who will bring the sacrifice of praise into the house of the Lord. For I will cause the captives of the land to return as at the first,' says the Lord"* (Jeremiah 33:11). All He created is very good: *"Then God saw everything that He had made, and indeed it was very good. So the evening and the morning were the sixth day"* (Genesis 1:31). The world that God created during the six days of Creation is very good. Notice how God evaluated His labor on each day of Creation. He pronounced a righteous judgment over it: *"God saw that it was good"* (Genesis 1:12).

When a person follows the guidelines set by the Creator, the world remains good. And even when man pollutes the world physically or spiritually, God grants him the gift of repentance. Mankind longs for a good life, good health, good business, and a good year; but what is good? Success is often ephemeral (lasts a short time) and prosperity corrupting, while setbacks and adversity often set the stage for advancement and triumph. Only God knows what is truly, objectively good for man. [334]

According to the Midrash, TET (ט) alludes to mud (טיט), and is symbolic of physical matter, from which man's body was created. Man is in the Almighty's hand as clay is in the potter's hand (Jer. 18:6). [335]

COMPOSITE LETTER (ט)

When TET (ט) is formed from ZAYIN (ז) and KAF (כ), it has a composite numerical value of 27, which produces the *mispar katan* of nine. Nine is the final unit of single integers before the tens begin. Alternatively, the Hebrew letter TET (ט) can be formed from ZAYIN (ז) and NUN (נ), which forms the Hebrew word (זן), the maternal nourishment that a baby receives during its nine months of gestation. [336]

FULFILLMENT OF TET (ט)

The fulfillment of the Hebrew Living Letter TET (ט) is either TET-YUD-TAV (טית) or TET-TAV (טת). When fulfilled as (טת), its *mispar katan* is 13 (9+400= 409=13). Recall that 13 is the same number as the Hebrew word *echad* (אֶחָד), which signifies the unity of God.[337] The second fulfillment of the alternate spelling of (טית) has a final *mispar katan* of 9, the same value as the numerical value and ordinal value of the sacred letter TET (ט).[338]

QUANTUM 9

Atomic Number 9—Fluorine (F). Quantumly speaking, fluorine is a chemical element that has the atomic number **9**, the atomic weight **18.998**, and the atomic symbol **F**. A single Fluorine atom has 9 protons, 9 electrons, and 10 neutrons.

Fluorine is the lightest member of halogen elements. Since fluorine is a halogen, its valency is one. This means that fluorine lacks one electron; and therefore, is an electron recipient that acts as an oxidizing agent. Always remember that electron acceptors are always oxidizing agents and electron donors are always reducing agents.[339]

Its chemical activity can be attributed to its extreme ability to attract electrons. Since Fluorine is the most electronegative element, it is extremely reactive with other elements, except for light inert gases. Atomic groupings rich in fluorine are often negatively charged due to fluorine being the most electronegative element.[340]

Fluorine's chemical activity can also be attributed to the small size of its atoms. The small size of the fluorine atom makes it possible to pack a relatively large number of fluorine atoms or ions around a central atom where it forms many stable complexes—for example, hexafluorosilicate $(SiF_6)^{2-}$ and hexafluoroaluminate $(AlF_6)^{3-}$.[341]

In nature and at its standard state, fluorine is a pale yellow gas with an irritating odor that's very toxic when inhaled. When fluorine is cooled, it becomes a yellow liquid.[342] Not only is fluorine found in the air, but in the earth's crust. Fluorine is the thirteenth most abundant element in the earth's crust. It is always found in a combined state with other elements.[343]

No one cared to industrially produce fluorine until World War II when people realized that uranium hexafluoride has nuclear properties and can be a source of energy. Since then, industrial production of fluorine rose exponentially.[344]

9 IN SCRIPTURE

TET (ט) is 9. The number nine represents the nine gifts of the Spirit. "*⁴ There are diversities of gifts, but the same Spirit. ⁵ There are differences of ministries, but the same Lord. ⁶ And there are diversities of activities, but it is the same God who works all in all. ⁷ But the manifestation of the Spirit is given to each one for the profit of all: ⁸ for to one is given the word of wisdom through the Spirit, to another the word of knowledge through the same Spirit, ⁹ to another faith by the same*

Spirit, to another gifts of healings by the same Spirit, ¹⁰ to another the working of miracles, to another prophecy, to another discerning of spirits, to another different kinds of tongues, to another the interpretation of tongues. ¹¹ But one and the same Spirit works all these things, distributing to each one individually as He wills" (1 Corinthians 12:4-11).

The number nine also represents the nine fruits of the Spirit. "²² But the fruit of the Spirit is love, joy, peace, longsuffering, kindness, goodness, faithfulness, ²³ gentleness, self-control. Against such there is no law. ²⁴ And those who are Christ's have crucified the flesh with its passions and desires. ²⁵ If we live in the Spirit, let us also walk in the Spirit" (Galatians 5:22-25).

Additionally, the number nine represents the nine plants in the bridal garden. "¹² A garden enclosed is my sister, my spouse, a spring shut up, a fountain sealed. ¹³ Your plants are an orchard of pomegranates with pleasant fruits, fragrant henna with spikenard, ¹⁴ spikenard and saffron, calamus and cinnamon, With all trees of frankincense, myrrh and aloes, With all the chief spices—¹⁵ A fountain of gardens, a well of living waters, and streams from Lebanon" (Song of Solomon 4:12-15).

TET (ט) is 9. Christ died at the ninth hour of the day (3 PM in Jerusalem), as the Lamb of God who takes away the sins of the world; thus, making the way of salvation for everyone: "³⁴ And at the NINTH hour Jesus cried out with a loud voice, saying, 'Eloi, Eloi, lama sabachthani?' which is translated, 'My God, My God, why have You forsaken Me?' ³⁵ Some of those who stood by, when they heard that, said, 'Look, He is calling for Elijah!' ³⁶ Then someone ran and filled a sponge full of sour wine, put it on a reed, and offered it to Him to drink, saying, 'Let Him alone; let us see if Elijah will come to take Him down.' ³⁷ And Jesus cried out with a loud voice, and breathed His last. ³⁸ Then the veil of the temple was torn in two from top to bottom. ³⁹ So when the centurion, who stood opposite Him, saw that He cried out like this and breathed His last, he said, 'Truly this Man was the Son of God!'" (Mark 15:34-39).

The Day of Atonement *(Yom Kippur)* is an audio-visual display of Messiah Yeshua's atoning sacrifice on the Cross. *Yom Kippur* is the only annual feast of the Lord when His people worship and fast for one day. The Hebrews claim this is the holiest day of the year. *Yom Kippur* begins at sunset on the ninth day of the seventh Hebrew month: "²⁶ And the LORD spoke to Moses, saying: ²⁷ 'Also the tenth day of this seventh month shall be the Day of Atonement. It shall be a holy convocation for you; you shall afflict your souls, and offer an offering made by fire to the LORD. ²⁸ And you shall do no work on that same day, for it is the Day of Atonement, to make atonement for you before the LORD your God. ²⁹ For any person who is not afflicted in soul on that same day shall be cut off from his people. ³⁰ And any person who does any work on that same day, that person I will destroy from among his people. ³¹ You shall do no manner of work; it shall be a statute forever throughout your generations in all your dwellings. ³² It shall be to you a sabbath of solemn rest, and you shall afflict your souls; on the NINTH day of the month at evening, from evening to evening, you shall celebrate your sabbath'" (Leviticus 23:26-32).

TWISTED & ROLLED TOGETHER

Beyond the primary meaning of a snake or to surround, Gesenius tells us that there is a secondary meaning for TET (ט). TET is believed to be something twisted or rolled together, which is supported by both Arabic and Hebrew words.

First, note that if you desire to have greater clarity about the prophetic Kingdom Day in which we now live, study the phrase "third day" in Scripture. For those of you unfamiliar with this concept, meditate upon if *"with the Lord one day is as a thousand years" (2 Peter 3:8),* then we have entered the third day since Jesus Christ walked this earth.

Let's simply look at how Scripture shows us that on the third day, the tribe of Zebulun was instructed by God to bring the tribal offering for the dedication of His Altar: *"11 Then the Lord said to Moses, 'Let them present their offerings, one leader each day, for the dedication of the altar.' 24 On the third day it was Eliab the son of Helon, leader of the sons of Zebulun" (Numbers 7:11,24).* The dedication ceremony for God's Altar is a picture of the way God's people can enter His Dwelling Place. Understand that when we dedicate something, we devote it to the worship of a divine being, set it apart for sacred uses, and commit to a goal or way of life. The Brazen Altar in God's Tabernacle represented the hearts of the people. The word "heart," in Ancient Hebrew, is a word picture that tells us that the heart is what controls the family or the heart controls that which is on the inside.

If the Israelite's hearts were not right in trying to obey and walk humbly before their God, their sacrifices meant nothing. If we are to dedicate our hearts like Zebulun, what does that mean? Jacob prophesied over his son: *"Zebulun shall dwell at the seashore; and he shall be a haven for ships" (Genesis 49:13).* The name "Zebulun" means dwelling place. It also means flashing light. Zebulun was to be a keeper of the lights of the harbor. In Ancient Hebrew, the word "dwell" means the work of returning (to our primordial state, as in the Garden of Eden) and it also means to have a house. If you research the Hebrew meaning of the word "dwell," you see the word *"shakan"* (pronounced shaw-kan). *Shakan* is a primary root word, and its meaning is akin to transmutation through the idea of lodging. When we behold Yeshua, we shall become like Him (i.e., we become like the Messiah by holding Him dear). I looked up the word "transmutation" in my dictionary. It means to change or alter in form or nature, especially to a higher form. For example, conversion of base metals into gold or silver. A synonym for the word transmutation is the word "transform."

When Zebulun presented their offerings for the dedication of the altar (Num. 7:24), Eliab the son of Helon was chosen in Moses' day. Eliab's name firstly means "God of his father."[345] So what does his father's name mean? Helon means to be strong or a force such as an army, wealth, virtue, or strength.[346] Don't forget the Hebrew word for "strength" in *"Love God with all your strength" (Deuteronomy 6:5)* means to love vehemently; wholly, speedily, especially when repeated; diligently, especially, exceedingly, greatly, louder and louder, mightily, utterly; to rake together; a poker for turning and gathering embers; a firebrand (i.e., a torch). Specifically, this force is a dance, or it means to writhe in pain (i.e., to have fellowship in His sufferings).

Figuratively this force or strength means to wait as in *"be still and know I am God,"* to bear, to bring forth, or to travail (birth). It also means to twist or whirl in a circular, spiral manner. This is interesting because the root meaning of Eliab's name is the Hebrew word *"uwl,"* which so far has been an unused root and means to twist or roll the body together.

The definition of "twist" in the *World Book Dictionary* includes turning with a winding motion, revolving, interweaving, connecting closely together, or associating intimately. As a noun, "twist" means a spiral line or pattern, a spin or twirl, or a cord, thread, or strand formed by twisting fibers, yarns, etc. This is the thread that the Lord is using to knit His people together. The thread is each individual part, you and I, working properly according to the leading of the Spirit of the Living God. It's a heavenly strand of DNA: *"But speaking the truth in love, we are to grow up in all aspects into Him, who is the head, even Christ, from whom the whole body, being fitted and held [knitted] together by what every joint supplies, according to the proper working of each individual part, causes the growth of the body for the building up of itself in love"* (Ephesians 4:15-16 *NASB Addition mine*).

I had a vision where I saw Yeshua differently than ever before. He had a chiseled appearance like an Academy Awards Oscar statute. He was entirely dark, in fact, black in appearance, except for royal blue highlights. I believe that the color black symbolizes the secret things hidden in darkness as well as Yeshua being the Definitive One, and royal blue represents revelation. Below Yeshua's chest was a square opening like a door. I saw nothing below His waist. I came to Him in the vision and fitted perfectly into the very chamber that was below His heart, just like a dovetail joint in a chest of drawers. We were connected in an embrace—the two becoming one. As soon as we were joined together, we started to spin in a circular motion—to twist. The circular spin rolled into an upward spiral going higher and higher. I then saw a sinister figure that I knew was a principality. We whirled right past him—higher and higher—so high above the principality that he no longer knew where we were. The scene then changed to Yeshua in a courtroom—the Judge. I saw Him straighten His arm and point His finger; then I saw that I was tucked in the chamber under His heart, which looked like a doorway. When Yeshua pointed, I saw that I pointed as well, perfectly synchronized with Him. Yeshua is "the Judge." In Ancient Hebrew "judge" is a picture of the door to life, taking us higher to be one with Him. "Eden" in Ancient Hebrew means to see the judge or to see the door of life—Yeshua. We will behold Jesus as we offer our hearts as a daily sacrifice by spending intimate time with Him. Before a bride gets married, she is consumed with her lover—thoughts, conversations, hopes, and dreams. We will become that Dwelling Place for the King. We will manifest what the Strong's definition of "dwell" says: to lodge, permanently stay, abide, have a habitation, lie down for rest or intimacy, ravish, and decease from any other purpose at all. Won't you make room for Him in your inn?

In this prophetic third day, Almighty God will strengthen us, as we make Him our dwelling place; and He Himself, will twist and roll His Body together. In other words: As we dwell with Him, He will twist and roll us together, as one.

QUANTUM QUARK (ט)

In summary, TET (ט) is the ninth Hebrew Living Letter. In Ancient Hebrew, TET (ט) means a snake or to surround. The redemption of TET (ט) for the Quantum 22 belongs to its number 9 and the surround sound of creation. The One New Man in Christ sits in the midst of the Father's perfect heart. Having the Heavenly Father's heart changes the atmosphere around a person. We can choose to hear His heart and join as one man in the Righteous Messiah to be pulled into His lights, sounds, and colors. We can choose to step into common union (communion) with Christ. We can choose to become the sons of Living God that have a higher, enhanced, living color frequency that is a surround sound. The One New Man in the Messiah who sits in Abba's heart is fully awake and aware of their surroundings. They are Christ Conscious.

The whole universe is touched by the mysteries of sound; and its partners: frequency, energy, vibration, light, color, etc. The human body is literally a symphony of sound. Every organ, every bone, every tissue, every cell has its own resonant frequency, its own sound. Every organ has its own frequency-based spectrum that varies according to environment, physical fitness, mentality, etc. When an organ is out of tune, the entire body is affected. A less-than-optimal frequency leads to states of disease and disintegration. In this, frequency plays an important role to identify a problem in the human body as well as having the potential to assist in fixing it.

Q: What are your thoughts about the human body being a symphony of sound?

Q. How can you co-labor with Christ to become a higher, enhanced living color frequency that's a surround sound?

י

YUD (י)—10
DIVINE POINT ENERGY

REDEMPTION OF YUD (י)

The redemption of the Messiah's Quantum 22 continues with the tenth letter of the Hebrew Alphabet. The pronouncement of the letter YUD (י) rhymes with "mode," and it has a numerical value of ten. YUD (י)'s pictograph is a hand, specifically a closed hand, which symbolizes a work, or a deed done. [347]

The work done by God during the six days of Creation is the first primordial work done; therefore, we can extrapolate that YUD (י) is additionally a symbol of Creation. A fundamental part of the creative energy that God used to create the Universe (the heavens and the earth) are the frequencies of the Hebrew letters and their associated numbers. Co-creating with Christ also involves the energies of His Hebrew letters and numbers.

Our gracious Heavenly Father showed a small group the creation facet for the redemption of YUD that He wants to highlight:

> We are traveling to the YUD (י) point. As we get closer and closer, there is stillness and peace while we are pulled to this intersection point. We see a very tall waterfall with two large rock formations on either side. The high waterfall is gentle; but it feels thunderous, especially at its base. We get the impression of His Great Name.
>
> The waterfall is very majestic and seems to be ascending higher. It gets taller as we get closer. We hear: "It's about perspective." The glorious letter YUD spins, as we behold it. There is a spinning sensation on our lips. The waterfall is like His heart ... gentle but strong. The sound is powerful. The sound goes throughout the land outward, like a continuous flow of His sound.
>
> We sense a joyful bubbling up, as it flows away. The spray from the waterfall is saturated with His sound, which affects everything around it. Somehow, it is changing things. We see the Hebrew Living Letters flowing down the waterfall. We see fresh green plants in the crevasses of the rock on both sides of the waterfall. It fits our lips very well, like a green sensation. It is a very misty place. The plants are likened to us.
>
> The voice of many waters is the voice of the Almighty, who speaks life and upholds life. The sacred letters—the building blocks of the Word of God—feed into us. Life is sustained and connected to the Living Word ... a fresh word.

"Deep calls to deep in the roar of Your waterfalls; all Your waves and breakers have swept over me" (Psalms 42:7 ₙᵢᵥ).

The Father's great love and compassion wash over us. The Father's love comes down and goes up. We meet in the intersection of the waterfall. We dance in the midst. We spin around, like the spinning YUD (י). The water gets more intense with more mist. The water washes down over us. It swirls and connects us. We enjoy the swirling.

Our hand touches a fish, and we get a connection to the quantum letter NUN (נ). The fish are joining in. They are wanting in on it. We also get ALEF (א), which is the letter that represents the Father. We understand the letter combination of ALEF-NUN-YUD. The Hebrew letter combination ALEF(א)-NUN(נ)-YUD(י) is one of the Names of God with the intent to expand our perspective, to discover hidden aspects and purposes behind every problem, to decipher unseen secrets, and reveal hidden truths.

We hear Abba Father say, "My precious Bride" in the swirling dance. We hear the sound of the living word in the water, just as the Word—Ten Commandments *Ketubah*—was given during Pentecost. We feel that we are in the center of many directions, and we can follow any direction. Following His sound is following the living word. There are many different tongues and languages here.

We seem to be in a cavern that acts like an invisible umbrella over us. When the water hits us, it turns to light. We expand more and more, as we soak up the living word. We are wearing a green YUD when we dance with the Father. Creation is laughing, as the green YUD garment transforms into a rainbow gown.

The fish become rainbows. They swim in the air. Dancing upright. Defying gravity. These rainbow fishes are witnesses to attest to us. The water goes around us, like a ribbon. The fish—creatures of the water—are going vertical with us. They move through the mist in the air. It looks like an iridescent rainbow, like rainbow trout. They are climbing vertically, not horizontally. We sense that these rainbow fish have to do with being fishers of men. It has to do with the end-time harvest of fish to be caught. We receive a diamond earring in a tear-drop shape. It is for reaching souls in despair.

As we spin in the essence of the Father's mightiness, the promises of God go out. There is a powerful bridal connection in the spin that goes higher and higher and gets stronger and stronger. His sound gets louder and moves out. There are big drops of water coming out of the spin that drops on all nations of the world. It is the Latter Rain. This Latter Rain transformation is connected to bridal hearts unifying with God Almighty. These rainbow fish turn into people at a certain point in their rising.

Why is the Bride wearing YUD on her lips? So, she can speak to creation. YUD is the beginning of Creation. The spiraling green YUD on her lips means: "There is power in her words. There is the power of life in her words." My Name YHVH (יהוה) is on your lips—the first point of Creation. From Zero Point Energy—YUD—that small point Creation begins. [348]

ANCIENT YUD (י)

YUD (י) is the smallest letter in the Hebrew Alef-Bet. It is barely larger than a dot and suspended in mid-air. Some say the shape of YUD resembles an apostrophe. YUD (י) is said to be the atom of the consonants, and the form by which all the other Hebrew letters begin and end. "The first dot with which the scribes first start writing a letter, or the last dot that gives a letter its final form—is a yod."[349]

Being the smallest of all letters, YUD (י) is a picture of humility. Take for example, when Jacob *(Ya'akov)* was renamed to Israel *(Yisrael)*, all that remained of his former name was the sacred letter YUD. An extra YUD (י) is the mark of humility in the text that says that Moses was the most humble man upon the face of the earth (Num. 12:3).

YUD represents Divine Point Energy, which can also be called Zero Point Energy. Zero Point Energy is a cubic centimeter of nothing… no temperature, no magnetic energy, no electromagnetic energy, and no radiation. Zero Point Energy is the God Particle that the Father of Lights used to speak us into this physical realm. This is where the creative energy is coming from. It is the energy He uses to cause you to exist. It is part of the glory of God. It is part of the glory of God above the speed of light. There is a quantum jump when you know that you are hooked up to Messiah Yeshua in oneness, as His Bride. This is the place of endless energy. Zero Point Energy is where Messiah Yeshua is pulling the energy to cause you and everything else to be. In the physical realm, you blink out of here, but your spirit doesn't blink on and off, because it is eternal. Your spirit has eternal life. It is your body that has a problem. Your body is made up of approximately 103 elements (from the Periodic Table of Elements) with the Voice of God speaking, causing it to be. There is a non-physical reality that you are. [350]

The point where this physical universe meets the spiritual is often called Zero Point. It is the point where matter and energy converge as one. It is also the point where time and space are the same thing. [351]

Since YUD (י) is used to form all other letters, it indicates God's omnipresence. Additionally, since Yeshua upholds all things by the word of His power and YUD is part of every Hebrew Living Letter, YUD (י) is considered to be the spark of the Spirit in everything. In other words, YUD is considered to be the starting point of the Presence of God in all things (Heb. 1:3).

"For with [His Name] YAH, HASHEM is the Rock of the Universe" (Isaiah 26:4). Through the study of Scripture and its nuances, many Hebrew Sages believe that God created the universe with the sacred letters YUD (י) and HEI (ה)—YAH (יָהּ).

The Hebrew word for "formed" (וַיִּיצֶר) in Genesis 2:7 has double YUDs, which symbolizes the dual nature of man—earthly and heavenly, mortal and immortal. [352]

YAH (יָהּ) is the shortened form of YHVH (יהוה). Not only is YUD the first Hebrew letter of YAH and YHVH, but it also begins the name of the Savior of the world—Yeshua (ישוע). [353]

Additionally, YUD (י) is the first letter for the four names given to the Hebrews in Scripture:

[1] Ya'akov (יַעֲקֹב)—Jacob.

[2] Yisrael (יִשְׂרָאֵל)—Israel.

[3] Yehudi (יְהוּדִי)—Jews.

[4] Yeshurun (יְשֻׁרוּן)—Jeshurun. [354]

SIMPLE LETTER (י)

YUD (י) is a simple letter with a numerical value of 10, which produces the final *mispar katan* of 1, as in the One and Only Creator of us all. [355]

FULFILLMENT OF YUD (י)

The fulfillment of the letter YUD (י) is YUD-VAV-DALET (יוד). These are all simple letters. The total numerical value of the hidden fulfillment of this expansion (וד) is ten, which is the same as the letter itself. [356]

QUANTUM 10

Atomic Number 10—Neon (Ne). Quantumly speaking, neon is a chemical element that has the atomic number **10**, the atomic weight **20.180**, and the atomic symbol **Ne**. A single neon atom has 10 protons and 10 neutrons, and 10 electrons in two shells.

Neon is an inert gas of the noble gases (Group 18 of the Periodic Table). Neon is the second lightest gas among the noble gases. Neon gas is colorless, odorless, tasteless, lighter than air, and non-flammable. All of the noble gases' indifference toward oxygen confers nonflammability upon them. [357]

Neon is more abundant in the cosmos than on Earth. It ranks fifth in cosmic abundance after hydrogen, helium, oxygen, and carbon. During cosmic nucleogenesis of the elements, large amounts of neon are built up from the alpha-capture fusion process in stars. Although neon is a very common element in the universe and our solar system, it is rare on Earth. [358]

Neon gas occurs in minute quantities in Earth's atmosphere as well as being trapped within rocks in the Earth's crust. [359] Neon is monatomic, making it lighter than the molecules of diatomic nitrogen and oxygen which form the bulk of Earth's atmosphere. A balloon filled with neon will rise into the air, but more slowly than a helium balloon. [360]

Neon is used in electric signs and fluorescent lamps. Since noble gases absorb and emit electromagnetic radiation in a much less complex way than other substances, they are used in discharge lamps and fluorescent lighting devices. If any of the noble gases are confined at a low pressure in a glass tube and an electrical discharge is passed through them, their gas will glow. [361] You will know neon is involved in low-voltage glow lamps, high-voltage discharge tubes, and advertising signs when you witness its distinctive reddish-orange glow. Its glow is fluorescent and can be seen from a distance. In the signage industry, neon signs are electric signs lighted by long luminous gas-discharge tubes that contain rarefied neon or other gases (argon/mercury). [362]

Helium-neon lasers have a red emission line due to the neon. Neon is also used in some plasma tubes and refrigerant applications, but few other commercial uses. As compared to hydrogen, the refrigerating capacity of neon is three times higher than liquid hydrogen per volume unit basis. Neon's refrigerating capacity is 40 times more than that of helium liquid. [363]

Since air is the only source for neon, it is considerably more expensive than helium. Commercially, neon is extracted by fractional distillation of liquid air. Liquid air is the air that has been cooled to very low temperatures (i.e., cryogenic temperatures), so that it condenses into a pale blue mobile liquid. Liquid air is stored in specialized containers (vacuum-insulated flasks are often used) to insulate it to keep the liquid air from becoming a gas at room temperature. Liquid air is often used for condensing other substances into liquid and/or solidifying them. [364]

10 IN SCRIPTURE

YUD (׳) is 10. The world was created with ten utterances. In the first chapter of Genesis the phrase *"and God said"* appears nine times. If you add the first three words of the Bible (also considered an utterance), you get a total of ten utterances.

YUD (׳) is 10. There were ten generations from Adam to Noah to make known how long-suffering the LORD God Almighty truly is.

YUD (׳) is 10. Abraham withstood ten trials, which can be counted in several different ways. One count is as follows: twice ordered to move (Gen. 12:1,10), twice in connection with his two wives (Gen. 12:11, Gen. 21:10), once his war with the kings (Gen. 14:13), once at the covenant between the pieces (Gen. 14:13), once being thrown in the furnace by Nimrod in Ur of the Chaldees (Gen. Rabbah 38:11), and once at the covenant of circumcision (Gen. 17:9). Some commentaries speak of a connection between Abraham's ten tests and the ten utterances with which the world was created. Abraham withstanding his ten trials proved that he was worthy of sustaining the world created by God's ten declarations.

YUD (׳) is 10. There were ten plagues in Egypt that precipitated the Exodus: blood, frogs, gnats, flies, livestock, boils, hail, locusts, darkness, and death of the firstborn. These ten plagues are also called the ten miracles. There were ten miracles performed at The Red Sea:

[1] The sea split.

[2] The water formed a tent over their heads.

[3] The land became firm (not muddy).

[4] It turned muddy when Egyptians tried to cross.

[5] The sea split into 12 strips so each tribe could travel separately.

[6] The water froze and became as hard as a rock.

[7] The water which became rock was actually many rocks beautifully arranged.

[8] The water remained clear so the tribes could see each other.

[9] The water, fit for drinking, leaked from all sides.

[10] After they finished drinking the water, the water immediately froze again. [365]

Some commentators equate the ten different verbs used to describe the death of the Egyptians in Exodus 15 with the ten plagues performed at the Red Sea:

[1] "he has thrown" (15:2),

[2] "he has cast" (15:4),

[3] "deeps cover them" (15:5),

[4] "they went down into the depths" (15:5),

[5] "dashes in pieces the enemy" (15:6),

[6] "You overthrow them that rise up against You" (15:7),

[7] "it consumes them like straw" (15:7),

[8&9] "the waters were piled up, the floods stood upright like a heap" (15:8), and

[10] "they sank as lead" (15:10).

YUD (י) is 10. Ten copper chariot lavers were used to wash the sacrifices in Solomon's Temple. Copper is key to the redemption of the human body. There is a battery that generates energy in each cell—the mitochondria. There are 40 quadrillion mitochondria in each human, which work to turn oxygen into water. This reaction of changing oxygen into water releases the energy that keeps a person alive. At the center of every mitochondrion are 50,000 atoms of copper. Without copper, energy production stalls and fatigue sets in. Please refer to the "Quantum Entanglement with the Quantum One" section of the *Water into Wine* chapter as well as the "Burnt Offerings and Ten Copper Chariot Lavers" section of the *DNA—The Multidimensional Code of Life* chapter.

YUD (י) is 10, and is the symbol of holiness according to the tithe (10%) of the herd or flock as well as the tithe of the land. "*[30] And all the tithe of the land, whether of the seed of the land or of the fruit of the tree, is the Lord's. It is holy to the Lord. [31] If a man wants at all to redeem any of his tithes, he shall add one-fifth to it. [32] And concerning the tithe of the herd or the flock, of whatever passes under the rod, the TENTH one shall be holy to the Lord. [33] He shall not inquire whether it is good or bad, nor shall he exchange it; and if he exchanges it at all, then both it and the one exchanged for it shall be holy; it shall not be redeemed*'" (Leviticus 27:30-33).

QUANTUM QUARK (י)

In summary, YUD (י) is the tenth Hebrew Living Letter. In Ancient Hebrew, YUD (י) is the smallest letter in the Hebrew Alphabet, and as such is a picture of humility. YUD (י) is said to be the atom of the consonants, and the form by which all the other Hebrew letters begin and end. It is the spark of the Spirit in everything. In other words, YUD (י) is considered to be the starting point of the Presence of God in all things. Not only is YUD (י) the first sacred letter of YAH (יָה) and YHVH (יהוה), but it also begins the name of the Savior of the world—Yeshua (ישוע). YUD (י) is the small point of Zero Point Energy where Creation begins. The Bride of Christ wears YUD (י) on her lips, as in the Name YHVH (יהוה), so she can speak His beautiful heart to creation.

Q: Why is YUD (י) being on the Bride of Christ's lips so important? Hint: Oneness.

Q. How do you focus on the Creator to co-create with Him?

KAF (כ)—11
CONFORMED TO HIS IMAGE

REDEMPTION OF KAF (כ)

The redemption of the Messiah's Quantum 22 continues with the eleventh letter of the Hebrew Alphabet. The pronouncement of the letter KAF (כ) has the sound of "k" as in kite. KAF (כ) has an ordinal value of 11 and a numerical value of 20. [366]

Almighty God is highlighting *kanaph* wings and His *kalah* Bride for the redemption of KAF (כ). The Ancient Hebrew Word Picture for *kanaph* is made up of three Hebrew letters: KAF, NUN, and PEY that looks like—כָּנָף. Having just thoroughly studied the word "throne" in Scripture, I also understood that the picture for KAF (כ) as a wing meant a kingdom covering or being enthroned. Therefore, *kanaph*'s Ancient Hebrew Pictograph conveys several things:

[1] *Kanaph* wings are the beginning of covered (enthroned, kingdom) life and action.

[2] *Kanaph* wings speak of kingdom life and action as well as open things up to it.

[3] *Kanaph* wings open things by speaking life.

Once a person is manifestly joined to the Father, the Son, and the Holy Spirit, as a New Living Creature, they still need to press on into greater levels of their high calling of God in Christ to be completely ONE not only with God Himself, but with others. This is a divine work that connects the Head—Jesus Christ who is the Messiah—to His Body. This work of connecting the Messiah's Head to His Body is wrought not by might, nor by power, but by His Spirit (Zech. 4:6) as well as at the intersection where His will and your will are one. We need to understand how powerful the crucified concept is of *"not My will, but thine, be done" (Luke 22:42)*. God's Kingdom comes where His perfect will is heard, understood, and done.

The corporate level of His new creation of the One New Man in Christ is portrayed several ways. One of the ways is the Living Creature's wings. Ezekiel 1:9 tells us that the four living creatures' wings touch one another as they move straight ahead without turning. To understand the significance of the New Living Creature's wings, we need to take a Hebraic peek. The most common (and first) occurrence of the word "wings" in Scripture is the Hebrew word *kanaph* (pronounced kaw-nawf').

The Strong's Concordance tells us that *kanaph* is an edge or extremity, specifically of a wing or an army. This definition got my attention, because I had just come off a third-heaven group experience where the Great Shepherd (i.e., Yeshua) led an excursion to take captivity captive for a group of pastors where the Lord divinely imparted some spiritual equipment to each one of us for our journey: a golden Melchizedek breastplate with triple-concentric circles in the midst, a golden shepherd rod, and a set of wings connected to our hearts. Significantly, He told me ahead of time that four people needed to be present for the type of thing that we all wanted to accomplish. So, when I saw the above definition for *kanaph*, I immediately received a greater revelation: *Kanaph* speaks of a precise wing of God's end-time Melchizedek Army flying by the Spirit with one another.

Next, I flipped over to my *Merriam-Webster's Collegiate Dictionary (10th Edition)* and looked up the definition for "wing":

> **wing** ~*noun* : one of the movable feathered or membranous paired appendages by means a creature is able to fly : power of flight : means of flight or rapid progress : a flank of an army : one of the offensive positions on either side of a center position [Christ] : a unit of the U.S. Air Force higher than a group and lower than a division : two or more squadrons of naval airplanes : a dance step marked by quick outward and inward rolling glide of one foot : insignia consisting of an outspread pair of stylized bird's wings which are awarded on completion of prescribed training to a qualified pilot or aircrew member. [367]

Then I dug even deeper by meditating upon the Ancient Hebrew Pictograph of the word *kanaph* (כָּנָף), which we have seen communicates *kanaph* wings are the beginning of enthroned and covered life and action as well as opening things up to kingdom life and action by speaking life.

When the wings of the Living Creature touch one another, it is a picture of a wing of God's end-time Melchizedek Army touching the heart of a matter—or in other words, a wing of royal priests hitting the bull's eye of the Father's perfect will. It portrays God's Kingdom moving straight ahead when a group of Crucified Ones flies by the Spirit with one another according to His throne. It's the place where His word goes forth and doesn't return void; but accomplishes exactly what is in the heart of the Father, here on earth as it is in heaven.

Let's point our wings to open things to God's kingdom life and action. [368] One of the ultimate points of God's glorious kingdom is His *kalah* bride. The Passion Translation reveals what was on Yeshua's heart and mind, as He spoke His last words on the Cross: *"It is finished, My Bride!"* (John 19:30 TPT). The Passion Translation gets their phrase *"It is finished, My Bride"* from the Hebrew word *kalah*, which has a homonym that means "fulfilled [completed]" and "bride." "This translation has combined both concepts." [369]

Even though the core of *kalah* is the Hebrew word for "bride," it has greater complexity and depth. *Kalah* evokes images of celebration, love, and commitment. It embodies the beauty and sacredness of the union between two individuals and families. Furthermore, the term *kalah* is saturated with spiritual significance. It represents the divine connection and blessings bestowed upon the couple as they embark on a new chapter of their lives. *Kalah* is not just a term for a bride; it is a reflection of love and the enduring bonds that unite us all. [370] *Kalah* is the *kanaph* wings attached to one's heart that first touch God and then one another.

The Lord gave us some divine insight into *kalah* during a *Mystic Mentoring* Group Ascension:

> Due to Yeshua's death, burial, and resurrection, our Pattern Son took off His garment of humanity and put on resurrection life—the garment of His *Kalah* Bride. In this Kingdom Day, His Bride is being clothed in holy rainbow fire and so is Messiah Yeshua's Body. Where He goes, we go. We are one: "I in you and you in Me." It's His favorite garment. The Wise Virgins with full lamps are filled with the Seven Spirits of God. [371] Behold, the Quantum 22 are falling like confetti. It's a restoration party for His *Kalah* Bride where His confetti letters cover her like a veil. She scoops up His Hebrew Living Letters in her heart and in her hands; then offers them back to her Beloved. [372]

ANCIENT KAF (כ)

KAF (כ) is the symbol of Crowning Accomplishment. The Hebrew word for crown is KEH-tehr (כֶּתֶר). Rabbi Munk speaks of three crowns: the crown of priesthood, the crown of kingship, and the crown of Torah (the Word of God Himself). There were three vessels in God's Temple that were topped with a golden crown all around. The crown atop the Golden Altar of Incense alludes to the crown of the priesthood (Exo. 30:3). The crown atop the Table of Showbread alludes to the crown of kingship (Exo. 25:24). The crown atop the Ark of the Covenant alludes to the ultimate crown of the Word of God—Torah (Exo. 25:11). God's original intention was that the entire nation should wear all three crowns.[373]

> "*3 And Moses went up to God, and the Lord called to him from the mountain, saying, 'Thus you shall say to the house of Jacob, and tell the children of Israel:* 4 *"You have seen what I did to the Egyptians, and how I bore you on eagles' WINGS and brought you to Myself.* 5 *Now therefore, if you will indeed obey My voice and keep My covenant, then you shall be a special treasure to Me above all people; for all the earth is Mine.* 6 *And YOU SHALL BE TO ME A KINGDOM OF PRIESTS AND A HOLY NATION." These are the words which you shall speak to the children of Israel'* " *(Exodus 19:3-6).*

At Sinai, all of Israel volunteered to accept God's Wedding Proposal unconditionally, saying, *"we will do and (then) we will understand" (Exodus 24:7).* This is when everyone received the crown of the Word of God. After the sin of the Golden Calf, the crowns of kingship and priesthood were no longer easily accessible. At the time, these crowns were lost to the holy nation as a whole; however, God preserved them. The priesthood was appointed to the lineage of Moses' brother Aaron while kingship was allotted to the House of David. [374]

A deposit of the crown of the Word of God has been given to all men: "*1 In the beginning was the Word, and the Word was with God, and the Word was God.* 2 *The same was in the beginning with God.* 3 *All things were made by him; and without him was not any thing made that was made.* 4 *In him was life; and the life was the light of men" (John 1:1-4).*

We are now in a *karios* time in this Kingdom Day when the door has been opened wide for whosoever will to become a king and priest after the Righteous Order of Melchizedek. *"⁶ And I looked, and behold, in the midst of the throne and of the four living creatures, and in the midst of the elders, stood a Lamb as though it had been slain, having seven horns and seven eyes, which are the Seven Spirits of God sent out into all the earth. ⁷ Then He came and took the scroll out of the right hand of Him who sat on the throne. ⁸ Now when He had taken the scroll, the four living creatures and the twenty-four elders fell down before the Lamb, each having a harp, and golden bowls full of incense, which are the prayers of the saints. ⁹ And they sang a new song, saying: 'You are worthy to take the scroll, And to open its seals; For You were slain And have redeemed us to God by Your blood out of every tribe and tongue and people and nation, ¹⁰ And have made us kings and priests to our God; And we shall reign on the earth'" (Revelation 5:6-10).*

Believers in our Lord and Savior Jesus Christ are commissioned into the Order of Melchizedek when they *"enter the Kingdom of God."*

"⁵ Jesus answered, 'Most assuredly, I say to you, unless one is born of water and the Spirit, he cannot ENTER THE KINGDOM OF GOD. ⁶ That which is born of the flesh is flesh, and that which is born of the Spirit is spirit'" (John 3:5-6).

"Strengthening the souls of the disciples, exhorting them to continue in the faith, and saying, 'We must through many tribulations ENTER THE KINGDOM OF GOD'" (Acts 14:22).

"²³ Then Jesus said to His disciples, 'Assuredly, I say to you that it is hard for a rich man to enter the kingdom of heaven. ²⁴ And again I say to you, it is easier for a camel to go through the eye of a needle than for a rich man to ENTER THE KINGDOM OF GOD'" (Matthew 19:23-24).

"²³ Then Jesus looked around and said to His disciples, 'How hard it is for those who have riches to ENTER THE KINGDOM OF GOD!' ²⁴ And the disciples were astonished at His words. But Jesus answered again and said to them, 'Children, how hard it is for those who trust in riches to ENTER THE KINGDOM OF GOD! ²⁵ It is easier for a camel to go through the eye of a needle than for a rich man to ENTER THE KINGDOM OF GOD'" (Mark 10:23-25).

"¹⁷ 'Assuredly, I say to you, whoever does not receive the Kingdom of God as a little child will by no means enter it' . . . ²⁴ᵇ He [Yeshua] said, 'How hard it is for those who have riches to ENTER THE KINGDOM OF GOD! ²⁵ For it is easier for a camel to go through the eye of a needle than for a rich man to ENTER THE KINGDOM OF GOD'" (Luke 18:17, 24b-25).

Since KAF's (כ) pictograph looks like the palm of a hand (anything contained in the palm of a hand) or a wing, it represents the place in the body where potential is actualized. The two quantum letters that spell KAF (כף) are the initial letters of two Hebrew words *koach* ("potential") and *poel* ("actual"); therefore, KAF hints at the latent power within the spiritual realm of the potential to fully manifest in the physical realm—the actual.

Grammatically, when a Hebrew word is prefixed with a KAF (כ), it carries the meaning being "like" or "as." When we "prefix" ourselves with the KAF (כ) of the Messiah, we are conformed to His image. We resemble Him, specifically, we resemble Yeshua in what we do.

Michael Munk tells us: KAF (כ) as (כּף) has a dual symbolism. "It stands for the palm of the hand serving as a container and at the same time as the measure of what it holds." "(כּף) also denotes productivity and accomplishment (Ibn Ezra to Psalms 73:13), which result through mental and physical efforts." The accomplishments through mental and physical efforts of KAF (כּף) are unlike YUD (י), which stands for hand but indicates power and possession. [375]

The Havdalah blessing at the conclusion of the Sabbath and festivals refers to the transition from holy activities (Sabbath) to normal activities (weekdays). To indicate the difference between the two, a person opens and closes their hand near the multi-wicked candlelight. The open hand symbolizes the work-free rest of the out-going Sabbath while the closed hand signifies the readiness for action and acquisition. [376]

SIMPLE LETTER (כ)

The sacred letter KAF (כ) belongs to the seven simple letters. It is used in the composition of seven letters TET (ט), KAF (כ), LAMED (ל), MEM (מ), SAMECH (ס), PEY (פ), and KOOF (ק).

FULFILLMENT OF KAF (כ)

The fulfillment of the letter KAF (כ) is KAF-FEY (כּף), which has a numerical value of 100 = 20+80 and a final *mispar katan* of 1. Its second fulfillment (פא כּף) also has a final *mispar katan* of one, which emphasizes the Unity of YHVH. [377]

QUANTUM 11

Atomic Number 11—Sodium (Na). Quantumly speaking, sodium is a chemical element of the alkali metal group that has the atomic number **11**, the atomic weight **22.990**, and the atomic symbol **Na**. A single sodium atom has 11 protons, 11 electrons, and 11 neutrons.

Sodium is a very soft silvery-white metal. It is the most common alkali metal and the sixth most abundant element on Earth. [378] Because sodium is extremely reactive, it never occurs in the free state in Earth's crust. However, one of sodium's most known and important compounds found in nature is common table salt—sodium chloride (NaCl). [379]

Sodium salts, particularly sodium chloride, are found almost everywhere in biological material. It is essential for all animals and some plants. Sodium is an essential element for life, as is potassium. Sodium and potassium are the two elements in the human body that needs to maintain a balance within the cell structure. The electrolyte balance between the inside of a cell and the outside is maintained by the "active transport" of potassium ions into the cell and sodium ions out of the cell. [380] Even though sodium is essential for life, it is rarely deficient in diets. High sodium intake is linked to hypertension—high blood pressure. [381]

When sodium chloride forms the mineral halite, it constitutes about 80 percent of the dissolved constituents of seawater. [382] The sodium content of the sea is approximately 1.05 percent, corresponding to a concentration of approximately 3 percent of sodium halides. Sodium has been identified in both the atomic and ionic forms in the spectra of stars, including the Sun. [383]

Most sodium compounds are prepared either directly or indirectly from sodium chloride, which occurs in seawater, in natural brines, and as rock salt. Large quantities of sodium chloride are employed in the production of other heavy (industrial) chemicals as well as being used directly for ice and snow removal, water conditioning, and food. [384]

There are many uses for sodium bicarbonate. As baking soda, it is used as a raising agent in baking. Sodium bicarbonate can be used to extinguish small grease or electrical fires by being thrown over a fire because the heating of sodium bicarbonate releases carbon dioxide. When mixed with water, sodium bicarbonate can be used as an antacid to treat heartburn and acid indigestion. It reacts with the body's stomach acid to produce salt, water, and carbon dioxide. Sodium bicarbonate is used in some mouthwashes, as a cleanser for teeth and gums. Some even brush their teeth with it. Sodium bicarbonate's cleansing qualities extend to soda-blasting, which removes paint and corrosion. [385]

Cattle can also receive the benefits of sodium bicarbonate, which acts as a buffering agent for their rumen when added as a supplement to their feed.

During the Manhattan Project, which fostered the development of the nuclear bomb in the early 1940s, the chemical toxicity of uranium was an issue. Uranium oxides were found to stick to cotton cloth, and they did not wash out with soap or laundry detergent. However, uranium would wash out with a 2% solution of sodium bicarbonate.

Millions of tons of sodium chloride, hydroxide, and carbonate are produced annually.

11 IN SCRIPTURE

KAF (כ) is 11. Eleven symbolizes disorder, chaos, and judgment in the Bible. The number 11 is used twenty-four times in Scripture and "11th" is found nineteen times. The number ten (10) represents ordinal perfection, law, and responsibility while eleven (11) represents the opposite. The irresponsible breaking of laws brings disorder, chaos, and judgment. Speaking of chaos and disorder, the Babylonians are notably connected to the number 11 in Scripture; thus, so are rebellion and witchcraft too.

KAF (כ) is 11. Genesis 11 sets the precedence for the number eleven. Man rebelled against God with Nimrod as their leader to build the Tower of Babel. God judged their trying to reach heaven through unlawful means by confusing their languages and disbursing the people from the plains of Shinar. These were the first seventy nations of the world.

KAF (כ) is 11. One of the last kings of Judah—Jehoiakim—ruled for 11 years from 609 to 598 BC. King Jehoiachin was his successor who only ruled three months before the Babylonians took control of Jerusalem. When King Nebuchadnezzar of Babylon-fame set up Zedekiah as his puppet ruler of Judea, he didn't anticipate Zedekiah's rebellion. After only 11 years, Nebuchadnezzar had to conquer Jerusalem again; but the second time he destroyed both the city of Jerusalem and its temple.

KAF (כ) is 11. Moses spoke to Israel before they were to enter the Promise Land during the eleventh month of the fortieth year. Moses reminded them how they could have reached their Promised Land from Horeb in eleven days instead of wandering in the wilderness for forty years. Israel had rebelled against the command of God by believing the spies' bad report, complaining in their tents, and refusing to go up and possess the land forty years earlier: *"¹ These are the words which Moses spoke to all Israel on this side of the Jordan in the wilderness, in the plain opposite Suph, between Paran, Tophel, Laban, Hazeroth, and Dizahab. ² It is ELEVEN days' journey from Horeb by way of Mount Seir to Kadesh Barnea. ³ Now it came to pass in the fortieth year, in the ELEVENTH month, on the first day of the month, that Moses spoke to the children of Israel according to all that the LORD had given him as commandments to them"* (Deuteronomy 1:1-3).

KAF (כ) is 11. The eleventh son of Jacob was favored and given a coat of many colors. Joseph dreamt of the sun, the moon, and the eleven stars—his brothers—bowing down to him (Gen. 37:9), which they did in Egypt many years later.

KAF (כ) is 11. The "eleventh hour" is a phrase that communicates the latest possible moment before it is too late. This phrase comes from a parable in Matthew 20 in which a few last-minute workers were hired long after the others, but they were paid the same wages.

KAF (כ) is 11. The eleventh foundation stone for the New Jerusalem is called "jacinth" in English and *"leshem"* in Hebrew. *"The foundations of the wall of the city were adorned with all kinds of precious stones: the first foundation was jasper, the second sapphire, the third chalcedony, the fourth emerald, the fifth sardonyx, the sixth sardius, the seventh chrysolite, the eighth beryl, the ninth topaz, the tenth chrysoprase, the ELEVENTH jacinth, and the twelfth amethyst"* (Revelation 21:19-20). Jacinth has another name "hyacinth." This precious stone comes in the colors of fiery judgment. Jacinth includes transparent red, orange, or yellow varieties of the gemstone zircon. Significantly, Jacinth was also included on the third row of the High Priest's Breastplate of Righteousness (Exo. 39:12).

KAF (כ) is 11. Eleven doesn't always speak of something evil, as the eleventh foundation stone of the New Jerusalem attests. Doug Addison reveals eleven is the number for transition. The eleventh hour is just before the start of a new day. In Deuteronomy 11:11, God promises Israel when they make it to the Promised Land, they will transition into a new season where God's blessing would flow from the beginning of the year to the end. *"But the land you are crossing the Jordan to take possession of is a land of mountains and valleys that drinks rain from heaven"* (Deuteronomy 11:11 $_{NIV}$). Not only does eleven symbolize transition but also new revelation. In Luke 11:11, Jesus shares a new revelation about our Heavenly Father where He is not mean or judgmental, but caring and kind. *"Which of you fathers, if your son asks for a fish, will give him a snake instead?"* (Luke 11:11 $_{NIV}$). [386]

Yeshua was about 33 years old when he died for the sins of the world. 33 is a multiple of 11. After Judas Iscariot betrayed Jesus and hung himself (Matt. 27:5), only eleven disciples remained. These eleven were rebuked by Yeshua for their unbelief and hardness of heart regarding His resurrection: "*⁹Now when He rose early on the first day of the week, He appeared first to Mary Magdalene, out of whom He had cast seven demons. ¹⁰ She went and told those who had been with Him, as they mourned and wept. ¹¹ And when they heard that He was alive and had been seen by her, they did not believe. ¹² After that, He appeared in another form to two of them as they walked and went into the country. ¹³ And they went and told it to the rest, but they did not believe them either. ¹⁴ Later He appeared to the ELEVEN as they sat at the table; and He rebuked their unbelief and hardness of heart, because they did not believe those who had seen Him after He had risen. ¹⁵ And He said to them, 'Go into all the world and preach the gospel to every creature'" (Mark 16:9-15).*

KAF (כ) is 11. The number eleven is especially connected to resurrection. In Revelation 11:11, we observe the two witnesses being resurrected: *"Now after the three-and-a-half days the breath of life from God entered them, and they stood on their feet, and great fear fell on those who saw them."*

John 11:11 reveals Yeshua telling His disciples that He will awaken Lazarus after being clinically dead for four days: *"This He said, and after this He said to them, 'Our friend Lazarus has fallen asleep; but I am going so that I may awaken him from sleep.'"*

Romans 11:11 puts an exclamation point to the number eleven's association with resurrection: *"¹¹ I say then, they did not stumble so as to fall, did they? Far from it! But by their wrongdoing salvation has come to the Gentiles, to make them jealous. ¹² Now if their wrongdoing proves to be riches for the world, and their failure, riches for the Gentiles, how much more will their fulfillment be! ... ¹⁵ For if their rejection proves to be the reconciliation of the world, what will their acceptance be but life from the dead? (Romans 11:11-12,15 _{NASB}).*

KAF (כ) is 11. The eleventh verse of Psalms 119 is near and dear to my heart due to it being the theme verse for my son memorizing Bible verses when He was younger: *"Your word I have hidden in my heart, That I might not sin against You."*

ROYAL LETTER (כ)

There are three Hebrew Living Letters that compose the Hebrew word for "king" (מֶלֶךְ). *Melech* is made up of KAF (כ), LAMED (ל), and MEM (מ). Notice how they appear in reverse alphabetic order, which represents the attribute of judgment. It is a monarch's responsibility to maintain law and order in their kingdom.

Significantly, when the kings of the dynasty of *David ha Melech* (King David) were anointed, special oil was smeared on their heads in the shape of the letter KAF (כ). As previously discussed, the letter KAF (כ) also represents a crown (כֶּתֶר). In particular, the three crowns of priesthood, kingship, and the Word of God.

QUANTUM QUARK (כ)

In summary, KAF (כ) is the eleventh Hebrew Living Letter. The number 11 in Scripture can symbolize disorder, chaos, and judgment. However, the other side of this two-edged sword is that 11 is the number for transition and resurrection. In Ancient Hebrew, KAF (כ) means a palm or wing. Since KAF's (כ) pictograph looks like the palm of a hand (anything contained in the palm of a hand), it represents the place in the body where potential is actualized.

The noise of the wings of the four living creatures are like the voice of the Almighty (Ezek. 1:24). The most common and first occurrence of the word "wings" in Scripture is the Hebrew word *kanaph* (כָּנָף), which speaks of the beginning of enthroned and covered life and action as well as opening things up to kingdom life and action by speaking life. When the wings of the Living Creature touch one another, it is a picture of a wing of God's Righteous End-time Melchizedek Army—His *Kalah* Bride—touching the heart of a matter.

Q: What is a potential in your life that you would like to see actualized?

Q. How can you point your wings to kingdom life and action?

ל

LAMED (ל)—12
SHEPHERD OF LOVE

REDEMPTION OF LAMED (ל)

The redemption of the Messiah's Quantum 22 continues with the twelfth letter of the Hebrew Alphabet. LAMED (ל) is pronounced *lah-med*. It has an ordinal value of 12 and a numerical value of 30. The ancient pictograph for LAMED (ל) is a rod or a staff, like a shepherd's staff.

We return to the first chapter of this book—*Three Spheres and the Hebrew Living Letters*—for the redemption of LAMED (ל). The first two paragraphs bring to light:

> "Like a comet, we ascend in Christ by the Blood of the Lamb to the intersection place where the Heavenly Father and the Son are one, where there is fullness of joy and love. Light expands from us and pulsates out. The vibrancy of the white light is a very high frequency. The brightest light appears to be coming down like some sort of pole. Then, it wraps around us, as a whirlwind.
>
> We behold three spherical lights in front of us. We understand that this is the source of the light. It is the Father, the Son, and the Holy Spirit. There is a horizontal beam behind the three spheres of light that is a soft lemony color. The pale lemon-yellow band of color appears to go behind, above, and below the three spheres. We understand that the pastel lemon color encompasses all three, which speaks of a particular type of unity. It is their goodness uniting them."

The LORD revealed that the pole of brightest light that wrapped around the group, like a whirlwind, was the LAMED (ל) Light before Creation, which is the goodness uniting the Three Spheres: The Father, The Son, and The Holy Spirit. YHVH first showed us the glorious letter LAMED (ל); then we saw it turn into a shepherd staff of light that wrapped around us. The Hebrew Living Letter YUD (י) was first in Creation itself, but LAMED (ל) was first before Creation.

The first five books of the Bible—the Torah—begins with the Hebrew letter BET (ב), as in *bereshit* (בְּרֵאשִׁית), and ends with LAMED (ל). Therefore, we can say that the entire Torah is contained between the letters LAMED-BET (לֵב)—the heart (Osios R'Akiva). The heart—

lev (לֵב)—is the center of a man's body, just as LAMED (ל) is the center of the Hebrew Alphabet. The Hebrew Pictograph for the word *"lev"* tells us that the heart is what controls the inside.

LAMED (ל) ending the Torah is all about the heart of the Word—the heart of Messiah Yeshua—from which everything flows. Many moons ago, I had a vivid vision during my quiet time before dawn. I saw Yeshua from the waist up with His arms outstretched. I heard Him say, "Come into My heart," as He crossed His arms over His chest before His image changed into a classic heart shape. Instantly, I was in a mauve-colored cavern, which was warm (just the right temperature), lit up, and so very clean. To one side, I beheld a small waterfall and heard: "Out of areas of brokenness My living waters flow." [387]

ANCIENT LAMED (ל)

LAMED (ל) is a unique and majestic quantum letter that towers above the other letters from its position in the heart of the Hebrew Alef-Bet. Since LAMED (ל) towers over the other letters from its central position, it is said to represent *melekh hamelakhim*—the King of Kings. It is not a coincidence that the three central letters of the Hebrew Alef-bet can spell out the Hebrew word *Melech* or *Melekh* (מלך), which means king. *"These will make war with the Lamb, and the Lamb will overcome them, for He is Lord of lords and KING OF KINGS; and those who are with Him are called, chosen, and faithful"* (Revelation 17:14).

> *"Now out of His mouth goes a sharp sword, that with it He should strike the nations. And He Himself will rule them with a rod of iron. He Himself treads the winepress of the fierceness and wrath of Almighty God. And He has on His robe and on His thigh a name written: KING OF KINGS and Lord of Lords"* (Revelation 19:15-16).

LAMED (ל) is the symbol of Teaching and Purpose. The Hebrew name for the letter itself—LAMED (למד)—comes from the root *lamad*, which means to teach or to learn. The first occurrence of *lamad* (למד) in Scripture is in Deuteronomy 4:1. *"Now, O Israel, listen to the statutes and the judgments which I TEACH you to observe, that you may live, and go in and possess the land which the Lord God of your fathers is giving you"* (Deuteronomy 4:1).

LAMED (למד) can be seen as an acronym for the phrase *lev meivin da-at*, which is translated as "a heart that understands knowledge." In other words, the goal of learning and teaching—LAMED (ל)—is to absorb the lessons into one's heart. And since LAMED ascends over the other quantum letters, it represents the prominence of learning and understanding to the heart. Just as the heart sustains the body, so does heartfelt learning of the Word of God sustain the spirit. Just as faith without works is dead (Jam. 2:20), so is learning without action: *"Even a child is known by his deeds, whether what he does is pure and right"* (Proverbs 20:11). Studying is not the ultimate goal; but rather, the actions that result from one's study.

"In the days of the Scriptures, God taught His ways to Israel through His prophets. Then, man could understand history and perceive it as the language of God's will. Without prophecy, it remains for man to search for God's hand in His conduct of events, for it is an article of faith that Divine Providence is always at work. In this sense, God is a Teacher, using events to show what is called for from man. This concept is implied in the large LAMED (ל) of "He cast them into another land" (Deut. 29:27). This enlarged LAMED (ל) alludes to the manner in which the Supreme Teacher watches over His charges, guiding and teaching them—sometimes in a merciful way, at other times with strict lessons. Even when a severe ordeal, such as *galus* [exile], is brought upon Israel, it serves an instructive purpose. The purpose of this suffering is to arouse Israel to change its faulty attitudes and to repent. (Shach). In fact, the large *lamed* indicates to the Jews that the length of the exile depends on them, for it will not end until its lesson is learned (Minchas Shai). Furthermore, R'Hirsch comments: Wherever the Jewish people are to be found away from their original homeland, they are there not by chance. The Director of history cast them there to teach the nations belief in One God." [388]

COMPOSITE LETTER (ל)

LAMED (ל) is composed of the letter KAF (כ) as its base and VAV (ו) as its neck. Together KAF and VAV have a numerical value of 26. Twenty-six is the same numerical value as His Sacred Name—YHVH. [389]

FULFILLMENT OF LAMED (ל)

The fulfillment of the sacred letter LAMED (ל) is LAMED-MEM-DALET (למד), which has a numerical value of 74 = 30+40+4 and a final *mispar katan* of 2. 74 = 7+4 = 11 where 1+1 =2. The second fulfillment of LAMED (למד) is LAMED, MEM, DALET with each spelled out, which has a *mispar katan* of 30. Thirty is the same value as the numerical value of the letter itself. [390]

QUANTUM 12

Atomic Number 12—Magnesium (Mg). Quantumly speaking, magnesium is a chemical element that has the atomic number **12**, the atomic weight **24.305**, and the atomic symbol **Mg**. The most common and stable type of magnesium atom contains 12 protons, 12 electrons, and 12 neutrons.

Magnesium is a shiny gray metal that has a low density, low melting point, and high chemical reactivity. After iron, oxygen, and silicon, magnesium is the fourth most common element on the Earth, making up 13% of our planet's mass. After sodium and chlorine, magnesium is the third most abundant element in seawater. It is also the eighth most abundant element in the Earth's crust. However, it does not occur freely in nature; but in combination with other elements in compounds such as sulfate (Epsom salts), oxide (magnesia), and carbonate (magnesite). [391]

Magnesium is a chemical element of the alkali-earth metals of Group 2 of the periodic table. It is the lightest structural metal, used widely in construction and medicine. [392] "The metal, which burns in air with a bright white light, is used in photographic flash devices, bombs, flares, and pyrotechnics; it is also a component of lightweight alloys for aircraft, spacecraft, cars, machinery, and tools. The compounds, in which it has valence 2, are used as insulators and refractories and in fertilizers, cement, rubber, plastics, foods, and pharmaceuticals (antacids, purgatives, laxatives)." [393]

I find it very interesting that magnesium is the eleventh most abundant element by mass in the human body, essential to all living cells, and some 300 enzymes. It makes sense that magnesium is a key element in human nutrition. In fact, magnesium is one of the critical elements of cellular life. "The Mg^{2+} ion is involved with the critically important biological polyphosphate compounds DNA, RNA, and adenosine triphosphate (ATP). Many enzymes depend on magnesium for their functioning. About one-sixth as plentiful as potassium in human body cells, magnesium is required as a catalyst for enzyme reactions in carbohydrate metabolism. Magnesium also is an essential constituent of the green pigment chlorophyll, found in virtually all plants, algae, and cyanobacteria. The photosynthetic function of plants depends upon the action of chlorophyll pigments, which contain magnesium at the center of a complex, nitrogen-containing ring system (porphyrin). These magnesium compounds enable light energy to drive the conversion of carbon dioxide and water to carbohydrates and oxygen and thus directly or indirectly provide the key to nearly all living processes." [394]

In the Universe, magnesium is produced in large aging stars by the sequential addition of three helium nuclei to a carbon nucleus. When these stars explode, as supernovas, much of their magnesium is expelled into the matter and radiation in the space between the star systems (interstellar medium) where it is available to be recycled into new star systems. [395]

12 IN SCRIPTURE

LAMED (ל) is 12, which is an extremely prominent number in Scripture. The number 12 can be found in 187 places in Scripture. The Book of Revelation alone has 22 occurrences. Twelve is considered a perfect number. 12 symbolizes God's power and authority as well as serving as a perfect governmental foundation, indicating completeness.

Jacob had twelve sons each a prince of the twelve tribes of Israel:

- [1] Reuben means to "see a son."
- [2] Simeon means "hated."
- [3] Levi means "loved."
- [4] Judah means "praised."
- [5] Dan means "vindicated" or "judge."
- [6] Naphtali means "prevailed."
- [7] Gad means "fortunate."

[8] Asher means "happy."

[9] Issachar means "wages."

[10] Zebulun means "dwell with honor."

[11] Joseph means "add to me."

[12] Benjamin's name meant "son of sorrows" when Rachel died; and then, he was renamed "son of my right hand" by his father Jacob.

Seven in Scripture illustrates the plan of God regarding the Messiah. For example, the seven sons of Leah tell the whole story of the Messiah. There is a proverb in the meaning of their names: The Messiah came as God's Son. He was hated by some but loved by others. We should praise Him, because He paid the wages for sin. Soon, He will come to dwell with us, and we will be His Bride [the only sister Dinah is the feminine version of "judge" with a bridal connotation]. The Messiah is the son of sorrows at the right hand of the Father.

LAMED (ל) is 12. The High Priest's Breastplate had twelve precious stones embedded in it, representing the 12 Tribes of Israel. There's a reason you don't hear a lot about the twelve Breastplate stones. The details surrounding them are quite enigmatic. The tribal names (and possible additions) written on these stones are particularly mysterious. There are over 30 various opinions on whose names were on which stone. Personally, I like Flavius Josephus' take because his is the oldest recorded account of the stones on the Breastplate, and he was an eyewitness that served in the Temple in Jerusalem as a priest as well. Josephus tells us that the tribes on the Breastplate were written in birth order. Don't forget that the Hebrews read from right to left.

God's invisible government is portrayed in the High Priest's Breastplate, which was connected to his shoulders and placed on his heart. God's government is upon Yeshua's shoulders—the Great High Priest of the Righteous Order of Melchizedek. Consider that Yeshua is the Head while His believers are His Body, which has shoulders. The background of the ancient picture of the Royal Priesthood of Believers includes the Breastplate of Judgment, sometimes called the Breastplate of Righteousness *(Urim v'Tumim)*. The Breastplate of Judgment on the High Priest's garments was made of a 28-strand colored wool yarn/linen/gold material that had been twisted and rolled together, then folded in half to form a pouch. The 28-strand Breastplate was shaped like a rectangle measuring 1 cubit by ½ cubit (approximately 18 inches by 9 inches). In this pouch-like pocket, Moses inserted a piece of parchment that is said to have had the Sacred Name YHVH written on it. It's a mystery what Moses actually wrote, or how he wrote it, but we do know that the only one at that time to have met God face to face was Moses. [396]

Moses was the only one who had the spiritual knowledge to write that Name on that holy piece of parchment, which was then inserted into the High Priest's Breastplate. This Name or Names inserted into the cleft of the Breastplate was called *Urim*, from the Hebrew word "light," because it would cause individual letters of the tribal names on the Breastplate to light up. This Name was also called *Tumim*, from the word "completeness," because if read in the proper order, these luminous letters presented complete and true answers to the questions of national import that the *Kohen Gadol* (High Priest) would ask of God. [397]

A Jewish Sage named Ramban gives an example of how this *Urim v'Tumim*/Breastplate process took place. "When God's people crossed the Jordan and had to undertake the conquest of the Land, the question arose which tribe should begin the war against the Canaanites. Phinehas, the *Kohen Gadol,* entered the Tabernacle and posed the question. The name Judah lit up and the letters HEI, LAMED, AYIN, and YUD. The *Kohen* had to know what this combination of letters represented, because they could be placed in several orders, thus forming different combinations of words. A Divine spirit gave him the wisdom to know that the message of the *Urim v' Tumim* was "The tribe of Judah shall go forth [to wage war]" (See Judges 1:1-2). [398]

LAMED (ל) is 12. The New Jerusalem has twelve gates where each is a pearl with an angel at each gate. Additionally, the names of the 12 Tribes of the children of Israel are written on these gates: *"⁹ᵇ 'Come, I will show you the Bride, the Lamb's wife.' ¹⁰ And he carried me away in the Spirit to a great and high mountain, and showed me the great city, the holy Jerusalem, descending out of heaven from God, ¹¹ having the glory of God. Her light was like a most precious stone, like a jasper stone, clear as crystal. ¹² Also she had a great and high wall with TWELVE gates, and TWELVE angels at the gates, and names written on them, which are the names of the TWELVE tribes of the children of Israel: ¹³ three gates on the east, three gates on the north, three gates on the south, and three gates on the west"* (Revelation 21:9b-13).

LAMED (ל) is 12. The New Jerusalem has twelve foundations with the names of the 12 Apostles on them: *"¹⁴ Now the wall of the city had TWELVE foundations, and on them were the names of the TWELVE apostles of the Lamb. ¹⁵ And he who talked with me had a gold reed to measure the city, its gates, and its wall. ¹⁶ The city is laid out as a square; its length is as great as its breadth. And he measured the city with the reed: TWELVE thousand furlongs. Its length, breadth, and height are equal. ¹⁷ Then he measured its wall: one hundred and forty-four [12×12] cubits, according to the measure of a man, that is, of an angel. ¹⁸ The construction of its wall was of jasper; and the city was pure gold, like clear glass. ¹⁹ The foundations of the wall of the city were adorned with all kinds of precious stones: the first foundation was jasper, the second sapphire, the third chalcedony, the fourth emerald, ²⁰ the fifth sardonyx, the sixth sardius, the seventh chrysolite, the eighth beryl, the ninth topaz, the tenth chrysoprase, the eleventh jacinth, and the TWELFTH amethyst. ²¹ The TWELVE gates were TWELVE pearls: each individual gate was of one pearl"* (Revelation 21:14-21 Additions mine).

Notice that the 12th foundation stone is amethyst. It radiates at the frequency of rest, peace, and government. *"Of the increase of His government and peace there shall be no end"* (Isaiah 9:6). The classic color of amethyst crystals is a royal purple, which emanates alpha waves that are neural oscillations in the frequency range of 8–12 Hz. This is the feeling you get just before you fall asleep and just before you wake up. [399] This frequency range bridges the gap between our conscious thinking and subconscious mind. In other words, alpha is the frequency range that helps us calm down when necessary and promotes feelings of deep relaxation. [400]

LAMED (ל) is 12. Twelve loaves of unleavened bread were placed on the Table of Showbread. This pure gold Table was placed near the north wall of the Wilderness Tabernacle. Twelve large loaves of *lechem hapanim* "show bread" were baked on Friday, then put on the Table on the Sabbath when the week-old loaves were removed and divided among Aaron's descendants.

Miraculously, all twelve loaves (each equivalent to the volume of 86.4 eggs) remained fresh all week. Like the Ark of the Covenant, the Table of Showbread had a crown (Exo. 25:23-30). [401] The 12 showbreads were divided into two stacks when they were placed on the Table with a large spoon full of frankincense on each stack. These loaves of showbread have been likened to the Sabbath because both are said to be an eternal covenant (Exo. 31:16; Lev. 24:8). Just as God provided food for the Sabbath in the Wilderness with a double portion of manna on Friday, so does the show bread symbolize God providing sustenance for His servants (Lev. 24:5-9). [402]

LAMED (ל) is 12. There are twelve thousand servants of God with the seal of the Living God on their foreheads from each of the 12 Tribes of Israel, which are 144,000 precious souls. Others can be incorporated sovereignly. Why I say this is I had an epiphaneous appearance from Yeshua and Gabriel on 4 December 1999 where I was told that the 144,000 covers the broadest spectrum possible within the confines of being from the 12 Tribes of Israel listed in Revelation Chapter 7: from full-blooded to one drop, from an elderly adult to a child, from male to female, from black to white, from singles to families to groups.

> "*¹ After this I saw four angels standing at the four corners of the earth, holding back the four winds of the earth so that no wind would blow on the earth, or on the sea, or on any tree. ² And I saw another angel ascending from the rising of the sun, holding the seal of the living God; and he called out with a loud voice to the four angels to whom it was granted to harm the earth and the sea, ³ saying, 'Do not harm the earth, or the sea, or the trees until we have sealed the bond-servants of our God on their foreheads. ⁴ And I heard the number of those who were sealed: 144,000, sealed from every tribe of the sons of Israel: ⁵ from the tribe of Judah, TWELVE thousand were sealed, from the tribe of Reuben TWELVE thousand, from the tribe of Gad TWELVE thousand, ⁶ from the tribe of Asher TWELVE thousand, from the tribe of Naphtali TWELVE thousand, from the tribe of Manasseh TWELVE thousand, ⁷ from the tribe of Simeon TWELVE thousand, from the tribe of Levi TWELVE thousand, from the tribe of Issachar TWELVE thousand, ⁸ from the tribe of Zebulun TWELVE thousand, from the tribe of Joseph TWELVE thousand, and from the tribe of Benjamin, TWELVE thousand were sealed'*" (Revelation 7:1-8 $_{NASB}$).

On Mount Zion stands the Lamb with 144,000 souls with the Lamb's Name and His Father's Name written on their foreheads. The new song that these 144,000 first fruits sing is the tune coming from their transformed DNA where their mortality has taken on immortality on Mount Zion (1 Cor. 15:53; Rev. 14:1).

Our DNA is driven and influenced by the all-pervading light frequency of the cosmos. It has a base frequency of 144,000 cycles per second (cps). $12 \times 12 = 144$ is a harmonic fractal of the cosmic fractal system that includes all in the living matrix of the interconnected web of life that exists throughout the cosmos. Researchers at first thought that our DNA twinned helices were like a radio transmitter, but we realize now it is more than this. Our DNA is 12-stranded, and it is a light antenna. [403]

It's the Lamb who facilitates the 144,000 arising, shining, and transforming, like Enoch (Phil. 3:20-21). These are the ones who follow the Lamb of God wherever He goes. Remember, they are only firstfruits, which means more (wise virgins) will follow in waves when Christ decides they are ready. *"¹ Then I looked, and behold, the Lamb was standing on Mount Zion, and with Him 144,000 who had His name and the name of His Father written on their foreheads. ² And I heard a voice from heaven, like the sound of many waters and like the sound of loud thunder, and the voice which I heard was like the sound of harpists playing on their harps. ³ And they sang a new song before the throne and before the four living creatures and the elders; and no one was able to learn the song except the 144,000 who had been purchased from the earth. ⁴ These are the ones who have not defiled themselves with women, for they are celibate. These are the ones who follow the Lamb wherever He goes. These have been purchased from mankind as first fruits to God and to the Lamb. ⁵ And no lie was found in their mouths; they are blameless"* (Revelation 14:1-5 NASB).

QUANTUM 30

Not only does LAMED (ל) have an ordinal value of 12, but a numerical value of 30. Therefore, let's explore the 30th element on the Periodic Table.

Atomic Number 30—Zinc (Zn). Quantumly speaking, zinc is a chemical element that has the atomic number **30**, the atomic weight **65.38**, and the atomic symbol **Zn**. A single zinc atom contains 30 protons in its nucleus, 30 electrons, and 34 neutrons in its most abundant isotope. It is the first element of the twelfth column in the Periodic Table, which is classified as a transition metal. This directly correlates to LAMED (ל) being both 12 and 30.

Under standard conditions, zinc is a hard and brittle metal with a bluish-white color. It becomes less brittle and more malleable above 100 degrees C. Zinc has relatively low melting and boiling points for a metal. It is a fair electrical conductor. When zinc comes into contact with the air it reacts with carbon dioxide to form a thin layer of zinc carbonate. This layer protects the element from further reaction. [404]

Zinc is not found in its elemental form, but it is found in minerals with other chemical elements in the Earth's crust where it is approximately the 24th [12 × 2] abundant element. Ocean water and air on Earth contain small traces of zinc. [405]

Several minerals are mined for zinc: sphalerite, smithsonite, hemimorphite, and wurtzite. Sphalerite has the highest percentage of zinc (~60%); therefore, it is the most mined. More than half of the zinc that is mined is used to galvanize steel, iron, and other metals. Galvanizing is when other metals are coated with a thin coating of zinc in order to prevent them from corroding or rusting. The other major use for zinc metals is making brasses and alloys for die-casting. [406]

Zinc plays an important role in biology. It is an essential trace element for humans, animals, plants, and for microorganisms. In the brain, zinc plays a key role in synaptic plasticity, which affects learning. [407] High concentrations of zinc can be found in the red blood cells of humans, which plays an important role in respiration by influencing CO_2 transport in the blood. Proper amounts of zinc in the pancreas aids in the storage of insulin while zinc in some enzymes aid in digesting proteins. After iron, zinc is the most abundant trace metal in all enzyme classes. [408]

Zinc is necessary for prenatal and postnatal development. Zinc deficiency affects about two billion people and is associated with many diseases in children: growth retardation, delayed sexual maturation, infection susceptibility, and diarrhea. [409] Although zinc is an essential requirement for good health, excess zinc can be harmful. Excessive absorption of zinc suppresses copper and iron absorption. [410]

Zinc oxide is an ultraviolet light absorber; therefore, it is used to prevent sunburn. It is also used as a dietary supplement, seed treatment, and photoconductor. Zinc has many other compounds that are used in various industrial and consumer applications, including: pesticides, pigments, dyes, fluxes, and wood preservatives, like a wooden shepherd's staff. [411]

30 IN SCRIPTURE

LAMED (ל) has a numerical value of 30. In Scripture, thirty is connected to the keepers of the shepherd's rod—the priests. The descendants of the first High Priest of Mercy—Aaron—were first dedicated to serving in God's House at thirty years old (Num. 4:3). John the Baptist began his ministry at 30 years old. He was of priestly descent. On his mother's side, she was a descendant of the daughters of Aaron and his father was also a priest. The Great Shepherd of our souls—Messiah Yeshua—also began his ministry at 30 years old: *"Now Jesus Himself began His ministry at about THIRTY years of age, being (as was supposed) the son of Joseph" (Luke 3:23).*

LAMED (ל) is 30. Thirty silver coins were paid to Judas to betray Yeshua, which means through extrapolation that the number 30 can represent the price Jesus paid on the Cross: *"³ Then Judas, His betrayer, seeing that He had been condemned, was remorseful and brought back the THIRTY pieces of silver to the chief priests and elders, ⁴ saying, 'I have sinned by betraying innocent blood.' And they said, 'What is that to us? You see to it!' ⁵ Then he threw down the pieces of silver in the temple and departed, and went and hanged himself. ⁶ But the chief priests took the silver pieces and said, 'It is not lawful to put them into the treasury, because they are the price of blood.' ⁷ And they consulted together and bought with them the potter's field, to bury strangers in. ⁸ Therefore that field has been called the Field of Blood to this day. ⁹ Then was fulfilled what was spoken by Jeremiah the prophet, saying, 'And they took the THIRTY pieces of silver, the value of Him who was priced, whom they of the children of Israel priced, ¹⁰ and gave them for the potter's field, as the Lord directed me'" (Matthew 27:3-10).*

HEART AND HEALING

The 32 threads of the *tzitzit* allude to the word *lev* (לֵב)—heart—with its numerical value of thirty-two. The word *tzitzit* (צִיצִית) is literally defined as "fringes," and refers to the strings attached to the corners of the *tallit* (טלית)—the Jewish prayer shawl. God commanded the

Jewish people to affix fringes to the corners of their clothing so that they would constantly remember Him and His commandments. *"Again the Lord spoke to Moses, saying, 'Speak to the children of Israel: Tell them to make tassels on the corners of their garments throughout their generations, and to put a blue thread in the tassels of the corners. And you shall have the tassel, that you may look upon it and remember all the commandments of the Lord and do them, and that you may not follow the harlotry to which your own heart and your own eyes are inclined, and that you may remember and do all My commandments, and be holy for your God. I am the Lord your God, who brought you out of the land of Egypt, to be your God: I am the Lord your God'"* (Numbers 15:37-41).

When this commandment began for the Jewish people, the common garment was a simple sheet of cloth where they affixed a fringe to each of its four corners. Over time, the *tzitzit's* eight strings and five knots on each corner of the *tallit* came to be a physical representation of the 613 commandments of Torah, because the numerical value of *tzitzit* (צִיצִית) adds up to 600 with the addition of 8 threads and 5 knots on each tassel to come up with 613.

The four corners of the *tallit* with its *tzitzit* are known as the wings, as in *"to you who fear My name the Sun of Righteousness shall arise with healing in His wings"* (Malachi 4:2). The *tzitzit* (צִיצִית) tassels on Yeshua's garment are what the woman with the issue of blood touched to get her healing: *"And suddenly, a woman who had a flow of blood for TWELVE years came from behind and touched the hem of His garment. For she said to herself, 'If only I may touch His garment, I shall be made well.' But Jesus turned around, and when He saw her He said, 'Be of good cheer, daughter; your faith has made you well.' And the woman was made well from that hour"* (Matthew 9:20-22).

In accordance with the Torah, *tekhelet* was used in ancient times to dye the *tzitzit* (צִיצִית) blue. *"Speak to the people of Isra'el, instructing them to make, through all their generations, tzitziyot on the corners of their garments, and to put with the tzitzit on each corner a blue thread. It is to be a tzitzit for you to look at and thereby remember all of Adonai's mitzvot and obey them, so that you won't go around wherever your own heart and eyes lead you to prostitute yourselves; but it will help you remember and obey all My mitzvot and be holy for your God"* (Numbers 15:38-40 CJB).

The secrets about the precious *tekhelet* blue dye were lost about 1300 years ago; because the mysterious sea creature—the Chillazon—that produced the blue dye was lost. "Over the past half-century, a convergence of research and discoveries by Rabbis, scientists, archaeologists, and others has led to the conclusive identification of the sea snail, *Murex trunculus*, as the authentic source of *tekhelet*."[412]

Notice that it is the blue thread that reminds God's people of His commandments: *"Make . . . tzitziyot on the corners of their garments, and to put with the tzitzit on each corner a blue thread. It is to be a tzitzit for you to look at and thereby remember all of Adonai's mitzvot and obey, so that you won't go around wherever your own heart and eyes lead you to prostitute yourselves"* (Numbers 15:38-39 CJB). This was to remind God's people of their Marriage Contract *(Ketubah)* with the Most High God, which was originally written on two sapphire cubes at Sinai—The Ten Commandments (Exo. 19).

QUANTUM QUARK (ל)

In summary, LAMED (ל) is the twelfth Hebrew Living Letter. The ancient pictograph for LAMED (ל) is a rod or a staff, like a shepherd's staff. The Torah begins with the Hebrew letter BET (ב), as in *bereshit* (בְּרֵאשִׁית), and ends with LAMED (ל). Therefore, we can say that the entire Torah is contained between the letters LAMED-BET (לֵב)—the heart. The heart—*lev* (לֵב)—is the center of a person's body, just as LAMED (ל) is the center of the Hebrew Alphabet. The Hebrew Pictograph for the word *"lev"* tells us that the heart is what controls the inside. LAMED (ל) ending the Torah is all about the heart of the Word—the heart of Messiah Yeshua, the Shepherd of Love—from which everything flows.

LAMED (ל) is a unique and majestic sacred letter that towers above the other letters from its position in the heart of the Hebrew Alef-Bet. Since LAMED (ל) towers over the other letters from its central position, it represents an important facet of the Good Shepherd. The Shepherd of our souls is also *melekh hamelakhim*—the King of Kings (1 Pet. 2:25; Rev. 19:16). It is not a coincidence that the three central letters of the Hebrew Alphabet can spell out the Hebrew word *melech* (מלך), which means "king."

Q: How can you better connect to the Good Shepherd of Love? Hint: Psalms 23.

Q. Why is the majesty of the King of Kings hidden within the kindness of the Chief Shepherd?

מ

MEM (מ)—13
FOUNTAIN OF DIVINE WISDOM

REDEMPTION OF MEM (מ)

The redemption of the Messiah's Quantum 22 carries on with the thirteenth letter of the Hebrew Alphabet. MEM (מ) is pronounced *mem* with the sound of "m" as in mom. This sacred letter has an ordinal value of 13 and a numerical value of 40. Its ancient pictograph looks like waves of water.

"We start our life being 99-percent water, and by the time we reach adulthood we are down to 70-percent. If we die of old age, we will probably be about 50-percent water. In other words, throughout our lives we exist mostly of water." [413] However, the vast majority of water in the human body is in the liquid crystalline state of water that exists in a gel-like state.[414]

Water "has a crystalline form—ice. Yet research reveals water has a second crystalline form—a liquid crystal. Although molecules remain mobile in liquid crystalline form, they tend to move together, like a school of fish. This is structured water—also referred to as organized water, hexagonal water, and liquid crystalline water. Liquid crystals are a unique phase of matter. Like solid crystals, the repeating pattern provides an efficient pathway for the smooth flow of energetic information. Liquid crystals store and transmit information just like solid crystals, yet they are flexible and many times more responsive." [415]

In a healthy human body, much of the water is in a liquid crystal state. Cell membranes and collagen work cooperatively with structured water to create an information network that reaches every cell. In fact, the presence of liquid crystal water is vital for cellular integrity. "It is the liquid crystalline organization of the human body that accounts for the instantaneous transfer of signals and other biological information." [416] We could say that your body is a liquid crystal and a crystal is basically an antenna.

Structured liquid crystal water is responsible for the stability of DNA and the maintenance of a strong electromagnetic field around it. Typically, youthful DNA is surrounded by structured water and has a much stronger electromagnetic field than the DNA of the elderly. [417] Behold, the redemption of MEM (מ) is God's holy priesthood's resurrected, immortal bodies will be perpetually youthful with self-sustaining impenetrable electromagnetic fields and perfectly structured water, which will be their eternal life state. Please note that an encoded picture of the overcoming Bride of Christ is the New Jerusalem. It is a holy crystalline city that's a six-sided gold cube (Rev. 21:9-16). Six (6) is always associated with mankind created on the sixth day.

For more than a century, scientists have noticed water has a variety of unusual properties that's don't fit within the class model of being a solid, liquid, or a gas. "Dr. Gerald Pollack, professor of bioengineering at the University of Washington, has proved significant evidence for water's capacity to form large zones of liquid crystalline/structured water" [418]

Liquid crystalline provides a mechanism to explain many unexplained mysteries of the human body. "Water's crystalline structure is based on tetrahedral geometry where oxygen atoms form the center of each tetrahedron. Under ideal circumstances, as water's tetrahedra join together, a repeating hexagonal pattern emerges with oxygen atoms forming the vertices of each hexagon. This is the reason liquid crystalline water has also been referred to as hexagonal water." [419] Structured water assembles itself into layers of offset hexagonal sheets with the formula H_3O_2 that behaves like a liquid crystal. Both light and sound can generate H_3O_2. [420]

There is a very interesting Living Creature connection between hexagonal water and the appearance of the wheels and their work being like unto beryl (Ezek. 1:16). When you read about the Living Creature (i.e., four or more living creatures) in Ezekiel 1, understand that it's talking about the corporate and righteous Order of Melchizedek who are leading their own inner Bride forth. [421] Not only does this apply to the kings and priests' bodies that are around 70-percent water, as adults; but to their spirits as well because *"the spirit of the Living Creature is in the wheels" (Ezekiel 1:20,21)*. This reality manifests during or right after the individuals (living creatures) become a corporate one "Living Creature" (Ezek. 1:19-20). The key is the hexagonal structure of both the crystalline water in people's bodies and the mineral beryl.

Add to this fact that researchers at the Oak Ridge National Laboratory (ORNL) squeezed water in combination with the mineral beryl to discover a multidimensional reality within our 3-D physical realm. ORNL put water molecules under extreme confinement in hexagonal ultra-small channels (5 angstroms across) to learn that the H_2O molecules delocalized and were simultaneously present in all six symmetrically equivalent positions. The reason scientists say that the water molecules were in all six symmetrically equivalent positions is that beryl is a mineral that forms hexagonal (six-sided) prisms. [422] Spiritually-speaking, this means that the corporate ascended spirit in the wheels of a group of kings and priests, who are dialed into the perfect will of the Heavenly Father, are in all six symmetrically equivalent positions (up, down, right, left, in and out) at the same time. This is a foreshadow of the potential of the six-sided golden cube of the New Jerusalem where God dwells with His bridal people.

To get the water in our bodies into the crystalline form is a complex process; however, one thing that we know is we need to be more full of light. Therefore, the resurrected eternal life body state will be a body completely full of light. We really are water beings dancing to the tune of frequency and light. Behold, the Light Bodies of His overcomers—the perpetually youthful Immortal Ones—will be clearly seen and known soon; but it's not by man's might or power; but by His Spirit (Zech. 4:6).

ANCIENT MEM (מ)

MEM (מ) has two forms (מ) and (ם). The closed MEM is only used at the end of a word. MEM (מ) is the symbol of the revealed and the concealed (Moses and Messiah). MEM (מ) stands for Moses and the revealed Torah; because Moses "started to reveal the infinite Torah on a level that man could perceive."[423] "Not everything was transmitted by Moses, however. The concealed will be revealed by the Messiah."[424]

In Proverbs, King Solomon reveals: *"It is the glory of God to conceal a matter, but the glory of kings is to search out a matter" (Proverbs 25:2)*. The first letter of the Book of Proverbs is an enlarged MEM (מ). "Before starting to write, King Solomon fasted forty days to attain the spiritual level that Moses had reached when he fasted forty days and received the Torah."[425]

As we have discussed, MEM (מ) represents water. The Hebrew word for water is *mayim*, which has a plural ending. God founded our physical world on the waters: *"The earth is the Lord's, and all its fullness, The world and those who dwell therein. For He has founded it upon the seas, And established it upon the waters" (Psalms 24:1-2)*.

Righteousness is equated to the security of life-giving water. Psalms 1:3 compares a righteous man to a tree securely planted beside channels of water (Psa. 1:1-3). After the LORD speaks of the fast that He chooses—to loose the bonds of wickedness, to undo heavy burdens, to break every yoke, to let the oppressed go free, etc.—God promises: *"The Lord will guide you continually, and satisfy your soul in drought, and strengthen your bones; you shall be like a watered garden, and like a spring of water, whose waters do not fail" (Isaiah 58:11)*.

The LORD brings water into the mix when He speaks of His people forsaking and neglecting Him. *"For My people have committed two evils: they have forsaken Me, the fountain of living waters, and hewn themselves cisterns—broken cisterns that can hold no water" (Jeremiah 2:13)*. How significant that life-giving water is a major theme in the Lord's vindication of Zion. *"10 'Rejoice with Jerusalem, and be glad with her, all you who love her; rejoice for joy with her, all you who mourn for her; 11 That you may feed and be satisfied with the consolation of her bosom, that you may drink deeply and be delighted with the abundance of her glory.' 12 For thus says the Lord: 'Behold, I will extend peace to her like a river, and the glory of the Gentiles like a flowing stream. Then you shall feed; on her sides shall you be carried, and be dandled on her knees. 13 As one whom his mother comforts, so I will comfort you; and you shall be comforted in Jerusalem' " (Isaiah 66:10-13)*.

COMPOSITE LETTER (מ)

The sacred letter MEM (מ) is composed of the letter KAF (כ) and VAV (ו). The VAV is attached at the back of its neck by one point to the forehead of the KAF. Together KAF (כ) and VAV (ו) have a numerical value of 26, which represents the Unity of God due to twenty-six being the same numerical value as His Sacred Name—YHVH.

FULFILLMENT OF MEM (מ)

The fulfillment of MEM (מ) is MEM-MEM (מם), which has a numerical value of 80 = 40+40 with a *mispar katan* of 8. [426]

QUANTUM 13

Atomic Number 13—Aluminum (Al). Quantumly speaking, aluminum is a chemical element that has the atomic number **13**, the atomic weight **26.982**, and the atomic symbol **Al**. A single aluminum atom contains 13 protons, 13 electrons, and 14 neutrons.

Aluminum is a silvery-gray metal. In standard conditions, aluminum is fairly soft and strong as well as being lightweight. Aluminum has a density lower than other common metals, approximately one-third the density of steel.[427] It has an affinity towards oxygen, therefore, aluminum forms a protective layer of oxide on its surface when exposed to air. Visually aluminum resembles silver in its color and in its ability to reflect light. It is soft, non-magnetic, and ductile. [428]

Aluminum is the third most abundant element, and the most abundant metal found in the Earth's crust. It makes up about 8% of the Earth's crust by weight and is generally found on Earth in minerals and compounds such as feldspar, beryl, cryolite, and turquoise. [429]

Chemically, aluminum is a post-transition metal in the boron group. Because of its chemical activity, aluminum never occurs in the metallic form in nature, but its compounds are present to a greater or lesser extent in almost all rocks, vegetation, and animals. Despite its prevalence in the environment, no living organism is known to use aluminum salts for its metabolism, but aluminum is well tolerated by plants and animals. [430]

In space, aluminum is the twelfth most abundant of all elements and third most abundant among the elements that have odd atomic numbers, after hydrogen and nitrogen. [431] The only stable isotope of aluminum, ^{27}Al, is the eighteenth most abundant nucleus in the Universe. It is created almost entirely after the fusion of carbon in massive stars that will later become Type II supernovas. This fusion creates ^{26}Mg, which upon capturing free protons and neutrons becomes aluminum. Some smaller quantities of ^{27}Al are created in hydrogen-burning shells of evolved stars where ^{26}Mg can capture free protons. Essentially all aluminum now in existence is ^{27}Al.[432]

Aluminum is almost always alloyed, which improves its mechanical properties, especially when tempered. For example, aluminum foils and aluminum beverage cans are alloys with 92% to 99% aluminum.

The major uses for aluminum are:[433]

> [1] Transportation (automobiles, aircraft, trucks, railway cars, marine vessels, bicycles, spacecraft, etc.). Aluminum is used because of its low density.

> [2] Packaging (cans, foil, frames, etc.). Aluminum is used because it is non-toxic, non-adsorptive, and splinter-proof.

> [3] Building and construction (windows, doors, siding, building wire, sheathing,

roofing, etc.). Since steel is usually cheaper, aluminum is used when lightness, corrosion resistance, or engineering features are important.

[4] Electricity-related uses (conductor alloys, motors, generators, transformers, capacitors, etc.). Aluminum is used because it is relatively cheap, highly conductive, has adequate mechanical strength and low density, and resists corrosion.

[5] A wide range of household items, from cooking utensils to furniture. Low density, good appearance, ease of fabrication, and durability are the key factors of aluminum usage.

[6] Machinery and equipment (processing equipment, pipes, tools). Aluminum is used because of its corrosion resistance, non-pyrophoricity, and mechanical strength.

[7] Portable computer cases. Currently rarely used without alloying, but aluminum can be recycled and clean aluminum has residual market value: The used beverage can (UBC) material can be used to encase the electronic components.

13 IN SCRIPTURE

MEM (מ) is 13. It appears to be a bipolar number that represents both good and evil. Thirteen is symbolic of rebellion and lawlessness in Scripture. The mighty hunter—Nimrod—was the thirteenth generation in Ham's lineage.

MEM (מ) is 13. The "valley of Hinnom," where infants were sacrificed to Molech, occurs thirteen times in Scripture. *"He [Ahaz] burned incense in the Valley of the Son of Hinnom, and burned his children in the fire, according to the abominations of the nations whom the Lord had cast out before the children of Israel" (2 Chronicles 28:3).*

MEM (מ) is 13. The dragon, which is symbolic of Satan, is found thirteen times in the Book of Revelation: *"So the great dragon was cast out, that serpent of old, called the Devil and Satan, who deceives the whole world; he was cast to the earth, and his angels were cast out with him" (Revelation 12:9).*

MEM (מ) is 13. In Romans 1, there are 23 characteristics of sinful people with a debased or reprobate mind. The thirteenth characteristic is they are haters of God.

MEM (מ) is 13. The enemy of God's people in Queen Esther's day—Haman—signed a decree on the thirteenth day of the first month that on the thirteenth day of the twelfth month, all Jews were to be killed in the Persian empire (Esth. 3:7-9).

MEM (מ) is 13. The other edge of this two-edged sword understands that "13" is the number connected to the King of Righteousness and King of Peace—Melchizedek. In God's Temple, there were thirteen *Songs of the Sabbath Sacrifice* sung to Melchizedek for thirteen consecutive Seventh-Day Sabbaths. All these *Sabbath Songs* focus on God as King and are heavenly-focused. The community which recited the *Sabbath Songs* was led through a progressive experience, which means that it was a corporate 13-week cycle. [434] This community was led to animate life and give spirit to the heavenly temple until the worshippers of the Most High God experienced the holiness of the *Merkabah* (God's Chariot Throne) and the Sabbath Sacrifice conducted by the angelic high priests. [435] In the book *Angelic Liturgy: Songs of the Sabbath Sacrifice* by Newsom, Charlesworth, Strawn and Rietz, we are told that "angelic praise is superior to human praise because of the angels' superior knowledge and understanding of divine mysteries. Above all, the *Sabbath Songs* focus on the role of the angels as priests in the heavenly temple, a function also attested in works such as *Jubilees,* the *Testament of Levi*, and *First Enoch*." [436]

Most of us have lost track of the fact that First-Century Christians thought of themselves as angels on Earth. Margaret Barker in her excellent book *Temple Theology* reveals that those who stood in God's council and learned wisdom have been born in the Holy of Holies. Those who entered the Holy of Holies and stood before the Throne of the Holy One become angels. They experience resurrection and become transformed into their angelic (cherubic) state. Recall that Yeshua Himself exhorted believers to become sons of light (John 12:36). This is the resurrected state of the High Priesthood of All Believers made after the Righteous Order of Melchizedek. *Luke 20:36* says, *"nor can they die anymore, for they are equal to the angels and are sons of God, being sons of the resurrection."* The intense preoccupation with angelic priests points to the individual and communal glorification of the Melchizedek Priesthood in and through the Thirteen Sabbath Songs.

MEM (מ) is 13. In the DALET section, we went over the "four living creatures" who have "four faces" each, "four sides" each, "four wings" each, and a total of four wheels within wheels. There are 13 references in Scripture to the "four living creatures," which is a picture of a wing of the Melchizedek Army flying as one with the Spirit of the Living God.

MEM (מ) is 13. Thirteen is connected to the Righteous Order of Melchizedek bringing the restoration of all things to Creation, as pictured in the sacred geometry of the 13 spheres of Metatron's Cube. Metatron's Cube is a symbol of creation, which includes all five Platonic solids hidden inside. Metatron's Cube symbolizes the underlying geometric patterns throughout the Universe. Recall that the five Platonic solids are: the tetrahedron (4-faces), the cube (6-faces), the octahedron (8-faces), the dodecahedron (12-faces), and the icosahedron (20-faces).

The thirteen circles of Metatron's Cube are the fundamental blueprint for all atomic structures. From this matrix of 13 spheres, all platonic solids can be created, which are the basic geometries of life at all levels of reality. Recall that when water is frozen, its molecules crystalize in a hexagonal pattern, just like Metatron's Cube. This is a clue to the geometric structure underlying the thing it is freezing—our reality of three-dimensional space. The thirteen spheres configuration is a cypher for the restoration of all things, which the kings and priests made after the Righteous Order of Melchizedek need to understand.

MEM's chariot—the MERKABAH (מרכבה)—is all about ascending to the Redeemer Himself—MASHIACH (מָשִׁיחַ)—to see as He sees and to be like He is with the purpose of manifesting these realities on earth, as it is in heaven. [437]

QUANTUM 40

Not only does MEM (מ) have an ordinal value of 13, but a numerical value of 40. Therefore, let's explore the 40th element on the Periodic Table.

Atomic Number 40—Zirconium (Zr). Quantumly speaking, zirconium is a chemical element that has the atomic number **40**, the atomic weight **91.224**, and the atomic symbol **Zr**. A single zirconium atom contains 40 protons, 40 electrons, and 51 neutrons.

Zirconium is a lustrous, greyish-white, soft, ductile, malleable metal that is solid at room temperature, though it is hard and brittle at lesser purities.[438] It is a metal that does not easily corrode, which means that it doesn't break down easily when exposed to oxygen or water. This means that it can be used to make many useful objects because we don't have to worry that it will need to be replaced frequently. This is a major reason that zirconium is used in nuclear reactors for atomic energy. Approximately 90% of the zirconium that is produced each year is used to make nuclear power.[439]

Due to zirconium's high melting point and chemical resistance, it is used in a variety of high-temperature applications as well as aggressive environments. In the space industry, materials fabricated from zirconium metal and ZrO_2 are used in space vehicles where resistance to heat is needed.[440] In the aeronautics industry, high-temperature parts, like combustors, blades, and vanes in jet engines, are increasingly being protected by thin ceramic layers composed of zirconia and yttria.[441]

In the biomedical field, zirconium-bearing compounds are used for dental implants, crowns, hip replacements, and middle ear ossicular chain reconstruction. In manufacturing, zirconium is added to other metals as a lining for furnaces that reach very high temperatures, and to make crucibles, which are containers in which other metals can be melted. In addition, zirconium is used to make bricks, scissors, knives, food packaging, and parts of microwave ovens.[442]

Although zirconium has no known biological role, it is widely distributed in nature and is found in all biological systems. Zirconium is commonly used in commercial products (e.g., deodorant sticks, aerosol antiperspirants, cosmetics) and also in water purification (e.g., control of phosphorus pollution, bacteria- and pyrogen-contaminated water).[443][444]

The name "zirconium" is taken from the name of the mineral zircon, the most important source of zirconium. The word is related to Persian zargun (zircon; zar-gun, "gold-like" or "as gold".)[445]

Zircon ($ZrSiO_4$) and cubic zirconia (ZrO_2) are cut into gemstones to be used in jewelry. Zircon is also used in dating of rocks. Thousands of years ago, Egyptians also used zircons in their jewelry. A Zircon is a combination of elements that includes zirconium.[446]

40 IN SCRIPTURE

The numerical value of MEM (מ) is forty. In Scripture, the number 40 symbolizes a period of testing, trial, or probation.

MEM (מ) is 40. God caused it to rain on Earth forty days and forty nights while righteous Noah and his family were encased in the Ark of His protection. Moses spent forty years in Pharoah's palace, forty years in the Midian desert; and forty years as a leader of Israel. Moses was also on Mount Sinai for forty days and forty nights on two separate occasions (Exo. 24:18; Exo. 34:1-28). Moses sent out 12 spies to investigate the Promise Land for forty days: *"And they returned from spying out the land after FORTY days" (Numbers 13:25)*. Because the Israelites rejected the good report about the Promise Land from Joshua and Caleb, they were cursed to wander in the Wilderness for forty years—one year for every day they spied out the land.

Additionally, everyone of age, except Joshua and Caleb, were told that they would die during their wanderings in the desert, so they would know God's rejection: *"³⁴According to the number of the days in which you spied out the land, FORTY days, for each day you shall bear your guilt one year, namely FORTY years, and you shall know MY rejection. ³⁵ I the Lord have spoken this. I will surely do so to all this evil congregation who are gathered together against Me. In this wilderness they shall be consumed, and there they shall die" (Numbers 14:34-35)*.

MEM (מ) is 40. God was still with His people. He still blessed all the people who wandered in the wilderness for forty years where they lacked nothing: *"For the Lord your God has blessed you in all the work of your hand. He knows your trudging through this great wilderness. These FORTY years the Lord your God has been with you; you have lacked nothing" (Deuteronomy 2:7)*. Take for instance, the children of Israel who were supernaturally fed manna for forty years until they began to inhabit the Promise Land (Exo. 16:35). They also received water (MEM) from a rock that followed them in the Wilderness: *"⁵And the Lord said to Moses, 'Go on before the people, and take with you some of the elders of Israel. Also take in your hand your rod with which you struck the river, and go. ⁶ Behold, I will stand before you there on the rock in Horeb; and you shall strike the rock, and water will come out of it, that the people may drink'" (Exodus 17:5-6)*. That rock was Christ: *"¹ Moreover, brethren, I do not want you to be unaware that all our fathers were under the cloud, all passed through the sea, ² all were baptized into Moses in the cloud and in the sea, ³ all ate the same spiritual food, ⁴ and all drank the same spiritual drink. For they drank of that spiritual Rock that followed them, and that Rock was Christ" (1 Corinthians 10:1-4)*.

KING DAVID AND THE SUBTERRANEAN WATERS

There are many descriptions of the Lord subduing the seas (Psa. 33:7; 74:13; 89:9; Jer. 5:22). Therefore, the stories about King David subduing the subterranean waters before building the Temple of God are a variation on this same theme.

There is a sea (firmament) surrounding the Temple, which is the place of God's Throne (Ark). Some believe that there's a foundation stone called *'eben sh'tiyyah* on which the Altar (the place of sacrifice) in Jerusalem stood. They also believe that it was the place from which all the waters of the earth began and were controlled. The waters under the earth were all gathered beneath the Temple, they believed, and it was necessary to ensure that sufficient water was released to ensure fertility, but not so much as to overwhelm the world with a flood. This is a ruling and reigning (throne room) position. The association of the Temple with the control of the water and the forces of chaos goes back to the earliest times. King David played a prominent role in controlling these subterranean waters.

The following is attributed to third-century AD rabbis, but allegorical stories such as this are much older. "Rabbi Johannan said...When David dug the Pits... the Deep arose and threatened the world. 'Is there anyone,' inquired David, 'who knows whether it is permitted to inscribe the Name upon a sherd and cast it into the Deep that its waves should subside?'... Ahitophel said, 'It is permitted.' David inscribed the Name upon a sherd, cast it into the deep and it subsided sixteen thousand cubits. When he saw that it subsided to such a great extent, he said, 'The nearer it is to the earth the better the earth can be watered,' and he uttered the fifteen Songs of the Ascents and the Deep reascended fifteen thousand cubits and remained at one thousand cubits."[447]

The psalmist could write *"The LORD sits enthroned over the flood; the LORD sits enthroned as king forever" (Psalms 29:10),* because the association is there.

This foundation stone is the Rock of Israel which was the beginning of Creation. It is the fixed point from which all life was formed. *"¹ These are the last words [uttered through divine inspiration] of David: The words of David son of Jesse, and the words of the man who was established on high, the anointed one of the God of Jacob, and the pleasing composer of the songs of Israel: ² The spirit of HASHEM spoke through me; His word is upon my tongue. ³ The God of Israel has said— THE ROCK OF ISRAEL has spoken to me—'[Become a] ruler over men; a righteous one, who rules through the fear of God, ⁴ like the morning light when the sun shines—a morning without clouds, from the shine out of the rain, grass [sprouts forth] from the earth" (2 Samuel 23:1-4 $_{Tanach}$).*

The Hebrew word for "men" in *"Become a ruler over men"* refers to the quintessential man (woman) for whom God created the universe... the one who accepts the Word of God (Torah). The Rock of Israel is Messiah Yeshua, which followed them in the Wilderness and from which they drank (1 Cor. 10:4). This brings us to the connection between the Hebrew Living Letters and water. Recall that one of the conditions present before the first creative proclamation: *"Let there be light!"* was *"the Divine Presence hovered upon the surface of the waters" (Gen. 1:2).* These primordial waters are connected to the water of the Word (Eph. 5:26), which contains the infinite creative possibilities of His glorious letters.

On 7 September 2019, Yeshua stared deeply into my eyes. Then He looked down at Himself. I was greatly moved when I saw Him open His robe. In the midst of Messiah Yeshua is a mysterious place where the Hebrew letters swirl. Within Yeshua, the letters that make up the essence of the Word of God, live and move and have their being.

It is astounding that the Hebrew Living Letters make up the structure of physical matter, which are what I call "hyper-quantum" due to existing both on a spiritual level and a hidden physical level. All physical matter is composed of collectives of Hebrew letters, even though we typically can't see them. It's most similar to the molecular level of physical matter. Through Hebrew letters, spiritual light descends and lives in the world.

> "Let the One Love of the Blood be a blanket expression of the Father's love. We are transforming into ancient mysteries. We are each a Word of God—a sound—made visible. Let Wisdom and the Hebrew letters teach us. Merging with the power of the Blood is the protoplasm of Creation—the quantum letters of the Hebrew Alphabet. There is life in the Hebrew letters. There's an exchange of love within. Behold, the Hebrew Living Letters are the elemental material of the Word of God in creation. Behold, the quantum letters spiral into the DNA of Creation, which is the protoplasm of the Universe. The Father's Love is the spiritual foundation of the earth and all of creation. It is the heavenly love that holds it all together in Christ. Step into the middle of His amazing DNA-Strand where His love is transformative. Intertwine with the burning heart of love in the Hebrew Living Letters."[448]

QUANTUM QUARK (מ)

To summarize, MEM (מ) is the thirteenth Hebrew Living Letter. MEM (מ) in ancient Hebrew means water. The redemption for MEM (מ) is the total restoration of the crystalline waters within the human body, so we can become Light Beings of the New Jerusalem.

MEM (מ) is 13, which appears to be a bipolar number that represents both good and evil. Thirteen is symbolic of rebellion and lawlessness in Scripture. The mighty hunter—Nimrod—was the thirteenth generation in Ham's lineage. The other edge of this two-edged sword understands that "13" is the number connected to the King of Righteousness and King of Peace—Melchizedek. In God's Temple, there were thirteen *Songs of the Sabbath Sacrifice* sung to Melchizedek for thirteen consecutive Seventh-Day Sabbaths. All these *Sabbath Songs* focus on God as King and are heavenly-focused. This community was led to animate life and give spirit to the heavenly temple until the worshippers of the Most High God experienced the holiness of the *Merkabah* (God's Chariot Throne) and the Sabbath Sacrifice conducted by the angelic high priests.

Q: According to Scripture, how can you assist God filling your body completely with pure light? (Hint: Study "light" in the Bible.)

Q. What was the focus of the thirteen *Songs of the Sabbath Sacrifice* on the day of rest?

ב

NUN (נ)—14
KINGDOM LIFE & ACTION

REDEMPTION OF NUN (נ)

The redemption of the Messiah's Quantum 22 continues with the fourteenth letter of the Hebrew Alphabet. NUN (נ) is pronounced *noon*. This sacred letter has an ordinal value of 14 and a numerical value of 50. Its ancient pictograph looks like a fish darting through water. In Aramaic, NUN means fish which is a symbol of productiveness and abundance. [449] When Jacob blessed Joseph's children with *"they shall multiply" (Genesis 48:16)*, he was referring to the fruitful, rapid, plentiful multiplying of Joseph's lineage, like fish.

Let's consider the mysterious overflowing provision of 153 fish. What is the secret of 153 fish? *"³ Simon Peter said to them, 'I am going fishing.' They said to him, 'We are going with you also.' They went out and immediately got into the boat, and that night they caught nothing. ⁴ But when the morning had now come, Jesus stood on the shore; yet the disciples did not know that it was Jesus. ⁵ Then Jesus said to them, 'Children, have you any food?' They answered Him, 'No.' ⁶ And He said to them, 'Cast the net on the right side of the boat, and you will find some.' So they cast, and now they were not able to draw it in because of the multitude of fish. ⁷ Therefore that disciple whom Jesus loved said to Peter, 'It is the Lord!' Now when Simon Peter heard that it was the Lord, he put on his outer garment (for he had removed it), and plunged into the sea. ⁸ But the other disciples came in the little boat (for they were not far from land, but about two hundred cubits), dragging the net with fish. ⁹ Then, as soon as they had come to land, they saw a fire of coals there, and fish laid on it, and bread. ¹⁰ Jesus said to them, 'Bring some of the fish which you have just caught.' ¹¹ Simon Peter went up and dragged the net to land, full of large fish, ONE HUNDRED AND FIFTY-THREE; and although there were so many, the net was not broken" (John 21:3-11).*

Why 153 large fish? Mathematically speaking, 153 is the 17th triangular number $1 + 2 + 3 \ldots + 17 = 153$.

153 is also the sum of the first five factorials. The factorial of a number is the product of all the positive integers less than and equal to any given number. For example, $5! = 5 \times 4 \times 3 \times 2 \times 1 = 120$. When we sum all the totals of $1! + 2! + 3! + 4! + 5!$, we get a grand total of 153.

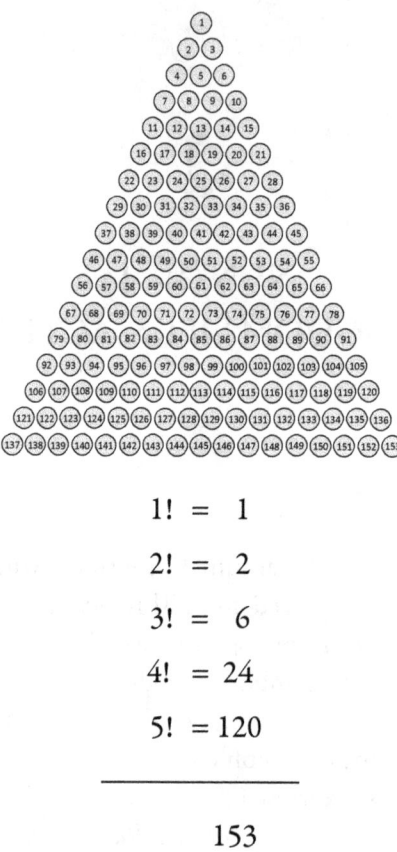

$$1! = 1$$
$$2! = 2$$
$$3! = 6$$
$$4! = 24$$
$$5! = 120$$
$$\overline{}$$
$$153$$

The secret of the 153 fish resides with the King of Kings who calls the fishermen (His disciples) to cast their net (the Gospel of the Kingdom) on the right side to bring the people in and out of the nations (sea of humanity).

The phrase "the sons of God" in Hebrew is equivalent to 153. *"For the anxious longing of the creation waits eagerly for the revealing of the SONS OF GOD" (Romans 8:19$_{NASB}$)*. Today, we usually hear the phrase "All creation waits for the manifestation of the sons of God," which is true and accurate, but more specifically this verse is speaking about the restoration of all things is waiting for a royal priesthood to arise, shine and rule the earth, as co-heirs in Christ. All of creation groans for the ones who will be delivered from the bondage of corruption into the glorious liberty of the children of God (Rom. 8:21-22). We truly are what we are waiting for. The ones who have entered and inherited the Kingdom of God (John 3:5; 1 Cor. 6:9-10; Gal. 5:19-21) are *"equal to angels and are SONS OF GOD, being sons of the resurrection" (Luke 20:36)*.

The goal of the One New Man in Christ is to come into the fullness of the Righteous Order of Melchizedek. Those who pursue this fullness are led into the third and highest corporate manifestation in the Body of Christ, which is His beautiful Bride. This is also the 100-percent fullness of the Heavenly One New Man in the Messiah (i.e., the Righteous Metatron Messiah Yeshua). "*²⁶ For you are all SONS OF GOD through faith in Christ Jesus. ²⁷ For as many of you as were baptized into Christ have put on Christ. ²⁸ There is neither Jew nor Greek, there is neither slave nor free, there is neither male nor female; for you are all one in Christ Jesus" (Galatians 3:26-28).* Please don't get hung up on gender specific terms; because both men and women are called to be the Sons of God, the Bride of Christ, and the One New Man.

Eleven times throughout the Torah, the number 153 connects to the word *HaPesach*, the Passover. Augustine observed that 10 is the number of the law and 7 that of the Spirit, which when added equals 17, and all the numbers from 1 to 17 when added equal 153, and thus the number practically represents all the elect. We have only scratched the surface of the depths of the meaning of the 153 fish.

ANCIENT NUN (נ)

The first mention of the word "nun" in Scripture is the reference to Joshua, the son of Nun: *"So the Lord spoke to Moses face to face, as a man speaks to his friend. And he would return to the camp, but his servant JOSHUA THE SON OF NUN, a young man, did not depart from the Tabernacle" (Exodus 33:11).* Joshua succeeded Moses and was able to enter the Promised Land, as the Son of Life. *"¹ After the death of Moses the servant of the Lord, it came to pass that the Lord spoke to JOSHUA THE SON OF NUN, Moses' assistant, saying: ² 'Moses My servant is dead. Now therefore, arise, go over this Jordan, you and all this people, to the land which I am giving to them—the children of Israel. ³ Every place that the sole of your foot will tread upon I have given you, as I said to Moses'" (Joshua 1:1-3).*

NUN (נ) is the symbol of Faithfulness, Soul, and Emergence. NUN (נ) has two forms. The bent (נ) is used at the beginning or middle of a word while the elongated (ן) is used at the end of a word. NUN (נ) stands for the reliable and faithful one while NUN (ן) denotes continuity. The most faithful one is God Himself. *"Your mercy, O Lord, is in the heavens; Your faithfulness reaches to the clouds" (Psalms 36:5).* ⁴⁵⁰ God's faithfulness can be emulated by humans. A person may be like a bent NUN whose heart is humble, serving in veneration and awe. Or they may be like the erect NUN who serves God with a steadfast heart of love, full of unwavering faith. ⁴⁵¹

"On the Day of Judgment, the person's own soul will testify before God" (Rashi), as it is written: The soul of man is the candle of HASHEM, searching all the inward parts (Prov. 20:27). ⁴⁵²

SIMPLE LETTER (נ)

NUN (נ) is one of the seven letters that are used to form other Hebrew Living Letters.

FULFILLMENT OF NUN (נ)

NUN (נ) is one of letters whose fulfillments (נון) begin and end with the same letter: VAV (ואו), MEM (מם), and NUN (נון). Notice that the fulfillment of NUN (נ) is NUN-VAV-END NUN (נון), which has a numerical value of 756 = 50+6+700 with a *mispar katan* of 9. The number 756 represents eternal truth.[453]

QUANTUM 14

Atomic Number 14—Silicon (Si). Quantumly speaking, silicon is a chemical element that has the atomic number **14**, the atomic weight **28.085**, and the atomic symbol **Si**. A single silicon atom contains 14 protons, 14 electrons, and most have 14 neutrons. Silicon is the second element in the fourteenth column of the Periodic Table. It is classified as a member of the metalloids, which have some properties in common with metals and some in common with non-metals.[454]

The properties that metalloids have in common are:

- Metalloids appear to be metal in appearance but are brittle.
- Metalloids can generally form alloys with metals.
- Some metalloids, like silicon and germanium, become electrical conductors under special conditions. They are called semiconductors.
- Metalloids are solids under standard conditions.
- Metalloids are mostly non-metallic in their chemical behavior.[455]

Silicon is the eighth most abundant element in the universe by mass, the most abundant metalloid on Earth, and the second most abundant element in the Earth's crust after oxygen.[456]

Silicon is a hard, brittle crystalline solid with a blue-grey metallic luster. In its amorphous form, it looks like brown powder. Even though silicon makes up approximately 28% of the Earth's crust, it's not generally found on Earth in its free form. However, silicon is found in almost all rocks, in sand, in clay, and in soils, combined with oxygen as silica (silicon dioxide, SiO_2) or with oxygen and metals as silicate minerals.[457] Silica is also found in quartz, granite, flint, talc, diorite, mica, and asbestos as well as in some gemstones: opals, agates, and amethysts.[458]

Most silicon is used industrially without being purified, which means the vast majority of its uses are in structural compounds: clays, sand, stucco, concrete, and most kinds of building stone.

"Silica is used in the form of sand and clay for many purposes; as quartz, it may be heated to form special glasses. Silicates are used in making glass, enamels, and ceramics; sodium silicates (water glass) are used in soaps, wood treatment, types of cements, and dyeing."[459] Additionally, Silica is used for contact lenses, breast implants, fiberglass, and glass wool for thermal insulation.

Silicon occurs in many plants and some animals. Although silicon is readily available in the form of silicates, very few organisms use it directly. Some plants accumulate silica in their tissues and require silicon for their growth, like rice. [460]

There is some evidence that silicon is important to human health for one's nails, hair, bone, and skin tissues. [461] Silicon is needed for the synthesis of elastin and collagen, of which the aorta (the largest and main artery in the human body) contains the greatest quantity. Silicon has been considered an essential element. Nevertheless, it is difficult to prove its essentiality; because silicon is very common and therefore deficiency symptoms are difficult to reproduce. [462]

"Semiconductors have had a monumental impact on our society. You find semiconductors at the heart of microprocessor chips as well as transistors. Anything that's computerized or uses radio waves depends on semiconductors. Today, most semiconductor chips and transistors are created with silicon. You may have heard expressions like 'Silicon Valley' and the 'silicon economy,' and that's why -- silicon is the heart of any electronic device."[463] In order to use silicon extensively as a semiconductor in solid-state devices, hyper-pure silicon is selectively doped with tiny amounts of boron, gallium, phosphorus or arsenic to control its electrical properties. [464]

"Silicon is one of the most useful elements to mankind. Most is used to make alloys including aluminum-silicon and ferro-silicon (iron-silicon). These are used to make dynamo and transformer plates, engine blocks, cylinder heads, and machine tools and to deoxidize steel.

Silicon is also used to make silicones. These are silicon-oxygen polymers with methyl groups attached. Silicone oil is a lubricant and is added to some cosmetics and hair conditioners. Silicone rubber is used as a waterproof sealant in bathrooms and around windows, pipes, and roofs." [465]

14 IN SCRIPTURE

NUN (נ) is 14. Fourteen is a multiple of 7, which implies a double portion of spiritual perfection. Yeshua's ancestors are listed in three sets of 14 generations between Abraham to Joseph, the husband of Mary. *"So all the generations from Abraham to David are FOURTEEN generations, from David until the captivity in Babylon are FOURTEEN generations, and from the captivity in Babylon until the Christ are FOURTEEN generations" (Matthew 1:17)*:

Abraham - Isaac - Jacob - Judah - Perez - Hezron - Ram - Amminadab - Nahshon
Salmon - Boaz - Obed - Jesse - King David

King David - Solomon - Rehoboam - Abijah - Asa - Jehoshaphat - Joram - Uzziah
Jotham - Ahaz - Hezekiah - Manasseh - Amon - Josiah (captivity)

Josiah (captivity) - Jeconiah - Shealtiel - Zerubbabel - Abiud - Eliakim - Azor
Zadok - Achim - Eliud - Eleazar - Matthan - Jacob - Joseph

Henry Gruver taught that it takes fourteen generations to prove anyone comes from pure blood. Therefore, three sets of 14 prove One with completely pure blood.

NUN (נ) is 14. In the Book of Proverbs, the phrase *"the fear of the Lord"* occurs fourteen times (Proverbs 1:7,29; 2:5; 8:13; 9:10; 10:27; 14:26,27; 15:16,33; 16:6; 19:23; 22:4; 23:17). It is wise to meditate on these incredible truths.

NUN (נ) is 14. The 14th of the first Biblical month (Nisan/Aviv) is the forever feast of Passover; therefore, fourteen represents deliverance and salvation. Approximately 430 years earlier, God made two covenant promises to Abraham on the beginning (night) of the 14th day of the first month. The first covenant promise concerned his physical seed, Isaac and his descendants. The second promise concerned those who would shine like the stars of heaven.

The binding of Isaac is known as the *Akedah Yitzhak* (Genesis 22). Post-Temple rabbis invented a seven-month Babylonian *Rosh haShana* (New Year's) tradition to replace the older and original one. Post-Temple rabbis instituted the man-made tradition of reading the *Akedah* on Tishri 2 (the second day of the rabbinically mandated new year in the Fall), which is done to this day. This was to hide the fact that Messiah Yeshua's death on the Cross, as the Lamb of God, was the fulfillment of the *Akedah*. The original place of the *Akedah* in the calendar was Nisan 14 when it actually happened.

"During the time Talmudic teaching was developing (40-500 AD), the entire life of Abraham's son of promise was intertwined with the Passover story. The near-sacrifice of Isaac was elsewhere compared with the sparing of the Israelite firstborn sons in the tenth plague—both occurring on Nisan 14."[466] "In the Mishnah (Megillah 3:5), the Torah reading for Rosh Hashanah (Tishrei 1, no second day) was not Genesis 22, the *Akedah* story; or Genesis 21, the birth of Isaac; but Leviticus 23, the command to blow the shofar as a memorial on Yom ha-Teruah."[467] The Book of Jubilees (dated around 150 BC, considered Scripture by Ethiopian Jews) tells of the *Akedah* taking place in Nisan (Jub. 17:15–18:19)."[468]

"According to Rabbi Kaunfer, the connection of the *Akedah* and Pesach was deliberately broken after the destruction of the Temple, in an effort to erase its powerful association with Yeshua's sacrifice: The selection of Genesis 22 as the reading for the second day of Rosh Hashanah reflected a conscious decision by certain of the Rabbis to move the *Akedah* away from its original calendrical home: Passover. This transfer was completed in order to distance the story of the *Akedah* with a time of the year that was increasingly associated with another martyr/sacrifice narrative, that of Jesus. The transfer of the Torah reading to Tishrei represented but one strategy on the part of the Rabbis to combat the Christological associations with the *Akedah*."[469]

"The *Akedah* may be exiled six months away from Nisan, but it still carries its original Nisan message. The Amidah prayer for Rosh Hashana refers to Isaac being bound *"for his seed,"* while the Musaf service begs God to grant us justification by remembering *"the son who was bound"* and "the merit of the innocent one"—without naming Isaac. Instead, another name is spoken. The name that generations of Hebrew-speaking rabbis have avoided with the euphemism, *ha-ish ha-hu*—"That man."[470] The name is invoked only once, in a silent whisper, during the first shofar blowing on the second day. It's printed in the tiniest type size possible for Hebrew prayer books. But it bypasses all of church history by honoring Him with an elegant Midrashic title unknown to Christians: '*Yeshua, Sar Ha-Panim.*'"[471]

Sar Ha-Panim means Prince of the Face (i.e., face to face), or Prince of the Presence. "May it be your will before you that at the sounding of the *Tekiah, Shevarim, Teruah*, that we are blowing, that it will be woven into the curtain by the appointed one, (Tartiel) according to the name that you received by Elijah (remembered for Good) Yeshua The Prince of the Presence, and the Prince Metatron, and that you will fulfill upon us mercy. Blessed are you lord of mercy." (*Machzor Rabah Rosh Hashanah*, pg. 168). [472]

QUANTUM 50

Not only does NUN (נ) have an ordinal value of 14, but a numerical value of 50. Therefore, let's explore the 50th element on the Periodic Table.

Atomic Number 50—Tin (Sn). Quantumly speaking, tin is a chemical element that has the atomic number **50**, the atomic weight **118.71**, and the atomic symbol **Sn**. A single tin atom contains 50 protons, 50 electrons, and 69 neutrons. Tin is the fourth element in the 14th column of the Periodic Table. Thus, tin is connected to the numbers 14 and 50, which are both intimately associated with the Hebrew letter NUN (נ).

Tin is classified as a post-transition metal. It is a soft, silvery-gray metal with a bluish tinge, which is very malleable (can be pounded into sheets) and polished to shine. Tin is resistant to corrosion from water. This allows it to be used as a plating material to protect other metals. [473] Since tin is malleable, non-toxic, and ductile, it is used to plate steel cans. These were the tin cans that used to contain food and to tie on the back of one's car when married. Pure tin is too weak to be used alone, but its many alloys include soft solder wire, pewter, bronze, and low-temperature casting alloys. [474]

"White tin has a body-centered tetragonal crystal structure, and gray tin has a face-centered cubic structure. When bent, tin makes an eerie, crackling cry, as its crystals crush each other."[475] The scream that comes from the breaking of tin's crystal structure is called a "tin cry."

Tin is about the 50th most abundant element in the Earth's crust, primarily in the ore cassiterite as dioxide: stannic oxide SnO_2. It is generally not found in free form. [476]

"The origins of tin are lost in antiquity. Bronzes, which are copper-tin alloys, were used by humans in prehistory long before pure tin metal itself was isolated. Bronzes were common in early Mesopotamia, the Indus Valley, Egypt, Crete, Israel, and Peru. Much of the tin used by the early Mediterranean peoples apparently came from the Scilly Isles and Cornwall in the British Isles, where tin mining dates to at least 300–200 BCE. Tin mines were operating in both the Inca and Aztec domains of South and Central America before the Spanish conquest. The symbol Sn for tin is an abbreviation of the Latin word for tin, stannum." [477]

In 2018, just under half of all tin produced was used in solder wire. Solder is a mixture of tin and lead that is used to join pipes and electronic circuits. The rest of tin's production was divided between tin plating, tin chemicals, brass and bronze alloys, and niche uses. [478] Other applications for tin include metal alloys such as bronze and pewter, the production of glass using the Pilkington process, toothpaste, and in the manufacture of textiles. [479]

50 IN SCRIPTURE

NUN (נ) is 50. In Scripture, fifty represents the jubilee year, which occurs after every seventh Sabbath year—every 50 years. It is an economic, cultural, communal, and environmental reset. Leviticus 25 speaks of a jubilee when the land and people rest, and all those who are in slavery are set free to return to their communities. *"8 And you shall count seven sabbaths of years for yourself, seven times seven years; and the time of the seven sabbaths of years shall be to you forty-nine years. 9 Then you shall cause the trumpet of the Jubilee to sound on the tenth day of the seventh month; on the Day of Atonement you shall make the trumpet to sound throughout all your land. 10 And you shall consecrate the FIFTIETH year, and proclaim liberty throughout all the land to all its inhabitants. It shall be a Jubilee for you; and each of you shall return to his possession, and each of you shall return to his family. 11 That FIFTIETH year shall be a jubilee to you; in it you shall neither sow nor reap what grows of its own accord, nor gather the grapes of your untended vine" (Leviticus 25:8-11).* The essence of "jubilee" is an interruption of the status quo. It is a reset emphasizing people and relationships above all other material things. Jubilee is a way of life for God's redeemed and liberated people, which accentuates mercy and justice. Jubilee is an expression of God's desire for all of creation to flourish in a broken world. Jubilee is radical. It is counterculture to the world of greed and materialism. It is a model for living well in a community according to God's righteous standards, so we can all thrive as individuals and as a community.

In Luke 4, at the beginning of Yeshua's earthly ministry, He proclaimed that He is the one foretold by the prophet Isaiah: *"16 So He came to Nazareth, where He had been brought up. And as His custom was, He went into the synagogue on the Sabbath day, and stood up to read. 17 And He was handed the book of the prophet Isaiah. And when He had opened the book, He found the place where it was written: 18 'The Spirit of the Lord is upon Me, Because He has anointed Me to preach the gospel to the poor; He has sent Me to heal the brokenhearted, To proclaim liberty to the captives and recovery of sight to the blind, to set at liberty those who are oppressed; 19 To proclaim the acceptable year of the Lord.' 20 Then He closed the book, and gave it back to the attendant and sat down. And the eyes of all who were in the synagogue were fixed on Him. 21 And He began to say to them, 'Today this Scripture is fulfilled in your hearing'" (Luke 4:16-21).* When Yeshua said, *"Today this Scripture is fulfilled,"* He was announcing that jubilee, liberty, and freedom are not just every 50 years; but even now, today, every day in Christ.

NUN (נ) is 50. Fifty clasps are used repeatedly in the Wilderness Tabernacle for oneness' sake. There were fifty gold clasps made to couple the curtains together, so that the tabernacle may be one (Exo. 26:6). Fifty bronze clasps were made to couple the curtains of goats' hair together for the tent over the Tabernacle, that it might be one (Exo. 36:14-18). These bronze clasps in the Tabernacle point to unity within God's Dwelling Place enabling His people to reach the world. The goats' hair tent was the covering for the Tabernacle, which touched the tabernacle itself. This is a picture of the harvesters in the Body of Christ who reach out to bring the people in. Harvesters can also be called fishers of men: *"Then Jesus said to them, 'Follow Me, and I will make you become fishers of men'" (Mark 1:17).*

PENTECOST

Since NUN (נ) has an ordinal value of fourteen and a numerical value of fifty, it can represent both Passover and Pentecost. We turn our focus to Pentecost due to NUN (נ) being fifty, which is a number associated with Pentecost. The term "Pentecost" comes from the Greek Πεντηκοστή *(Pentēkostē)*, which means "fiftieth." It refers to the Biblical festival of *Shavuot* celebrated on the fiftieth day after Passover. Pentecost is also known as the "Feast of Weeks" and the "Feast of 50 Days."

Fifty days after the first Passover, the children of Israel arrived at Mount Sinai whereupon Moses received the *Ketubah* (Torah with the Ten Commandments). Fifty days after Yeshua's crucifixion on the 14th of the first month, God poured out His tongues-of-fire Spirit (Acts 2).

The Babylonian-inspired institutional church perverts the Biblical mandate for Pentecost by saying that it is 50 days from the mixed pagan practice of Easter, which gives the Babylonian Queen of Heaven credit for Christ's Resurrection. The origins of Easter are so grievous I hate to even discuss the details. People dismiss the fact that infants were sacrificed on the Babylonian Queen of Heaven's altar at a sunrise service on Easter Sunday where the eggs of Ishtar were dyed with the blood of the sacrificed infants.

According to the "Melchizedek Section" of the *Dead Sea Scrolls* (11Q13), the messenger who brings good news, who announces salvation is the one of whom it is written to proclaim the year of the LORD's favor, the day of vengeance of our God; to comfort all who mourn (Isa. 61:2) [480] can be interpreted as Melchizedek (Crucified Ones made after the Righteous Order of Melchizedek) instructs about all the periods of history for eternity, the divine Biblical marks of time and in the statutes of the truth. In this Kingdom Day, we will restore His ancient foundations in more ways than one. One of His restored ways is we will worship our Heavenly Father in Spirit and in Truth at His appointed times and in His redemptive ways.

It is significant that God's people first received the Ten Commandments on Pentecost; then His devout ones in the Upper Room received the Holy Spirit on this same feast day. Recall that God gave Israel the Torah, which includes the Ten Commandments (Exo. 20:1-21). At the time, God was forging a betrothal contract—a *Ketubah* with Israel. The written betrothal contract represented "the Book of the Covenant," because marriage is a covenant. The Book of the Covenant spelled out the mutual obligations of God and Israel just as the *Ketubah* spells out the obligations between husband and wife. Exodus 19:8 reveals that Israel accepted God's marriage proposal by corporately declaring: *"All that the Lord has spoken we will do."* In Hebrew, it's *Na'aseh V'Nishmah*, which literally means we agree to do it even before we have listened.

We know about the oneness of Israel due to the Hebrew word for "camp" in Exodus 19:2, which is *chanah*. *Chanah* is singular while Israel is plural. By this, we understand that the necessary oneness requirement for marriage was fulfilled at Sinai. A biblical wedding ceremony requires that the marriage be consummated under a wedding canopy *(chupah)*. In Exodus 19:17, Moses brought all the people out of the camp to meet God. They stood at the nether part of the mount. The word "nether" in Hebrew implies that the people stood underneath the mountain.

This means that God provided Mount Sinai as a *chupah* that Israel stood underneath when the wedding took place.

Every biblical wedding requires two witnesses. They are called the friends of the bridegroom. One witness is assigned to the groom and one is assigned to the bride. In Exodus 19:17, Moses is seen as one of the two witnesses. The friends of the bridegroom's job are to escort the bride to meet the groom under the *chupah* (Mount Sinai). For the *Ketubah* to be a legal contract between the husband and wife, it must be signed by the two witnesses—the friends of the bridegroom. Unfortunately, when Moses returned from being with God on Mount Sinai, he did not sign the *Ketubah*. Instead, Moses broke the two Cubes of the Covenant due to witnessing the peoples' sin of the Golden Calf (Exo. 32:19).

With the once-for-all sacrifice of the Lamb of God on the Cross, the Word of God Himself became the *Ketubah* for whosoever believes in Him. How important that God's people fulfill their side of the *Ketubah*. The picture revealed in Scripture is that we need to quit Golden Calf worship, so we can be married to the King of Kings and the Lord of Lords. Then, fifty days later, we can receive the fullness of the outpouring of the Holy Spirit. Receiving the fullness of the outpouring of the Holy Spirit means that a person has been totally immersed and saturated with the Seven Spirits of God within their body, soul, and spirit to the point where they are crowned.

Rabbi Moshe Weissman wrote in the Midrash: "In the occasion of *Matan Torah* [the giving of the Torah], the *Bnai Yisrael* [children of Israel] not only heard Hashem's Voice but actually saw the sound waves as they emerged from Hashem's mouth. They visualized them as a fiery substance. Each commandment that left Hashem's mouth traveled around the entire Camp and then to each Jew individually, asking him, 'Do you accept upon yourself this Commandment with all *halochot* [Jewish law] pertaining to it?' Every Jew answered 'Yes' after each commandment. Finally, the fiery substance which they saw engraved itself on the *luchot* [tablets]." [481]

The fiery ticker tape on Sinai also happened in the upper room during *Shavuot*. During Pentecost, the people were also one when God poured out His tongues-of-fire Holy Spirit. *"And when the day of Pentecost was fulfilled, while they were assembled together, suddenly there came a sound from heaven as of a rushing mighty wind and it filled all the house where they were sitting. And there appeared to them tongues which were divided like flames of fire; and they rested upon each of them" (Acts 2:1-3 Lamsa's Aramaic).*

QUANTUM QUARK (נ)

In summary, NUN (נ) is the fourteenth Hebrew Living Letter. The picture of NUN (נ) in ancient Hebrew looks like a fish darting through water. It symbolizes action and life. In Aramaic, NUN means fish, which is a symbol of productiveness and abundance found in the mysterious catch of 153 large fish. 153 in Scripture is connected to the King of kings who calls the fishermen to cast their nets on the right side of the boat to redemptively bring people out of the sea of humanity. 153 stands for the "sons of God." All creation is waiting for these *"sons of the resurrection" (Luke 20:36)*.

NUN (נ) is 14, which is a multiple of 7. Fourteen implies a double portion of spiritual perfection. Yeshua's ancestors are listed in three sets of 14 generations between Abraham to Joseph, the husband of Mary (Matt. 1:17). It takes fourteen generations to prove anyone comes from pure blood. Therefore, three sets of 14 prove One with completely pure blood.

Q: What thoughts come to mind when you contemplate the number "153" in Scripture?

Q. Why was the proof of ALEF-TAV (את)'s pure blood important?

SAMECH (ס)—15
ENDLESS GLORY

REDEMPTION OF SAMECH (ס)

The redemption of the Messiah's Quantum 22 continues with the fifteenth letter of the Hebrew Alphabet. SAMECH (ס) is pronounced *sah-mekh*. This quantum letter has an ordinal value of 15 and a numerical value of 60. SAMECH (ס) means a prop or support or aid in ancient Hebrew. Its pictograph looks like a shield, and it symbolizes support as well as a slow twisting (like a plant being changed with a prop) as well as turning aside (like a plant).

The Hebrew sages say that SAMECH (ס) represents the endless and ever-ascending spiral of God's glory in the universe. The circular form of SAMECH (ס) speaks of the end of something intertwined in its beginning and vice versa. The trumpeting voice of Almighty God is like a breeze moving circularly in and out His shofar in the shape of the Hebrew Living Letter SAMECH (ס). The inherent unity between the end and the beginning before the Fall is connected to the glorious transcendent light of the Most High God, which encompasses every point of reality.

> In the supernatural shelter of SAMECH (ס), Yeshua brings the fiery torch of His Baptism of Fire to His pure worshipers, so they can unite with Him, being crucified in union with Christ in the Cross (Matt. 3:11; Gal. 2:20). There is a righteous king and priest aspect of Christ's Cross with the infinity cross in the midst of His *Merkabah*. The satanic Order of Melchizedek tries to duplicate, but they can't because being crucified with Christ requires death to one's own selfish ambitions and ways. Christ's Cross in the midst of His *Merkabah* is a higher level of consecration and dedication of His Baptism of Fire. There are no short-cuts. We are His Tabernacle, which is made solid by union with His Cross (Col. 1:24). When the fire of His Baptism comes down from on high, so does the white garments of the righteous Melchizedek Priesthood that are worn in the Holy of Holies. [482]

His endless cycles are reflected here on earth in the seasons as well as the Biblical Year being marked by the Feasts of the Lord. A glorious picture of the redemption of SAMECH (ס) is the restoration, redemption, and resurrection of the Messiah's Body, as portrayed in the Biblical Feast of Unleavened Bread with its connection to manna and Tabernacles. The Feast of

Unleavened Bread, the Feast of Tabernacles, and manna itself all started on the 15th day of their respective month; hence, their tie to the 15th Hebrew Living Letter SAMECH (ס).

Let's highlight the Feast of Unleavened Bread first. In Hebrew, it is called *Hag Ha Matzah*. For one full week, from the 15th through the 21st of the first month *(Nisan/Aviv)*, no leaven may be found in God's people's homes. As soon as Purim ends, a new Biblical time begins. For the next month, the Jewish people scrupulously cleanse their homes of *chametz* (leaven which represents sin). On the evening of the 13th of Nisan, the Father goes through His home by the light of a candle, carrying a feather, a wooden spoon, and a white linen cloth. Once the Father finds the last remnants of leaven in his house, he scoops them into the wooden spoon with the feather. He then wraps it in the white linen and removes it from the house. In the morning, the bundle of leaven is burnt in a fire. The search for leaven is called *Bedikat Chametz*.

> *"7 Therefore purge out the old leaven, that you may be a new lump, since you truly are unleavened. For indeed Christ, our Passover, was sacrificed for us. 8 Therefore let us keep the feast, not with old leaven, nor with the leaven of malice and wickedness, but with the UNLEAVENED BREAD of sincerity and truth" (1 Corinthians 5:7-8).*

Messiah Yeshua participated in *Bedikat Chametz* when He cleansed the Temple a few days before Passover. *"13 Now the Passover of the Jews was at hand, and Jesus went up to Jerusalem. 14 And He found in the temple those who sold oxen and sheep and doves, and the money changers doing business. 15 When He had made a whip of cords, He drove them all out of the temple, with the sheep and the oxen, and poured out the changers' money and overturned the tables. 16 And He said to those who sold doves, 'Take these things away! Do not make My Father's house a house of merchandise!' 17 Then His disciples remembered that it was written, 'Zeal for Your house has eaten Me up.' 18 So the Jews answered and said to Him, 'What sign do You show to us, since You do these things?' 19 Jesus answered and said to them, 'Destroy this temple, and in three days I will raise it up.' 20 Then the Jews said, 'It has taken forty-six years to build this temple, and will You raise it up in three days?' 21 But He was speaking of the temple of His body. 22 Therefore, when He had risen from the dead, His disciples remembered that He had said this to them; and they believed the Scripture and the word which Jesus had said" (John 2:13-22).*

We know that Messiah Yeshua literally rose from the dead in three days: *"For as Jonah was three days and three nights in the belly of the great fish, so will the Son of Man be three days and three nights in the heart of the earth" (Matthew 12:40)*. Remember Messiah Yeshua (Jesus Christ) is the Head of the multi-membered Body of Christ (Col. 1:15-20). Just as our Heavenly Father raised up Yeshua on the third day, so will He raise up the Messiah's Body: *"But, beloved, do not forget this one thing, that with the Lord one day is as a thousand years, and a thousand years as one day" (2 Peter 3:8)*. According to a Second Peter 3:8 timetable, it has been more than two thousand years since Yeshua came to earth and died for our leaven. *"After two days He will revive us; on the third day He will raise us up, that we may live in His sight" (Hosea 6:2)*. Also, according to a Second Peter 3:8 timetable, it has been more than six thousand years since Adam was created, which signals the Seventh Day Transfiguration of Man. *"1 Now after six days Jesus took Peter,*

James, and John his brother, led them up on a high mountain by themselves; ² and He was transfigured before them. His face shone like the sun, and His clothes became as white as the light" (Matthew 17:1-2).

This Transfigured One in its fullness is the Unleavened Body (Bread) of Christ. *"And He took bread, gave thanks and broke it, and gave it to them, saying, 'This is My body which is given for you; do this in remembrance of Me"* (Luke 22:19). Messiah Yeshua is the *matzah*—the unleavened bread eaten at Passover and Unleavened Bread: *"I am the Bread of Life, he who comes to Me will never hunger"* (John 6:35).

On the 14th of the first Biblical month (Nisan), Passover is celebrated by His people with a Seder meal. The main purpose is to tell our children about our redemption from slavery in Egypt (i.e., the world). We are instructed to eat the Seder meal reclining, which is in the manner of free men. The Seder's SAMECH (ס) fifteen steps progressively reveal our Beloved Messiah as the Lamb of God who takes away the sin of the world.

The *matzah* is bread without leaven, which is a literal picture of a man—One New Man in Christ—without sin. During the Passover Seder, the *matzah* is broken; and the middle *matzah* is wrapped in linen and hidden away from the eyes of the children. At the end of the Seder, the child who finds the middle *matza,* which is called the *Afikoman* (refreshment which comes after—dessert) receives a reward, just as the one who searches for the Messiah with a sincere heart finds eternal life.

The second highlight for the redemption of SAMECH (ס) is the Feast of Tabernacles due to this festival starting on the fifteenth day of the seven month. *"³³ Then the Lord spoke to Moses, saying, ³⁴ 'Speak to the children of Israel, saying: "The FIFTEENTH day of this seventh month shall be the Feast of Tabernacles for seven days to the Lord"'"* (Leviticus 23:33-34). Besides representing the Feast of Tabernacles, there were 15 "Songs of Decrees" sung by the Levites at the Feast of Tabernacles during the festive drawing of water.

"At the morning service on each of the seven days of the Feast of Tabernacles *(Sukkot)* a libation of water was made together with the pouring out of wine (Suk. iv. 1; Yoma 26b), the water being drawn from the Pool of Siloam in a golden ewer of the capacity of three logs. It was borne in solemn procession to the water gate of the Temple, where the train halted while on the shofar was blown *teki'ah, teru'ah, teki'ah.* The procession then ascended the *kebesh,* or slanting bridge to the altar, toward the left, where stood on the east side of the altar a silver bowl for the water and on the west another for the wine, both having snout-like openings, that in the vessel for the wine being somewhat the larger. Both libations were poured out simultaneously (Suk. iv. 9)."[483]

"On the night of the first day of the Feast of Tabernacles the outer court of the Temple was brilliantly illuminated with four golden lamps, each containing 120 logs of oil, in which were burning the old girdles and garments of the priests (Shab. 21a; Yoma 23a). These lamps were placed on high pedestals which were reached by ladders; and special galleries were erected in the court for the accommodation of women, while the men below held torches in their hands, sang hymns, and danced. On the fifteen steps of the Gate of Nicanor stood the Levites, chanting the fifteen 'songs of degrees'(Ps. cxx.-cxxxiv.) to the accompaniment of their instruments, of which the most important was the *ḥalil,* or flute, although it was used neither on the Sabbath nor on the first day of the feast (Suk. v. 1). The illumination, which was like a sea of fire, lit up every nook and corner of Jerusalem, and was so bright that in any part of the city a woman could pick wheat from the chaff. Whosoever did not see this celebration never saw a real one (Suk. 53a)."[484]

Please note that the burning of the old girdles and garments of the priests during the Feast of Tabernacles hints at the mystery *"For this corruptible must put on incorruption, and this mortal must put on immortality" (1 Corinthians 15:53)*, which correlates the Baptism of Fire to this Feast. "Meanwhile on the steps of the inner court stood the Levites singing Ps. cxx.-cxxxiv., accompanied by various musical instruments. The celebration continued till cockcrow, when the two priests at the Nicanor gate sounded the signal, and the crowd departed, facing about, however, at the eastern gate, when the priests recited, 'Our forefathers in this place turned their backs on the altar of God and their faces to the east, worshiping the sun; but we turn to God' (comp. Ezek. viii. 15, 16; Suk. v. 1-4; Tosef., Suk. iv.)."[485] They repented for the most detestable practice in God's eyes that drives Him far from His Sanctuary, you and I.[486] Remember that the coming of the Son of Man is from the east—the eastern gate (Matt. 24:27).

Any one of the 15 psalms from Psalms 120 to Psalms 134 were also sung by Hebrew pilgrims on their way to Jerusalem for the three great pilgrimage feasts. The other possibility is that Psalms 120-134 were sung while ascending Mount Zion, or while ascending the steps of the Temple.[487] The 15 Songs of Degrees are also called Gradual Psalms, Songs of Steps, Pilgrim Psalms, and songs for going up to worship, which are also called Songs of Ascents.

There is an exact one-to-one correspondence for each of the fifteen Songs of Ascents to each of the fifteen steps which separated the Court of Israel from the Court of Women. "That is why the Tosefta has the Levites greet the ministering *kohanim* in the last Song of Ascents. For the last song would have been sung from the top of the stairway from where the Levites would see the *kohanim* in the temple *ulam* or portico."[488] The medieval rabbi David Kimhi states the same thing: "These Songs of Ascents are fifteen [in number], and they say that the Levites used to recite them on the fifteen steps which were on the Temple Mount, and which separated the Court of Israel from the Court of Women. For they used to ascend by these steps from the Court of Women to the Court of Israel, singing one song on each step."[489] Additionally, David Kimhi (RaDaK)'s commentary on Psalms reveals that these Pilgrim Songs sung by the Levites standing on the fifteen steps were also sung by those that "ascend" from captivity.[490]

The third highpoint for the redemption of SAMECH (ס) is the manna that began to fall on the fifteenth day of the second month when the Lord told Moses in *Exodus 16:4: "I will rain bread from heaven for you."* The manna in the Wilderness was a foreshadowing of the true Bread of Life that rains down from heaven—Messiah Yeshua. The redeemed mature Body (Bread) of the Messiah attaches to this reality. *"31 Our fathers ate the MANNA in the wilderness; as it is written: 'He gave them BREAD out of heaven to eat.' 32 Jesus then said to them, 'Truly, truly, I say to you, it is not Moses who has given you the BREAD out of heaven, but it is My Father who gives you the true BREAD out of heaven. 33 For the BREAD of God is that which comes down out of heaven and gives life to the world.' 34 Then they said to Him, 'Lord, always give us this BREAD.' 35 Jesus said to them, 'I am the BREAD of Life; the one who comes to Me will not be hungry, and the one who* believes in Me will never be thirsty. . . . ⁵⁸ *This is the BREAD that came down out of heaven, not as the fathers ate and died; the one who eats this BREAD will live forever'* " *(John 6:31-35,58 NASB)*.

The central tenet of manna is *"they shall be taught of God" (John 6:45)*. Let's dig a little deeper for our resurrected body connection. *"And not only they, but ourselves also, which have the firstfruits of the Spirit, even we ourselves groan within ourselves, waiting for the adoption, to wit, the redemption of our body" (Romans 8:23 KJV)*.

Manna portrays the reality of the restoration, redemption, and resurrection of the Messiah's Body when:

[1] we have been humbled and tested by the Word to know what's in our hearts, to see whether we will walk in His instruction (Exo. 16:4).

[2] we have been humbled by God through Him feeding us manna, so we actually know and understand that man does not live by bread alone, but by everything that proceeds out of the mouth of the LORD (Deut. 8:3).

[3] we have been disciplined as sons to keep His commandments, to walk in His ways, and to fear Him (Deut. 8:5-6).

Enoch gives us an example of this transfigured Prince of the Presence. He is the first fruit of firstfruits of the perfectly pure and righteous Metatron, which is the 100-percent mature Head of Christ—Messiah Yeshua—attached to the 100-percent mature Body of Christ. This is the undifferentiated state of Messiah Yeshua. This type of Enochian transformation is a renewal at a totally different level. It's a sub-atomic transfiguration. It includes the renewal of the perfect substance of Creation—the Hebrew Living Letters. The Righteous Metatron is made up of people who have repeatedly denied themselves, taken up their cross, and followed Him daily to the point of being made into the exact same image as the Messiah.

ANCIENT SAMECH (ס)

SAMECH (ס) is the symbol of support, protection, and memory. It is a closed-rounded quantum letter, which represents divine support. Man's confident reliance on God's support is a mainstay of both Judaism and Christianity. It is so fundamental that King Solomon summarizes all of his teachings with the words: *"Here now is my final conclusion: Fear God and obey his commands, for this is everyone's duty. God will judge us for everything we do, including every secret thing, whether good or bad" (Ecclesiastes 12:13-14 NLT)*. [491]

Consequently, the middle of the sacred letter SAMECH (ס) has no support; but it miraculously was kept in place and suspended. [492] The perimeter of SAMECH (ס) can represent God—the Protector—while the center of SAMECH (ס) symbolizes what He protects. [493] *"¹ I love you, Lord; you are my strength. ² The Lord is my rock, my fortress, and my savior; my God is my rock, in whom I find protection. He is my shield, the power that saves me, and my place of safety" (Psalms 18:1-2 NLT)*. Protection is symbolized by the numerical value of SAMECH (ס), which is 60. As King Solomon states: *"SIXTY mighty ones round about it [his bed], of the mighty ones of Israel. All gripping the sword, learned in warfare. Each with his sword on his thigh"* (Song of Solomon 3:7). [494]

COMPOSITE LETTER (ס)

SAMECH (ס) is composed of KAF (כ) AND VAV(ו), which together have the same numerical value as God's Name YHVH—26. [495]

FULFILLMENT OF SAMECH (ס)

The fulfillment of the quantum letter SAMECH (ס) is SAMECH-MEM-KAF (סמך), which has a numerical value of 120 = 60+40+20 with a *mispar katan* of 3. The numerical value of the hidden fulfillment MEM-KAF (מך) is 60 = 40+20, which equals the numerical value of the original letter SAMECH (ס). [496]

QUANTUM 15

Atomic Number 15—Phosphorus (P). Quantumly speaking, phosphorus is a chemical element that has the atomic number **15**, the atomic weight **30.974**, and the atomic symbol **P**. A single phosphorus atom contains 15 protons, 15 electrons, and 16 neutrons. Phosphorus is the second element in the 15th column of the Periodic Table.

Phosphorus gets its name from the Greek word *"phosphoros,"* which means "bringer of light" due to the fact that this element glows in the dark. [497] Phosphorus is the twelfth most abundant element in the Earth's crust. [498] Phosphorus is classified as non-metal. Its high chemical reactivity assures that it does not occur in the free state (except in a few meteorites). Phosphorus always occurs as the phosphate ion. The principal combined forms in nature are phosphate salts. About 550 different minerals have been found to contain phosphorus, but, of these, the principal source of phosphorus is the apatite series in which calcium ions exist along with phosphate ions and variable amounts of fluoride, chloride, or hydroxide ions. [499]

"Only about 5 percent of the phosphorus consumed per year in the United States is used in the elemental form." [500] "Elemental phosphorus comes in various allotropes (different crystal structures) including white, red, violet, and black phosphorus." [501]

In living organisms, the role of phosphorus is essential. "Phosphorus is the sixth most abundant element of the human body." [502] Beans, nuts, eggs, fish, milk, and chicken are good sources of phosphorus in a human diet.

Phosphorus is a component of DNA, RNA, ATP, bones, and teeth. In organisms, the element phosphorus usually appears as phosphate. In its other forms phosphorus is usually very toxic. The ordinary allotrope, called white phosphorus, is a poisonous, colorless, semitransparent, soft, waxy solid that glows in the dark (phosphorescence) and combusts spontaneously in air, producing dense white fumes of the oxide P_4O_{10}; it is used as a rodenticide and a military smokescreen. Heat or sunlight converts it to the red phosphorus allotrope, a violet-red powder that does not phosphoresce or ignite spontaneously. Much less reactive and soluble than white phosphorus, it is used in manufacturing other phosphorus compounds and in semiconductors, fertilizers, safety matches, and fireworks. [503] In industry, phosphorus is primarily used in fertilizers; because it is a key element in the growth of plants.

15 IN SCRIPTURE

SAMECH (ס) is 15. The Israelites left Egypt to be free to worship the Most High God on the fifteenth day of the first month. *"They departed from Rameses in the first month, on the FIFTEENTH day of the first month; on the day after the Passover the children of Israel went out with boldness in the sight of all the Egyptians" (Numbers 33:3).*

As we have seen, SAMECH (ס) can represent The Feast of Unleavened Bread due to the festival starting on the fifteenth day of the first month (Lev. 23:6). We also considered how SAMECH (ס) can embody manna, because God first gave the bread from heaven on the fifteenth of the second month (Exo. 16:1-31). There is a three-fold cord woven between unleavened bread, the man without sin, and manna, which we will discuss more shortly. Additionally, SAMECH (ס), as 15, can also signify The Feast of Tabernacles due to this festival starting on the fifteenth day of the seven month (Lev. 23:34). Through the "fifteen" connections in Scripture, our three-fold cord becomes four: unleavened bread, the man without sin, manna, and God's Dwelling Place.

The four-fold "fifteen" connection in Scripture flows into the Tabernacle of God that's a city that is foursquare. What do we get when we combine unleavened bread, the man without sin, manna, and God's Dwelling Place? The Pure and Spotless Bride of the Messiah, which is the place where God truly dwells: *"¹ And I saw a new heaven and a new earth: for the first heaven and the first earth were passed away; and there was no more sea. ² And I John saw the holy city, new Jerusalem, coming down from God out of heaven, prepared as a bride, adorned for her husband. ³ And I heard a great voice out of heaven saying, Behold, the Tabernacle of God is with men, and He will dwell with them, and they shall be His people, and God himself shall be with them, and be their God. .... ¹⁶ And the city lieth foursquare, and the length is as large as the breadth: and he measured the city with the reed, twelve thousand furlongs. The length and the breadth and the height of it are equal" (Revelation 21:1-3,16).*

QUANTUM 60

Not only does SAMECH (ס) have an ordinal value of 15, but a numerical value of 60. Therefore, let's investigate the 60th element on the Periodic Table.

Atomic Number 60—Neodymium (Nd). Quantumly speaking, neodymium is a chemical element that has the atomic number **60**, the atomic weight **144.24**, and the atomic symbol **Nd**. A single neodymium atom contains 60 protons, 60 electrons, and 84 neutrons.

Neodymium is the fourth member of the rare-earth metals. The name "neodymium" comes from the Greek word *neos,* which means new, and *didymos,* which means twin. [504]

Neodymium is classified as a lithophile under the Goldschmidt classification. This means that neodymium is generally found combined with oxygen. Although it belongs to the rare-earth metals, neodymium is not that rare. Neodymium is the 27th most common element in the Earth's crust. [505]

"Neodymium is a ductile and malleable silvery-white metal. It oxidizes readily in the air to form an oxide, Nd_2O_3, which easily spalls, exposing the metal to further oxidation. The metal must be stored sealed in a plastic covering or kept in a vacuum or in an inert atmosphere." [506]

Neodymium compounds were first used to color glass. Its properties cause the color of the glass to change from purple to yellow to blue or green under different lighting conditions. We still use neodymium compounds in glass products today—to whiten the light in incandescent bulbs and to make goggles for welders and glass blowers. [507]

The real demands for neodymium are in products likely to be found in your pocket. Neodymium, iron, and boron (NIB) combine to create very powerful magnets. When alloyed with iron ($Nd_2Fe_{14}B$), neodymium makes the strongest type of permanent magnet—a ferromagnet. Tiny NIB magnets can be found in cell phones, earbuds, computer hard drives, and DVD/CD players. They can be found in microphones, loudspeakers, and headphones too. Larger NIB magnets are used in the electric motors of hybrid and electric vehicles as well as in some wind turbines. NIB magnets are also used to identify counterfeit money. Real paper money has tiny magnetic particles added to the inks when they are printed. [508]

60 IN SCRIPTURE

SAMECH (ס) is 60. In Scripture, sixty is associated with overcoming giants. In Numbers 21:33, Moses and company conquered sixty fortified cities in the kingdom of Og in Bashan. *"¹ Then we turned and went up the road to Bashan; and Og king of Bashan came out against us, he and all his people, to battle at Edrei. ² And the Lord said to me, 'Do not fear him, for I have delivered him and all his people and his land into your hand; you shall do to him as you did to Sihon king of the Amorites, who dwelt at Heshbon.' ³ So the Lord our God also delivered into our hands Og king of Bashan, with all his people, and we attacked him until he had no survivors remaining. ⁴ And we took all his cities at that time; there was not a city which we did not take from them: SIXTY cities, all the region of Argob, the kingdom of Og in Bashan. ⁵ All these cities were fortified with high walls, gates, and bars, besides a great many rural towns"* (Deuteronomy 3:1-5). The *Jewish Encyclopedia* reveals that Og was the Amorite king of Bashan, who reigned in Ashtaroth. He was one of the giants of the remnant of the Rephaim. [509]

SAMECH (ס) is 60. The length of Solomon's Temple was 60 cubits: "*¹And it came to pass in the four hundred and eightieth year after the children of Israel had come out of the land of Egypt, in the fourth year of Solomon's reign over Israel, in the month of Ziv, which is the second month* [manna month], *that he began to build the house of the Lord. ² Now the house which King Solomon built for the Lord, its length was SIXTY cubits, its width twenty, and its height thirty cubits*" (1 Kings 6:1-2 $_{Additions\,mine}$).

Significantly, God points out that the foundation of the House of God is 60 cubits in length as well: "*¹ Now Solomon began to build the house of the Lord at Jerusalem on Mount Moriah, where the Lord had appeared to his father David, at the place that David had prepared on the threshing floor of Ornan the Jebusite. ² And he began to build on the second day of the second month in the fourth year of his reign. ³ This is the foundation which Solomon laid for building the house of God: The length was SIXTY cubits (by cubits according to the former measure) and the width twenty cubits*" (2 Chronicles 3:1-3).

It is not a coincidence that King Nebuchadnezzar of Babylon fame made his golden image sixty cubits high: "*Nebuchadnezzar the king made an image of gold, whose height was SIXTY cubits and its width six cubits. He set it up in the plain of Dura, in the province of Babylon*" (Daniel 3:1).

SAMECH (ס) is 60. Sixty is the second step of three in the Parable of the Sower for those who hear the word, accept it, and bear fruit: "*some thirtyfold, some SIXTY, and some a hundred*" (Mark 4:20).

QUANTUM QUARK (ס)

To summarize, SAMECH (ס) is the fifteenth Hebrew Living Letter. SAMECH (ס) in ancient Hebrew looks like a shield or a prop, support, or aid. Its pictograph symbolizes support, a slow twisting (like a plant being changed with a prop) as well as turning aside (like a plant). The redemption for SAMECH (ס) focuses on the glorious restoration, redemption, and resurrection of the Messiah's Body, as portrayed in the Biblical Feast of Unleavened Bread with its connection to manna and Tabernacles. It is a subatomic transfiguration of the perfected Body of Christ facilitated by the flawless substance of His redeemed hyperquantum Hebrew Living Letters facilitated by the Lamb of God and His Cross.

The Hebrew sages say that SAMECH (ס) represents the endless and ever-ascending spiral of God's glory in the universe. His endless cycles are reflected here on earth in the seasons as well as the Biblical Year being marked by the Biblical Feasts. SAMECH (ס), as 15, maps to the fourfold "fifteen" connection in Scripture that flows into the Tabernacle of God that's a city that is foursquare. What do we get when we add up unleavened bread, the man without sin, manna, and God's Dwelling Place? The Pure and Spotless Bride of the Messiah, which is the place where God truly dwells (Rev. 21:3).

Q: What is the most intriguing aspect of the redemption for SAMECH (ס) for you?

Q. What are the key components to becoming part of the glorious Bride of the Messiah? Hint: Look to the fourfold connection in Scripture: unleavened bread, the man without sin, manna, and God's Dwelling Place.

ע

AYIN (ע)—16
SPIRITUAL LIGHT & SIGHT

REDEMPTION OF AYIN (ע)

AYIN (ע) represents the spiritual light mentioned in Genesis 1:3, which is different than the celestial light mentioned in Genesis 1:14-18. The divine light of God is greater than the light that emanates from the sun and stars, and its radiance can only be seen with one's inner eye given by the Holy Spirit.

The redemption of the Messiah's Quantum 22 continues with the sixteenth letter of the Hebrew Alphabet. AYIN (ע) is pronounced *ah-yeen*. This quantum letter has an ordinal value of 16 and a numerical value of 70. Its ancient pictograph is an eye.

ALEF- TAV (את) revealed what the redemption of AYIN (ע) looks like in this Kingdom Day through a profound and powerful Ascension in Christ. [510]

> As a small group focused solely on the perfect will of the Heavenly Father with no agenda, we experienced holy frequencies in the Father's House of the Blood of the Lamb and of the soft yellow atmosphere (Cloud of His Presence) of the Hebrew Living Letters. We beheld a circle on the floor in heaven that was put in place after The Fall, which is connected to the foundation of a human's soul. YHVH pierced the circle on heaven's floor with a pillar of pure uncreated light to release the completion of the restoration of all things. It freed the fullness of the Blood that speaks of better things (Heb. 12:24) in conjunction with His quantum letters being emancipated from the bondage of corruption (Rom. 8:21). God took the cap off a greater multidimensional reality that will shift our 3-D reality to 4-D and 5-D.

The mind's eye connection was singled out during our group ascension, specifically a right connection with our emotions in our mind's eye. The entire mind's eye of a person that is perfectly aligned and pure is linked to our pre-fallen multidimensional state. "Mind's eye" is the concept of humans having an eye in our mind. It's a mental image. Not all people have the same mental imagery ability, but they all have the potential since the first gift given when a person is born from above is to begin to spiritually, mentally, and physically see the Kingdom of God (John 3:3).

One of the seven seals of the human body—the pineal gland—produces what is called the mind's eye. [511] The pineal gland resembles a pinecone shape; hence, its name comes from the Latin *pinea* for pinecone. It is reddish grey in color and about the size of a grain of rice (5-8 mm). "The pineal gland is located in the epithalamus near the center of the brain, between the two hemispheres of the brain, tucked in a groove where the two halves of the thalamus join." [512] "The human pineal gland grows in size until about 1-2 years of age, remaining stable thereafter." [513] However, its weight increases gradually from puberty onwards. [514]

"The pineal gland consists mainly of pinealocytes, but four other cell types have been identified: interstitial cells, perivascular phagocyte, pineal neurons, and peptidergic neuron-like cells." [515] Unlike most of the brain, the pineal gland is not isolated from the body by the blood-brain barrier. [516] It has a profuse blood flow, second only to the kidney. [517]

The pineal gland produces melatonin. Melatonin has various functions in the central nervous system. One of the most important functions is to modulate sleep patterns. Due to the pineal gland being light sensitive, it possesses a circadian clock that directly regulates melatonin synthesis. [518] Melatonin is also essential to blood pressure and insulin sensitivity.

Common examples of mental images include daydreaming and mental visualization. Sometimes we visualize what we read in books. Sometimes a musician "sees" the music notes in their head when they hear a song. Sometimes an athlete pictures their winning performance ahead of time. Visual imagery is the ability to create mental representatives of persons, places, and things absent from a visual field. This ability is crucial to problem-solving, memory, and spacial reasoning. [519] Neuroscientists have found that imagery and perception share many of the same areas of the brain that function similarly. [520]

On 7 October 2024, while solely seeking the perfect will of the Heavenly Father, He confirmed that the pineal gland is the seat of the soul. The first person to suggest this was Rene Descartes (1596-1650) who regarded the pineal gland as the principal seat of the soul where our thoughts are formed. Descartes discussed the pineal gland in his first book—*Treatise of Man*—as well as his last book—*The Passions of the Soul*. Many of his assumptions were off; however, it appears that at least some of his theories about the pineal gland are on. [521]

The pineal gland has captured a large amount of cultural attention due to the notion of the New Age third eye as well as multiverse encounters, time travel, etc. [522] This is one of those cases when we must look at the root of something to see if it can be redeemed, because the "third eye" is part of the chakra system that's off. However, the pineal gland and how it functions has been created by God Himself; therefore, it can, and will be, redeemed to its pristine state by those who are perfect in unity with the Father, the Son, and the Holy Spirit (John 17:23).

There is some science behind the idea of the pineal gland as a "third eye" or "inner eye." What many call the third eye is a biological process that's part of the photo-neuro-endocrine system, which is composed of the retina, the central nervous system, and the pineal gland. [523] The photo-neuro-endocrine pineal gland is an integral part of the human brain. It offers information on the circadian rhythm which connects the outside world with a person's internal biochemical and physiological needs. [524]

"Rick Strassman, author and psychiatry professor of the University of Medicine, New Mexico, has theorized that the pineal gland is capable of producing N,N-dimethyl-tryptamine [DMT], an extremely powerful hallucinogen, derived from tryptophan, especially under certain stress conditions like the moment of birth, the process of giving birth, or the moments before death." [525] "This molecule might be the one responsible for the near-death experiences reported by patients resuscitated after cardiac-death, later research underlying a cascade effect in which serotonin and endogenic opioids are also involved in these hallucinatory experiences." [526] Rick Strassman has demonstrated that DMT exists in the pineal gland of rodents. [527]

"'The hallmark of the DMT effect is the feeling that what one is witnessing is more real than real,' Strassman noted." [528] I can personally testify that I have experienced several more real than real mental/spiritual pictures in my mind's eye during various ascensions in Christ. We will soon "see" how the pineal gland operates in its first created state, for those who shift into their fifth-dimension reality. They will have heightened pineal gland operations according to our original AYIN (ע) design.

ANCIENT AYIN (ע)

AYIN (ע) is the symbol of Sight and Insight. It is the sacred letter of perception and insight, for its name (עין) means eye. Therefore, from the name of this letter, we understand it encompasses the world of sight both physical and spiritual. "Man's outlook and perception—represented by AYIN—is considered the barometer of his character." [529]

The connection of AYIN with spiritual awareness is first found when Adam and Eve became aware of their sin. *"Then the eyes of both of them were opened, and they realized that they were naked" (Genesis 3:7).*[530] Adam and Eve received new insight. *Targum Yonasan* translates "the eyes of both were enlightened" with an awareness of shame. [531]

"Having eaten from the forbidden Tree of Knowledge their bodies came into conflict with their spirit. No longer could they master their senses."[532] "Sforno explains that after the fall of man, their eyes no longer aspired only to the spiritual but became agents of pleasure and temptations of the flesh." [533] "The eyes see, the heart desires and the person acts. By following his eyes and heart rather than God's will, man lost paradise" (R'Hirsch, Numbers 15:41). [534]

COMPOSITE LETTER (ע)

AYIN (ע) is formed using different combinations of the letters NUN (נ), ZAYIN (ז), VAV (ו), and YUD (י). The quantum letter AYIN (ע) has the numerical value of 70 which can be reduced to the *mispar katan* of seven. AYIN (ע)'s *mispar katan* of 7 connects this glorious letter to the holiness of Sabbath—the seventh day of the week—as well as the Seven Spirits of God. [535]

FULFILLMENT OF AYIN (ע)

The letter AYIN (ע) fulfills as the word (עין) 130=70+10+50, which has a *mispar katan* of 13. Thirteen is the same as *echad*, which signifies the Unity of God. [536]

QUANTUM 16

Atomic Number 16—Sulfur (S). Quantumly speaking, sulfur is a chemical element that has the atomic number **16**, the atomic weight **32.06**, and the atomic symbol **S**. A single sulfur atom contains 16 protons, 16 electrons, and 16 neutrons.

Sulfur is a nonmetallic element in Group 16 [AYIN(ע) is 16] of the Periodic Table. Sulfur gets its name from the Latin word *sulpur,* which is formed from a Latin root meaning "to burn."[537]

In cosmic abundance, sulfur ranks ninth or tenth among the elements. Sulfur occurs in the uncombined state as well as in combination with other elements in rocks and minerals that are widely distributed, although it is classified among the minor constituents of Earth's crust.[538] Sulfur can take the form of over 30 different allotropes (crystal structures). This is the most allotropes of any element. [539] Sulfur is found both in its native form and in metal sulfide ores. It is found in meteorites, in the ocean, in the earth's crust, in the atmosphere, and in practically all plant and animal life. [540]

Elemental sulfur is a bright or soft yellow, crystalline solid at room temperature. Sulfur is the fifth most abundant element by mass on Earth. Though sometimes found in pure, native form, sulfur on Earth usually occurs as sulfide and sulfate minerals.

Historically, sulfur is called brimstone, which means "burning stone."[541] "The history of sulfur is part of antiquity. The name itself probably found its way into Latin from the language of the Oscans, an ancient people who inhabited the region including Vesuvius, where sulfur deposits are widespread. Prehistoric humans used sulfur as a pigment for cave painting; one of the first recorded instances of the art of medication is in the use of sulfur as a tonic."[542]

Sulfur emits a blue flame when burned. It melts into a molten red liquid that forms a toxic gas sulfur dioxide when combined with oxygen (SO_2). Sulfur forms many different compounds, including the gas hydrogen sulfide, which has the infamous strong odor of rotten eggs. Hydrogen sulfide is flammable, explosive, and highly poisonous. Or in a word, dangerous.

Elemental sulfur can be found near hot springs, volcanic emissions, hydrothermal vents, and salt domes in many parts of the world. Today, almost all elemental sulfur is produced as a byproduct of removing sulfur-containing contaminants from natural gas and petroleum. [543] The greatest commercial use of the element is the production of sulfuric acid for sulfate and phosphate fertilizers, and other chemical processes. Sulfur is used to refine oil, extract minerals, and process water. It is also used to produce matches, car batteries, insecticides, and fungicides.

Sulfur is an essential element for all life. It is the eighth most abundant element in the human body by weight,[544] which is about equal in abundance to potassium, and slightly greater than sodium and chlorine. [545] The main dietary source of sulfur for humans is found in plant and animal proteins. [546] Biotin (B_7) and thiamine (B_1) are two vitamins that are organosulfur compounds crucial for life.

"The medicinal use of sulfur dates back to the time of Hippocrates (circa 500 BCE); and sulfur continues to be used primarily to treat acne, seborrheic dermatitis, rosacea, scabies, and tinea versicolor. Elemental sulfur and its various forms (e.g., sulfides, sulfites, and mercaptans) possess antimicrobial and antifungal properties in addition to acting as anti-inflammatory agents." [547]

16 IN SCRIPTURE

AYIN (ע) is 16. The territory given to the Tribe of Issachar contained sixteen cities and their villages (Josh. 19:17-23). Issachar's name means "reward or recompense." He was Leah's fifth son and Jacob's ninth. In *First Chronicles 12:32*, the Tribe of Issachar is described as men who *"had understanding of the times; to know what Israel ought to do."* The eye (AYIN) of their understanding was enlightened: *"17 That the God of our Lord Jesus Christ, the Father of glory, may give to you the spirit of wisdom and revelation in the knowledge of Him, 18 the eyes of your understanding being enlightened; that you may know what is the hope of His calling, what are the riches of the glory of His inheritance in the saints" (Ephesians 1:17-18).*

The Tribe of Issachar knew what to do and when to do it. Their perception was so keen that a whole nation waited for their insight so they could follow. The Tribe of Issachar understood chronological time through their study of the movements of the planets and stars. The Tribe of Issachar was responsible for calling the whole nation of Is-real together when the stars and moon aligned for Biblical Feast days. The Tribe of Issachar knew how to interpret the signs and wonders in the heavens associated with the sun, moon, and stars. [548]

The Tribe of Issachar not only understood chronological time, but they understood spiritual and political time as well. The sons of Issachar could discern what God was doing and when He was doing it. They knew when one move of God was ending and another beginning. [549] For instance, it was unusual for women to sit in authority of Israel. Nevertheless, God chose Deborah. The sons of Issachar knew it and went out to battle under her leadership (Jud. 5). The sons of Issachar also supported King David when he was not popular with King Saul. All the tribes were split on their support of David, except one—Issachar (1 Chr. 12:32). [550]

Given all this, it is no surprise that the sons of Issachar excelled in the understanding, wisdom, and knowledge of the Word of God. God filled the Tribe of Issachar with understanding and wisdom; and chose this tribe to be one of the three tribes who went out first when the nation moved: Judah, Issachar, and Zebulun. [551]

QUANTUM 70

Not only does AYIN (ע) have an ordinal value of 16, but a numerical value of 70. Therefore, let's explore the 70th element on the Periodic Table.

Atomic Number 70—Ytterbium (Yb). Quantumly speaking, ytterbium is a chemical element that has the atomic number **70**, the atomic weight **173.05**, and the atomic symbol **Yb**. A single ytterbium atom contains 70 protons, 70 electrons, and 103 neutrons.

Ytterbium is a rare-earth metal of the lanthanide series of the Periodic Table. A Swedish village is the origin of the name ytterbium. The mineral gadolinite (($Ce, La, Nd, Y)_2FeBe_2Si_2O_{10}$) was discovered in a quarry near the town of Ytterby, Sweden. It has been the source of a great number of rare earth elements. Today, ytterbium is primarily obtained through an ion exchange process from monazite sand (($Ce, La, Th, Nd, Y)PO_4$), which is a material rich in rare earth elements." [552]

"Ytterbium is the most volatile rare-earth metal. It is a soft, malleable silvery metal that will tarnish slightly when stored in air and therefore should be stored in vacuum or in an inert atmosphere when long storage time is required."[553] Ytterbium is readily dissolved by strong mineral acids. It also reacts slowly with cold water as well as oxidizing slowly in the air. [554]

In contrast with the other rare-earth metals, which usually have antiferromagnetic and/or ferromagnetic properties at low temperatures, ytterbium is paramagnetic at temperatures above 1.0 kelvin. [555] Paramagnetism is a form of magnetism whereby some materials are weakly attracted by an externally applied magnetic field, and form internal, induced magnetic fields in the direction of the applied magnetic field. In contrast with this behavior, diamagnetic materials are repelled by magnetic fields and form induced magnetic fields in the direction opposite to that of the applied magnetic field. [556]

Contrary to most other lanthanides, which have a close-packed hexagonal lattice, ytterbium crystallizes in the face-centered cubic system. Ytterbium has a significantly lower density (6.973 g/cm^3) than its neighboring lanthanide: thulium (9.32 g/cm^3) and lutetium (9.841 g/cm^3). Its melting and boiling points are also significantly lower than those of thulium and lutetium too. This is due to the closed-shell electron configuration of ytterbium ($[Xe]\ 4f^{14}\ 6s^2$), which causes only the two 6s electrons to be available for metallic bonding (in contrast to the other lanthanides where three electrons are available). [557]

Ytterbium has few uses. It can be alloyed with stainless steel to improve some of its mechanical properties and used as a doping agent in fiber optic cables. One of the ytterbium's isotopes is being considered as a radiation source for portable X-ray machines. [558] Memory devices and tunable lasers are some other uses as well as ytterbium being used as a less toxic industrial catalyst. [559]

70 IN SCRIPTURE

AYIN (ע) is 70. In Scripture, the number seventy occurs frequently. Seventy nations and languages were disbursed from the Tower of Babel. In the Book of Jubilees, seventy nations and languages in the world are based upon the seventy grandsons of Noah enumerated in Genesis 10, each of whom became the ancestor of a nation. [560]

AYIN (ע) is 70. Seventy bulls were sacrificed on God's Altar during the Feast of Tabernacles. These 70 bullocks are connected to the 70 nations. *Sukkot* is also called the Feast of the Nations (Num. 29:12-32).

The connection of the seventy bulls to the seventy nations is taken from Deuteronomy 32:8, Genesis 46:27, and Exodus 1:1-5. Please don't forget about the necessity of the nations of the world coming to Jerusalem to celebrate *Sukkot* (Tabernacles) and worship the King (Zech. 14:16-19). *Sukkot* is not just a harvest celebration, but the time when Solomon completed and dedicated the Temple in Jerusalem (2 Chr. 5:1-3). So will it be a time of Temple completion and dedication again. The 70 bulls sacrificed during *Sukkot* were according to the heavenly pattern, which always included all nations of the world in the plans for His Dwelling Place.

AYIN (ע) is 70. Seventy souls went to Egypt to be preserved in a time of great famine.

These seventy were the descendants of Jacob. Jacob was 130 years old when he made the journey. [561] *"¹ Now these are the names of the children of Israel who came to Egypt; each man and his household came with Jacob: ² Reuben, Simeon, Levi, and Judah; ³ Issachar, Zebulun, and Benjamin; ⁴ Dan, Naphtali, Gad, and Asher. ⁵ All those who were descendants of Jacob were SEVENTY persons (for Joseph was in Egypt already)" (Exodus 1:1-5).*

AYIN (ע) is 70. In the holy city of Jerusalem, which has seventy names, they built God's Temple, which has seventy pillars. There are seventy holy days in a Biblical Year—52 Sabbaths and 18 festivals, including the intermediate days of *Pesach* and *Sukkot*. [562]

FULL OF EYES

"The Cherubim of God's Chariot Throne are made up of people after the Righteous Order of Melchizedek whose aim is the purification of their human nature into its exalted state. Originally, we were not created as human beings. To be human is a fallen state. We were originally created as living beings (i.e., living souls). *"And the LORD God formed man of the dust of the ground, and breathed into his nostrils the breath of life; and man became a living soul" (Genesis 2:7 $_{KJV}$)*. The quest is one of self-conquest in order to return to being a living soul with a resurrected body and a quickening spirit. The soul that conquers oneself is the one who has the All-Consuming Fire sitting on their throne. Self-conquest is an arduous, painful, and trying quest. One can say its path is paved with the fire of many trials. Even though the flaming sword that guards the way to the Tree of Life is singular, this sword of fire wielded by the Cherubim is both your sword as well as God's sword." [563]

"Right after Revelation 19 reveals that the Bride of Christ is made ready by the righteous acts of the saints and how blessed everyone is who has been called to the Marriage Supper of the Lamb, we see the One who the Bride will be wed to, as one. He is called *"The Word of God"* as well as *"Faithful and True" (Revelation 19:13,11)*. We are also told that *"His EYES were as a flame of fire" (Revelation 19:12)*.

> *"And I beheld, and, lo, in the midst of the throne and of the four beasts, and in the midst of the elders, stood a Lamb as it had been slain, having seven horns and seven EYES, which are the Seven Spirits of God sent forth into all the earth" (Revelation 5:6 $_{KJV}$)*.

His eyes of a flame of fire are the torches before the throne that allows His New Living Creature full of eyes to be in the midst and round about His Throne: *"⁵ And out of the throne proceeded lightnings and thunderings and voices: and there were seven lamps of fire burning before the throne, which are the Seven Spirits of God. ⁶ And before the throne there was a sea of glass like unto crystal: and in the midst of the throne, and round about the throne, were four beasts FULL OF EYES before and behind" (Revelation 4:5-6 $_{KJV}$)*. Therefore, the Word of God that we are instructed to take up is a sword connected to the Seven Spirits of God. These are His seven eyes that are alive and powerful. These seven eyes can also rightly divide one's soul and spirit as well as judge the thoughts and attitudes of one's heart (Heb. 4:12)." [564]

These Seven Spirits of God are both His eyes and His flames of fire. They are sent into all of our earth, so His royal priesthood can be purified in order to rule and reign on earth (both within and without). There is a trans-dimensional garden east of Eden whose entrance is guarded with a flaming sword that turns every way, which means that there's no getting around this sword of fire that keeps the way of the Tree of Life (Gen. 3:24). [565]

His eyes of a flame of fire are searching to and fro seeking those who are His. His Beloved Bride will certainly be His, and she will epitomize: *"Love the LORD your God with all your heart and with all your soul and with all your strength" (Deuteronomy 6:5).* [566]

Notice that *Revelation 4:6* shows us that the four beasts (i.e., four living creatures) are *"FULL OF EYES before and behind."* [567] Ezekiel 1:18 also tells us that the rings on the wheels are full of eyes: *"As for their rings, they were so high that they were dreadful; and their rings were FULL OF EYES round about them four" (Ezekiel 1:18 KJV).* Fundamentally, let's understand that all these eyes have several dimensions to them. One dimension is that the Order of Melchizedek is His eyes. Ezekiel 10:12 goes on to tell us: *"And their whole body, and their backs, and their hands, and their wings, and the wheels, were FULL OF EYES round about, even the wheels that they four had."* The Corporate Order of Melchizedek's whole body—our backs, our hands, our wings, and our wheels (spirits)—are full of eyes. Think of every cell in your body as an eye. Our cells have intelligence, sight, and insight.

The wheel within the wheel beside each living creature is full of eyes. We are told that the spirit of the living creatures is in the wheels (Ezek. 1:20-21). Mysteries of mysteries... Somehow, our spirits have something similar to the cellular sight of our bodies, which gives us spiritual sight and insight.

WINDOW OF THE SOUL

AYIN (ע) represents an eye. A person's eyes are one of the most intricately designed organs in the human body. The human brain is the only part that is more complex. It takes one nerve fiber to carry sound to the brain while more than a half of million nerve fibers are used to carry visual pictures from the optic nerve to the brain.

The eye is called the window to the soul. Not only do eyes take pictures of what's on the outside, but eyes reveal what's on the inside of a person as well.

In the Sermon on the Mount, Messiah Yeshua taught His disciples the eye is the light of the soul. *"The lamp of the body is the EYE. If therefore your EYE is good, your whole body will be full of light. But if your EYE is bad, your whole body will be full of darkness. If therefore the light that is in you is darkness, how great is that darkness!" (Matthew 6:22-23).*

Notice the difference of the good eye and the evil eye: *"The EYE that mocks at his father, and despises the old age of his mother, the ravens of the valley shall pick it out and the young vultures shall eat it" (Proverbs 30:17 Lamsa's Aramaic).* Personally, I like to focus on the Bride of the Messiah's Holy Spirit eyes: *"Behold, you are fair, my love! Behold, you are fair! You have dove's EYES" (Song of Solomon 1:15).*

"You have ravished my heart, My sister, my spouse; You have ravished my heart With one look of your EYES, With one link of your necklace. How fair is your love, My sister, my spouse! How much better than wine is your love, And the scent of your perfumes Than all spices!" (Song of Solomon 4:9-10).

Kings and priests made after the Righteous Order of Melchizedek are Bridal Candidate Word Warriors advancing their kingdoms of God within. Through Psalms 19, the Kingdom of God advances: *"⁷ The law of the Lord is perfect, converting the soul; The testimony of the Lord is sure, making wise the simple; ⁸ The statutes of the Lord are right, rejoicing the heart; The commandment of the Lord is pure, enlightening the EYES"* (Psalms 19:7-8).

Listen to how the man after God's own heart, who is explicitly connected to the Order of Melchizedek in Psalms 110:4, sets nothing wicked before his eyes and keeps his eyes on the faithful. *"¹ I will sing of mercy and justice; To You, O Lord, I will sing praises. ² I will behave wisely in a perfect way. Oh, when will You come to me? I will walk within my house with a perfect heart. ³ I will set nothing wicked before my EYES; I hate the work of those who fall away; It shall not cling to me. ⁴ A perverse heart shall depart from me; I will not know wickedness. ⁵ Whoever secretly slanders his neighbor, Him I will destroy; The one who has a haughty look and a proud heart, Him I will not endure. ⁶ My EYES shall be on the faithful of the land, That they may dwell with me; He who walks in a perfect way, He shall serve me. ⁷ He who works deceit shall not dwell within my house; He who tells lies shall not continue in my presence"* (Psalms 101:1-7).

QUANTUM QUARK (ע)

In summary, AYIN (ע) is the sixteenth Hebrew Living Letter. AYIN (ע) in ancient Hebrew means an eye. AYIN (ע)'s pictograph of an eye symbolizes seeing, understanding, experiencing, and being seen. The redemption for AYIN (ע) focuses on the mind's eye of a person as it relates to the pineal gland in the human body. When a person is born from above, they begin to redemptively see the Kingdom of God spiritually, mentally, and physically (John 3:3). Since the pineal gland has been created by God, it will be redeemed and restored to its pristine state by those who are perfect in unity with the Father, the Son, and the Holy Spirit (John 17:23). The science behind the idea of the pineal gland as an "inner eye" is the biological process that's part of the photo-neuro-endocrine system.

The AYIN (ע) is the sacred letter of perception and insight, for its name (עין) means eye. Therefore, from the name of this letter, we understand it encompasses the world of sight both physical and spiritual. The connection of AYIN with spiritual awareness is first found when Adam and Eve became aware of their sin (Gen. 3:7). Adam and Eve received new insight. *Targum Yonasan* translates "the eyes of both were enlightened" with an awareness of shame. Having eaten from the forbidden Tree of Knowledge their bodies came into conflict with their spirit. No longer did they have mastery over their senses.

Q: What are your thoughts about God restoring your pineal gland to its pristine state?

Q. How can one regain mastery of their senses?

PEY (פ)—17
KINGDOM SPEECH

REDEMPTION OF PEY (פ)

The redemption of the Messiah's Quantum 22 continues with the seventeenth letter of the Hebrew Alphabet. PEY (פ) is pronounced *pay*. This quantum letter has an ordinal value of 17 and a numerical value of 80. Its ancient pictograph is a mouth.

The first mention of "mouth" in Scripture is in Genesis 4:11 where it tells us that the earth has a mouth. Add to this that God defended Moses by saying that He spoke with Moses mouth to mouth. "*⁶And He said, 'Hear now My words: If there is a prophet among you, I the Lord will make Myself known to him in a vision And I will speak to him in a dream. ⁷But it is not so with My servant Moses; He is entrusted and faithful in all My house. ⁸With him I speak MOUTH TO MOUTH* [directly], *Clearly and openly and not in riddles; And he beholds the form of the Lord. Why then were you not afraid to speak against My servant Moses?'*" *(Numbers 12:6-8 _{Addition mine})*.

In the Hebrew Alphabet, PEY (פ) follows AYIN (ע), which suggests there's a priority of seeing and understanding before the verbal expression of any insight. PEY (פ) encompasses the terms "word," "expression," "speech," "decree," "declaration," etc.

The redemption of PEY (פ) has to do with speaking the Word of God and words of life. Consider that your words have the power to change your DNA because spoken words are frequencies. The vibrational behavior of DNA has been studied by a Russian biophysicist and molecular biologist Pjotr Garjajev and his colleagues. They refer to human DNA as a biological internet, being far superior to the artificial one we've created. Their research explains phenomena such as self-healing through positive affirmations. These Russian researchers investigated the 90% of DNA that's labeled "junk DNA."

Pjotr Garjajev's team contained both linguists and geneticists. The Russian linguists found that our genetic code follows the same rules as all human languages. This makes sense since the Father of Lights spoke creation into existence. The linguists compared the rules of syntax (how we put words together to form phrases and sentences), semantics (the study of meaning in language forms), and the basic rules of grammar. They found that the alkaline in our DNA follow the rules of grammar, just like human languages. So, human language is simply a reflection of our DNA, and our DNA has been made in the same image as God (Gen. 1:26).

How does speaking words change your DNA? It's simple. Just say words that reflect what you want to see and watch the results to see how your reality changes. This is part of our original design of being co-creators in Christ as well as the concept behind Deuteronomy 30 directing God's people to choose this day blessings or curses, life or death: *"14 The word is very near you, in your mouth [PEY] and in your heart, that you may do it. 15 See, I have set before you today life and good, death and evil, 16 in that I command you today to love the Lord your God, to walk in His ways, and to keep His commandments, His statutes, and His judgments, that you may live and multiply; and the Lord your God will bless you in the land which you go to possess"* (Deuteronomy 30:14-16).

In the experiment, researchers found that DNA will always react to language-modulated laser rays and even to radio waves, as long as the proper frequencies are used. This explains the power behind positive words and thinking. The Russian researchers used this knowledge to work on devices that can change DNA through suitable radio waves and light frequencies to repair genetic defects. Garjajev's research group used this method with chromosomes damaged by dangerous frequencies, such as x-rays, to repair them. Unlike the western way of cutting out single genes from the DNA, Russians found a way to transfer all the information without any of the disharmonies. All of this was possible merely using frequency and language, which shows just how powerful our words and frequencies are. [568] [569]

The most powerful way that people can reprogram their DNA is through the power of the Word of God, using their words. Thoughts are powerful too, but the spoken word kicks it up a notch or two. Take for instance, when you speak a Bible verse out loud, as you meditate upon it and come into agreement with it, you realign yourself to your original design, as it was in the garden east of Eden (Gen. 2:8). "When you are speaking the Word of God over yourself, and changing your mind, the word can change your subatomic cellular structure and this in turn changes your family genetics." [570]

ANCIENT PEY (פ)

PEY (פ) stands for the mouth. It is the instrument of speech. When God created man, *"He blew into his nostrils the soul of life; and man became a living being" (Genesis 2:7).* [571] The Hebrew Sage Onkelos interprets the phrase "living being" as a speaking spirit. The mouth makes a person able to fulfill the ultimate purpose of Creation—to sing praises of the Almighty and speak His Word. As it is written: *"The dead do not praise God" (Psalms 115:17).*

"Through speech, man can articulate the soul's insights and concepts, and communicate them to others; therefore, intelligent speech is the basis of all humanity and civilization."[572] "Anything that happens, whether spiritual, emotional, or physical, can be transformed into action only after it finds expression in words whether or not they are verbalized." [573] To carry out the highest *mitzvah* (commandment, good deed), the study of God's Word includes teaching precepts with speech: *"teach them thoroughly to your children and speak to them" (Deuteronomy 6:7).* [574]

Everyone is supposed to learn God's word and teach it. "*¹⁰ Create in me a clean heart, O God, And renew a right and steadfast spirit within me. ¹¹ Do not cast me away from Your presence And do not take Your Holy Spirit from me. ¹² Restore to me the joy of Your salvation And sustain me with a willing spirit. ¹³ Then I will teach transgressors Your ways, And sinners shall be converted and return to You" (Psalms 51:10-13 ₐₘₚ).*

COMPOSITE LETTER (פ)

PEY (פ) is formed using the letters KAF (כ) and YUD (י). The letter PEY (פ) represents the Ten Commandments—the ten (י) which are housed within a box-like Ark, which pictorially is represented by the letter (כ) lying on its side. ⁵⁷⁵ PEY (פ) can also represent a bird that sits in a nest, which can be an allegory for God's *Shekinah* Glory dwelling in the Temple.

The sacred letter PEY (פ) can be written in its fulfilled form in three different ways: (פא) or (פה) or (פי). Both the fulfillments of (פא) and (פה) have a *mispar katan* of nine, which represents absolute truth. ⁵⁷⁶

FULFILLMENT OF PEY (פ)

The *mispar katan* of the fulfillment of (פה) is 13. Thirteen represents both *echad* (unity of the Godhead) and the Righteous Order of Melchizedek. ⁵⁷⁷

QUANTUM 17

Atomic Number 17—Chlorine (Cl). Quantumly speaking, chlorine is a chemical element that has the atomic number **17**, the atomic weight **35.45**, and the atomic symbol **Cl**. A single chlorine atom contains 17 protons, 17 electrons, and 17 neutrons for the ^{37}Cl isotope. Chlorine is a nonmetallic element. It is the second lightest member of the halogen elements, which is Group 17 of the Periodic Table, which directly corresponds to PEY (פ) being 17.

Chlorine gets its name from the Greek word *chloros*, which means "greenish yellow." Chlorine is a toxic, corrosive, greenish-yellow gas that is irritating to a person's eyes and to their respiratory system. ⁵⁷⁸ In fact, chlorine was used as a chemical weapon in World War I on 22 April 1915 by the German Army at Ypres, Belgium. ⁵⁷⁹ ⁵⁸⁰ Inhaling chlorine gas may cause pulmonary edema—an excessive buildup of fluid in the lungs that can lead to breathing difficulties—as well as severe burns and ulcerations, beyond eye and skin irritation. ⁵⁸¹ Because it is denser than air, chlorine tends to accumulate at the bottom of poorly ventilated spaces. ⁵⁸²

Just as chlorine can be toxic, so can the words of a person's mouth: "*¹⁶ But to the wicked God says: 'What right have you to declare My statutes, Or take My covenant in your mouth, ¹⁷ Seeing you hate instruction And cast My words behind you? . . . ¹⁹ You give your mouth to evil, And your tongue frames deceit. ²⁰ You sit and speak against your brother; You slander your own mother's son'*" *(Psalms 50:16-20).*

According to Mike Adams and Eric Coppolino, 2-3-7-8 tetrachlorodibenzo-p-dioxin (2-3-7-8 TCDD) is the most toxic molecule ever created by man. It consists of a pair of benzene rings, 2 oxygen atoms, and 4 chlorine atoms. The four chlorine atoms on its ends create a chemical armor where this molecule is practically indestructible. 2-3-7-8 TCDD dioxin suffers the same problems as DDT, which was banned in the 1960s because it was a Persistent Bioaccumulative Toxin (called PBT by the EPA). Dioxin became notorious in the 1980s; because of being used in the chemical weapon Agent Orange as well as several dioxin dumps (like Love Canal) being discovered in the United States. [583] [584]

Chlorine is an extremely reactive element and a strong oxidizing agent. Among the elements, it has the highest electron affinity and the third-highest electronegativity behind oxygen and fluorine. Because of its great reactivity, all chlorine in the Earth's crust is in the form of ionic chloride compounds, which includes table salt.

Chlorine is the second-most abundant halogen (after fluorine) [585] and the twenty-first most abundant chemical element in Earth's crust. However, these crustal deposits are dwarfed by the huge reserves of chloride in seawater. The most common compound of chlorine is salt, like the salt in seawater. [586] Salt's chemical formula is sodium chloride (NaCl), which represents a 1:1 ratio of sodium and chlorine ions.

The element chlorine has multiple applications. It is used to sterilize drinking water, disinfect swimming pools, and manufacture several commonly used products, such as paper, textiles, medicines, paints, and plastics, particularly PVC. Moreover, chlorine is used in the development and manufacturing of materials that make vehicles lighter, from seat cushions and seat covers to tire cords and bumpers. [587]

17 IN SCRIPTURE

PEY (פ) is 17. The number 17 in Scripture reveals the judgment and deliverance of God. It is a number specially associated with Noah. On the seventeen of the second month, the fountains of the great deep were broken up and the windows of heaven opened, which started the Great Deluge worldwide. "*⁷ So Noah, with his sons, his wife, and his sons' wives, went into the ark because of the waters of the flood. ⁸ Of clean animals, of animals that are unclean, of birds, and of everything that creeps on the earth, ⁹ two by two they went into the ark to Noah, male and female, as God had commanded Noah. ¹⁰ And it came to pass after seven days that the waters of the flood were on the earth. ¹¹ In the six hundredth year of Noah's life, in the second month, the SEVENTEENTH day of the month, on that day all the fountains of the great deep were broken up, and the windows of heaven were opened. ¹² And the rain was on the earth forty days and forty nights" (Genesis 7:7-12)*.

On the seventeen of the seventh month, Noah's Ark rested on the mountains of Ararat. *"¹ Then God remembered Noah, and every living thing, and all the animals that were with him in the ark. And God made a wind to pass over the earth, and the waters subsided. ² The fountains of the deep and the windows of heaven were also stopped, and the rain from heaven was restrained. ³ And* the waters receded continually from the earth. At the end of the hundred and fifty days the waters decreased. ⁴ Then the ark rested in the seventh month, the SEVENTEENTH day of the month, on the mountains of Ararat. ⁵ And the waters decreased continually until the tenth month. In the tenth month, on the first day of the month, the tops of the mountains were seen" (Genesis 8:1-5).

There is a connection in Scripture between the words of one's mouth and deep waters. "⁴ The words of a man's mouth are deep waters; The wellspring of wisdom is a flowing brook. ⁵ It is not good to show partiality to the wicked, Or to overthrow the righteous in judgment. ⁶ A fool's lips enter into contention, And his MOUTH calls for blows. ⁷ A fool's MOUTH is his destruction, And his lips are the snare of his soul. ⁸ The words of a talebearer are like tasty trifles, And they go down into the inmost body. ... ²⁰ A man's stomach shall be satisfied from the fruit of his MOUTH; From the produce of his lips he shall be filled. ²¹ Death and life are in the power of the tongue, And those who love it will eat its fruit" (Proverbs 18:4-8,20-21).

PEY (פ) is 17. Joseph was seventeen when he brought a bad report about his brothers—the sons of Bilhah (Dan and Naphtali) and the sons of Zilpah (Gad and Asher), which led to his enslavement and being brought to Egypt. "¹ Now Jacob dwelt in the land where his father was a stranger, in the land of Canaan. ² This is the history of Jacob. Joseph, being SEVENTEEN years old, was feeding the flock with his brothers. And the lad was with the sons of Bilhah and the sons of Zilpah, his father's wives; and Joseph brought a bad report of them to his father" (Genesis 37:1-2). Notice how Genesis 37 begins with *"this is the history of Jacob;"* and then, goes right into the story of Joseph.

The story of Joseph is the longest story in the first five books of the Bible (Torah). Joseph was a shadow and type of the Messiah. Jacob said to Joseph: Come son, I will send you down to your brothers (Gen. 37:13). Messiah Yeshua only came to do the will of the Heavenly Father (John 5:19). There are many parallels between Joseph's story and Yeshua's story. In one parallel, Joseph was cast down into a pit. He descended before ascending to a place of honor. Messiah Yeshua first descended before He ascended on high (Eph. 4:9-10). [588]"

Joseph received the longest blessing from Jacob in Genesis 49. In the midst of the blessing of sons, Jacob calls out Yeshua's name. *"For Thy salvation I wait, O Lord" (Genesis 48:18)*. In Hebrew, it literally says: *"for Yeshua I wait."* Yeshua is mentioned more than 300 times in the *Tanach* (Jewish Bible, which is the Old Testament). [589]

QUANTUM 80

Not only does PEY (פ) have an ordinal value of 17, but a numerical value of 80. Therefore, let's explore the 80th element of the Periodic Table.

Atomic Number 80—Mercury (Hg). Quantumly speaking, mercury is a chemical element that has the atomic number **80**, the atomic weight **200.59**, and the atomic symbol **Hg**. A single mercury atom contains 80 protons, 80 electrons, and 120 neutrons. It is located in the 12th column of the Periodic Table.

Hg comes from the word *hydragyrum*, which comes from the Greek word *hydragyros* meaning "water" and "silver." Another name for mercury is quicksilver ("living silver") due to its shiny liquid properties. [590] The Bible speaks of the pure words of YHVH being like living silver: *"The words of the LORD are pure words, like silver tried in a furnace of earth, Purified seven times" (Psalms 12:6).* Never forget that winsome words are spoken at just the right time, and they are as appealing as apples gilded in gold surrounded by silver (Prov. 25:11).

Mercury is named after the planet Mercury. Mercury is a heavy, silvery-white metal that's liquid at room temperature. It is the only known metallic element that is in the liquid state at standard temperature and pressure (STP). The only other element that is a liquid under these conditions is halogen bromine, although metals such as caesium, gallium, and rubidium melt just above room temperature. Compared to other metals, mercury is a poor conductor of heat, but a fair conductor of electricity. [591]

Mercury expands as its temperature is raised; hence, its use in thermometers. A basic thermometer is a small glass tube with some mercury inside. As the temperature rises, the mercury expands. Markings on the tube tell what the temperature is, based on how high the mercury rises. Mercury is also good for thermometers because it does not cling to glass, so the glass remains clear and easy to read. Mercury is used in thermometers, barometers, manometers, sphygmomanometers (blood pressure monitors), float valves, mercury switches, mercury relays, and many other instruments. Laboratory instruments took advantage of mercury's properties as a very dense, opaque liquid with a nearly linear thermal expansion. [592]

Gaseous mercury is used to make fluorescent lamps, mercury-vapor lamps, and some "neon sign" type advertising signs. An electric charge traveling through mercury vapor, or gas, will make the gas glow. Additionally, mercury is used in other electrical products as well.

Mercury used to be used for preserving wood, developing daguerreotypes (the first publicly available photographic process, widely used during the 1840s and 1850s), silvering mirrors, anti-fouling paints (discontinued in 1990), herbicides (discontinued in 1995), interior latex paint, handheld maze games, cleaning, and road leveling devices in cars. Mercury compounds have been used in antiseptics (antimicrobial), laxatives, antidepressants (treatment for major depression and anxiety), and antisyphilitics (elemental mercury chemotherapy was effective in the very early stages of syphilis). However, many died from medically-directed mercury poisoning in their desperate attempts to rid themselves of the excruciating, disfiguring, and humiliating disease. [593]

The element mercury and PEY (פ) are both connected to the number 80. Just as mercury is highly toxic, so can be spoken words. The Book of James speaks of the toxicity of an unbridled tongue (James 1:22-26). Psalms 5 reveals that *"the boastful shall not stand in Your sight"* (vs. 5), *"You shall destroy those who speak falsehood"* (vs. 6), and *"there is no faithfulness in their mouth; their inward part is destruction; their throat is an open tomb; They flatter with their tongue"* (vs. 9).

Due to concerns about the element's toxicity, mercury thermometers, and sphygmomanometers are being largely phased out in clinical environments. Mercury is very dangerous. It can make people very sick if it is swallowed, touched, or even breathed. Mercury poisoning can result from exposure to water-soluble forms of mercury (such as mercuric chloride or methylmercury), by inhalation of mercury vapor, or by ingesting any form of mercury. Mercury poisoning is intensified with lead co-exposures. Concurrent mercury and lead exposures are considered one risk factor for autism. [594]

Most mercury is found in a red ore called cinnabar. This is a rock that contains a compound, or mix, of mercury and sulfur. Scientists can separate the mercury from the sulfur after the ore is mined. Mercury also occurs on its own in nature in isolated drops and occasionally in larger fluid masses. This natural mercury is usually found with cinnabar, near volcanoes or hot springs. [595]

80 IN SCRIPTURE

PEY (פ) is 80. The land rested for eighty years when Israel conquered Moab. "*²⁹And at that time they killed about ten thousand men of Moab, all stout men of valor; not a man escaped. ³⁰So Moab was subdued that day under the hand of Israel. And the land had rest for EIGHTY years*" (Judges 3:29-30).

PEY (פ) is 80. Notice how the number "80" is connected to valiant warriors. For example, Jehu appointed eighty men to destroy the worshipers of Baal. "*So they went in to offer sacrifices and burnt offerings. Now Jehu had appointed for himself EIGHTY men on the outside, and had said, 'If any of the men whom I have brought into your hands escapes, whoever lets him escape, it shall be his life for the life of the other'*" (2 Kings 10:24).

PEY (פ) is 80. Eighty valiant priests withstood King Uzziah when he pridefully took on the priest's role. "*¹⁶But when he was strong his heart was lifted up, to his destruction, for he transgressed against the Lord his God by entering the temple of the Lord to burn incense on the altar of incense. ¹⁷So Azariah the priest went in after him, and with him were EIGHTY priests of the Lord— valiant men. ¹⁸And they withstood King Uzziah, and said to him, 'It is not for you, Uzziah, to burn incense to the Lord, but for the priests, the sons of Aaron, who are consecrated to burn incense. Get out of the sanctuary, for you have trespassed! You shall have no honor from the Lord God'*" (2 Chronicles 26:16-18).

SPEAKING AND HEART OVERFLOW

From the overflow of one's heart, a person speaks. "*The [intrinsically] good man produces what is good and honorable and moral out of the good treasure [stored] in his heart; and the [intrinsically] evil man produces what is wicked and depraved out of the evil [in his heart]; for his MOUTH speaks from the overflow of his heart*" (Luke 6:45 $_{AMP}$). King David's heart overflowed with the goodness of the Lord. "*¹My heart is overflowing with a good theme; I recite my composition concerning the King; My tongue is the pen of a ready writer. ²You are fairer than the sons of men; Grace is poured upon Your lips; Therefore God has blessed You forever*" (Psalms 45:1-2). Wisdom, justice, and truth flowed from within David's heart outward though the words of his mouth. "*²⁷Depart from evil and do good; And you will dwell [securely in the land] forever. ²⁸For the Lord delights in justice And does not abandon His saints (faithful ones); They are preserved forever,*

But the descendants of the wicked will [in time] be cut off. ²⁹ The righteous will inherit the land And live in it forever. ³⁰ The MOUTH of the righteous proclaims wisdom, And his tongue speaks justice and truth. ³¹ The law of his God is in his heart; Not one of his steps will slip" (Psalms 37:27-31 $_{AMP}$).

The mouth of the righteous flows with joy and wisdom while the wicked speak of perverse things and come to nothing. "³⁰ The [consistently] righteous will never be shaken, But the wicked will not inhabit the earth.³¹ The MOUTH of the righteous flows with [skillful and godly] wisdom, But the perverted tongue will be cut out.³² The lips of the righteous know (speak) what is acceptable, But the MOUTH of the wicked knows (speaks) what is perverted (twisted)" (Proverbs 10:30-32 $_{AMP}$).

The wise enjoy the good fruit of their mouth. They reap the abundant life that they sow. "¹ A wise son heeds and accepts [and is the result of] his father's discipline and instruction, But a scoffer does not listen to reprimand and does not learn from his errors. ² From the fruit of his MOUTH a [wise] man enjoys good, But the desire of the treacherous is for violence. ³ The one who guards his MOUTH [thinking before he speaks] protects his life; The one who opens his lips wide [and chatters without thinking] comes to ruin" (Proverbs 13:1-3 $_{AMP}$).

The tongue of the wise speaks knowledge. Soft, gentle, and thoughtful answers are the way of the Tree of Life. Harsh, painful, and careless words crush the spirit and stir up anger. "A soft and gentle and thoughtful answer turns away wrath, But harsh and painful and careless words stir up anger. ² The tongue of the wise speaks knowledge that is pleasing and acceptable, But the [babbling] MOUTH of fools spouts folly. ³ The eyes of the Lord are in every place, Watching the evil and the good [in all their endeavors]. ⁴ A soothing tongue [speaking words that build up and encourage] is a tree of life, But a perversive tongue [speaking words that overwhelm and depress] crushes the spirit" (Proverbs 15:1-4 $_{AMP}$).

The Kingdom of God advances when the words of your mouth and the thoughts of your heart are pleasing and acceptable in God's sight. "Let the words of my MOUTH and the meditation of my heart be acceptable in your sight, O Lord, my rock and my redeemer" (Psalms 19:14 $_{ESV}$).

Out of the heart comes evil thoughts, murder, adultery, sexual immorality, theft, false witness, and slander. These are the ones that don't show that they love God by keeping His commandments. "¹⁷ Do you not see that whatever goes into the MOUTH passes into the stomach and is expelled? ¹⁸ But what comes out of the MOUTH proceeds from the heart, and this defiles a person. ¹⁹ For out of the heart come evil thoughts, murder, adultery, sexual immorality, theft, false witness, slander. ²⁰ These are what defile a person. But to eat with unwashed hands does not defile anyone" (Matthew 15:17-20 $_{ESV}$).

The heart of the wise causes his speech to be discrete with sound judgment. Thus, their gracious words strengthen, heal, nourish, and encourage people's bodies and souls. "²⁰ Whoever gives thought to the word will discover good, and blessed is he who trusts in the Lord. ²¹ The wise of heart is called discerning, and sweetness of speech increases persuasiveness. ²² Good sense is a fountain of life to him who has it, but the instruction of fools is folly. ²³ The heart of the wise makes his speech judicious and adds persuasiveness to his lips. ²⁴ Gracious words are like a honeycomb, sweetness to the soul and health to the body" (Proverbs 16:20-24 $_{ESV}$).

QUANTUM QUARK (פ)

To summarize, PEY (פ) is the seventeenth Hebrew Living Letter. PEY (פ) in ancient Hebrew means a mouth. PEY (פ) follows AYIN (ע), which suggests there's a priority on seeing and understanding before the verbal expression of any insight. From the overflow of one's heart, a person speaks (Luke 6:45). The wise enjoy the good fruit of their mouth. They reap the abundant life that they sow (Prov. 13:1-3). The Kingdom of God advances when the words of your mouth and the thoughts of your heart are pleasing and acceptable in God's sight (Psa. 19:14).

The redemption of PEY (פ) focuses on the power of the Word of God when spoken to dictate to your DNA healing and wholeness. When we speak the Word of God and change our mind, we can change our subatomic cellular structure.

Q: What good words overflow from your heart?

Q. What is your favorite Bible verse to speak out loud?

צ

TSADIK (צ)—18
RIGHTEOUS ONES

REDEMPTION OF TSADIK (צ)

The proper name of the eighteenth letter of the Hebrew Alphabet is TZADDI or TSADE, but it is commonly called TSADDIK or TSADIK (צ). The alphanumeric TSADIK (צ) has an ordinal value of 18 and a numerical value of 90. TSADE's ancient pictograph is a fishhook. Today, it is called TSADIK, which is translated as a righteous one—God and/or man. The pictograph for TSADIK (צ) symbolizes to pull forward, something inescapable, desire, trouble, and a harvest. [596]

The facet that's being emphasized for the redemption of TSADIK (צ) in *Quantum 22* is the phrase "in which righteousness dwell." It is found in *2 Peter 3:13*— *"Nevertheless we, according to His promise, look for new heavens and a new earth in which righteousness dwell."* The Greek word for "righteousness" in this verse is *dikaiosynē* (δικαιοσύνη). In the broad sense, it is the righteous state of a person as they ought to be—the condition acceptable to God. This includes integrity, virtue, purity of life, rightness, correctness of thinking, feeling, and acting. In a narrower sense, it's righteous justice, or in other words, the virtue which gives each their due. [597]

The *dikaiosynē* righteousness, that dwells within the new heavens and a new earth, is the same righteousness spoken of in the following verses:

[1] *"Blessed are those who hunger and thirst for RIGHTEOUSNESS, For they shall be filled"* (Matthew 5:6).

[2] *"Blessed are those who are persecuted for RIGHTEOUSNESS' sake, For theirs is the kingdom of heaven"* (Matthew 5:10).

[3] *"But seek first the kingdom of God and His RIGHTEOUSNESS, and all these things shall be added to you"* (Matthew 6:33).

[4] *"So that as sin reigned in death, even so grace might reign through RIGHTEOUSNESS to eternal life through Jesus Christ our Lord"* (Romans 5:21).

The *"new heavens and a new earth in which righteousness dwell"* spoken of in Second Peter 3:13 is the dwelling place that God is building in this Kingdom Day: *"⁹ᵇ Come, I will show you the Bride, the Lamb's wife . . .* ¹ *Now I saw a new heaven and a new earth, for the first heaven and the first earth had passed away. Also there was no more sea.* ² *Then I, John, saw the holy city, New Jerusalem, coming down out of heaven from God, prepared as a bride adorned for her husband.* ³ *And I heard a loud voice from heaven saying, 'Behold, the tabernacle of God is with men, and He will dwell with them, and they shall be His people. God Himself will be with them and be their God'" (Revelation 21:9b,1-3).*

Notice that righteousness dwells in the Lamb of God's Wife, as portrayed in the fine linen given her to wear that's made of the righteous acts of the saints: "⁶ *And I heard, as it were, the voice of a great multitude, as the sound of many waters and as the sound of mighty thunderings, saying, 'Alleluia! For the Lord God Omnipotent reigns!* ⁷ *Let us be glad and rejoice and give Him glory, for the marriage of the Lamb has come, and His wife has made herself ready.'* ⁸ *And to her it was granted to be arrayed in fine linen, clean and bright, for the fine linen is the RIGHTEOUS acts of the saints" (Revelation 19:6-8).*

> Picture, if you will (by the Spirit of the Living God), gently rolling around in these soft ethereal particles from the color red-to-blue to red-to-blue to red-to-blue, which is the essence of the revelation of His love. After being simply coated in and out, a royal purple poof of color reveals a pear-shaped Bride before a gateway.

Let's consider a pear for a moment. The flesh of the pear fruit contains stone cells. Notice how apples float in water while pears sink. They are called stone cells, because their hard cell walls fill nearly all the cell's volume. The word for "stone" in Hebrew is a contraction of two words *av* (father) and *ben* (son). Therefore, a stone is a beautiful picture of the Father and Son being so integrated that you can't seem to figure out where one begins and another ends. It's their essence of oneness, as proclaimed in John 17:21. And so it will be with the Bride of Christ.

Yeshua is the living stone described in Isaiah 28:16 and 1 Peter 2:4. The white stones of Revelation 2:17 are also living stones of First Peter 2:5, which no man knows the new name written on them, except the recipient. One dimension of this white living stone picture coincides with the watch word for Israel—*The Shema* (Deut. 6:4)—which communicates that a person only truly knows when one hears, understands, and does. This is equivalent to loving the Lord our God with all our heart, soul, and strength (Deut. 6:5).

Therefore, in No Man's Land, there is a place where the Pear Bride manifests. These are people who daily yield to the crucifixion process, [598] which enables their Kingdom of Self to disappear and the Kingdom of God to appear. The perfect building block for the New Jerusalem—the new multidimensional heavens and a new earth in which righteousness dwell—are white living stones. The ultimate holy temple, which is the Lamb's wife, is made up of triumphant overcomers who have divinely become incorporated into the oneness matrix with the Father and the Son by His Spirit, and who continually aim at manifesting His righteousness, holiness, and truth in their lives.

The incredible chapter of John 17 articulates that those that are *"not of the world"* receives and truly understands the words which the Father gave to Yeshua as well as make the Father's name known to men. How did Yeshua make the Father's name known to men? According to John 17:6, Yeshua manifested the Father's name to men; and then, they kept God's Word: *"Jesus therefore answered and was saying to them, 'Truly, truly, I say to you, the Son can do nothing of Himself, unless it is something He sees the Father doing; for whatever the Father does, these things the Son also does in like manner' " (John 5:19 $_{NASB}$).*

"And by this we know that we come to know Him, if we keep His commandments . . . whoever keeps His word, in him the love of God has truly been perfected. By this we know that we are in Him: the one who says he abides in Him ought himself to walk in the same manner as He walked" (1 John 2:3,5-6 $_{NASB}$). Keeping God's Word to the degree where you abide in Him where the love of God is truly perfected within you is where I AM That I AM is. We know that Christ's Church has been, and is continuing to be, sanctified by Yeshua's journey to the cross and His cleansing her by washing of water with the Word (Eph. 5:26). I say "continuing to be sanctified" because like most works of the Spirit in an individual's life, the Spirit's work is a two-way street. Just because a truth, like the cross, exists doesn't mean that it's a reality in your life. Our part is to appropriate that spiritual truth—align our thoughts and deeds to the truth conveyed in God's Word.

Our destiny, as God's people, is to walk in the same glory that the Father gave to Yeshua. How did Yeshua glorify the Father on earth? John 17:4 says through accomplishing the work the Father gave Him to do. *"²¹ Was not Abraham our father justified by works when he offered Isaac his son on the altar? ²² Do you see that faith was working together with his works, and by works faith was made perfect? ²³ And the Scripture was fulfilled which says, 'Abraham believed God, and it was accounted to him for RIGHTEOUSNESS.' And he was called the friend of God. ²⁴ You see then that a man is justified by works, and not by faith only. . . . ²⁶ For as the body without the spirit is dead, so faith without works is dead also" (James 2:21-26).*

ANCIENT TSADIK (צ)

TSADDIK (צ), "the Righteous One, refers to the Almighty, who is called 'The Righteous and Upright One' (Deut. 32:4), devoid of every conceivable injustice. True righteousness can only exist in God and is an integral part of Him." [599] "The term *TZADDIK* is also applied to human beings, who emulate God's righteousness by conducting themselves with integrity, truth, and justice (Tosefos Yom Tov, Berachos 7:3)."[600] The Divine *Tzaddik* sustains and protects the world. The human *tzaddik* does too. Just as the angels are God's messengers in heaven, so are *tzaddikim* His ambassadors on earth (Ramban). [601]

TSADIK, TZADIK, TZADDI, TSADDI, or TSADE (צ) primarily represents *tzaddikim* —righteous ones. TSADI or TSADE means to hunt. "We are not only hunting for fractured people; but also, our own soul's lost sparks. Redeemed sparks serve to elevate the consciousness of a soul of the Tsadik to higher levels."

The first mention of "righteous" in Scripture is in Genesis 7:1 when God declares Noah righteous. *"Then the Lord said to Noah, 'Come into the ark, you and all your household, because I have seen that you are RIGHTEOUS before Me in this generation' " (Genesis 7:1).*

The father of faith—Abraham—is intimately connected to the realities of righteous and righteousness. "*⁵ Then He brought him outside and said, 'Look now toward heaven, and count the stars if you are able to number them.' And He said to him, 'So shall your descendants be.' ⁶ And he believed in the Lord, and He accounted it to him for RIGHTEOUSNESS*" (Genesis 15:5-6). The possessor of heaven and earth—Abraham—is the quintessential example of a righteous man of faith who contrary to hope, in hope believed. "*¹⁶ Therefore it is of faith that it might be according to grace, so that the promise might be sure to all the seed, not only to those who are of the law, but also to those who are of the faith of Abraham, who is the father of us all ¹⁷ (as it is written, 'I have made you a father of many nations') in the presence of Him whom he believed—God, who gives life to the dead and calls those things which do not exist as though they did; ¹⁸ who, contrary to hope, in hope believed, so that he became the father of many nations, according to what was spoken, 'So shall your descendants be.' ¹⁹ And not being weak in faith, he did not consider his own body, already dead (since he was about a hundred years old), and the deadness of Sarah's womb. ²⁰ He did not waver at the promise of God through unbelief, but was strengthened in faith, giving glory to God ²¹ and being fully convinced that what He had promised He was also able to perform. ²² And therefore 'it was accounted to him for RIGHTEOUSNESS'*" (Romans 4:16-22).

The righteousness of the faith is not according to the law but by grace. Grace's purpose is obedience (Rom. 1:5). However, never forget that faith without works is dead (Jam. 2:14-26). Notice how Abraham exemplifies providing a space and place for the way of the Lord, which makes righteousness and justice available. "*¹⁷ And the Lord said, 'Shall I hide from Abraham what I am doing, ¹⁸ since Abraham shall surely become a great and mighty nation, and all the nations of the earth shall be blessed in him? ¹⁹ For I have known him, in order that he may command his children and his household after him, that they keep the way of the Lord, to do RIGHTEOUSNESS and justice, that the Lord may bring to Abraham what He has spoken to him*" (Genesis 18:17-19).

Righteousness and justice are not only the way of the LORD, but they are the foundation and habitation of God's Throne. "*¹⁴ RIGHTEOUSNESS and justice are the foundation of Your throne; Lovingkindness and truth go before You. ¹⁵ Blessed and happy are the people who know the joyful sound [of the trumpet's blast]! They walk, O Lord, in the light and favor of Your countenance! ¹⁶ In Your name they rejoice all the day, And in Your RIGHTEOUSNESS they are exalted*" (Psalms 89:14-16 $_{AMP}$).

> "*¹ The Lord reigns; Let the earth rejoice; Let the multitude of isles be glad! ² Clouds and darkness surround Him; RIGHTEOUSNESS and justice are the foundation of His throne. ³ A fire goes before Him, And burns up His enemies round about. ⁴ His lightnings light the world; The earth sees and trembles. ⁵ The mountains melt like wax at the presence of the Lord, At the presence of the Lord of the whole earth. ⁶ The heavens declare His RIGHTEOUSNESS, And all the peoples see His glory*" (Psalms 97:1-6).

COMPOSITE LETTER (צ)

TSADIK (צ) is formed using the letters NUN (נ) and YUD (י) where the neck of the letter NUN (נ) is bent forward, representing the humility of a righteous person. A righteous person draws down the *Shekinah* (שְׁכִינָה)—Dwelling Presence Glory—as they rightly connect to the Lord of Love, which is symbolized by the letter YUD (י), which is placed above the bent NUN (נ). [602]

FULFILLMENT OF TSADIK (צ)

TSADIK (צ) has a numerical value of 90, which breaks down to a *mispar katan* of 9; thereby, representing truth. [603] TSADIK (צ) fulfills as either (צדי) or (צדיק). When spelled as (צדי), it translates as "hunting." The numerical value of the fulfillment of (צדי) is 104, which is four times the numerical value of Hashem's Name (26 × 4). When TSADIK fulfills as (צדיק), it translates as "a righteous person," which has a *mispar katan* of 15. Fifteen is the same numerical value as God's Name YAH (יה). [604]

QUANTUM 18

Atomic Number 18—Argon (Ar). Quantumly speaking, argon is a chemical element that has the atomic number **18**, the atomic weight **39.95**, and the atomic symbol **Ar**. A single argon atom contains 18 protons, 18 electrons, and 22 neutrons. Argon is in Group 18 on the Periodic Table; therefore, both Argon and TSADIK are connected to 18.

Argon's name comes from the Greek word *argos* (ἀργόν), the neuter singular form of ἀργός, which means idle, lazy or inactive. Indeed, for more than a hundred years after the discovery of argon, chemists were unable to get it to combine with any other element. [605] Argon being inactive is also a reference to the fact that the element undergoes almost no chemical reactions. The complete octet (eight electrons) in the outer atomic shell makes argon stable and resistant to bonding with other elements.

Argon was the first noble gas to be discovered. [606] It is the third most abundant gas in Earth's atmosphere. [607] Argon gas is more than twice as abundant as water vapor, 23 times as abundant than carbon dioxide, and more than 500 times as abundant as neon. The argon gas in the Earth's atmosphere comes from the radioactive decay of potassium-40, which produces the stable argon-40. Over 99% of Earth's argon is argon-40. [608] Argon is the most abundant noble gas in Earth's crust. [609]

Under standard conditions, argon is an inert gas, which means that it typically doesn't react with other elements to form compounds. [610] This correlates to not diluting righteousness—TSADIK (צ)—through mixture as well as demonstrating the righteous conduct of patience and longsuffering. When argon is excited by a high-voltage electric field it glows a violet color. [611] The violet color or "purple" is a kingly hue that is associated with the Spirit of the Reverential Fear of the Lord.

Because argon is the most abundant and cheapest of the noble gases, it is often used when an inert gas is needed. One of the main applications of argon is for the gas inside incandescent lighting. Because argon won't react with the filament used by light bulbs even at high temperatures, it helps the filament to last longer and keeps the glass of the bulb from blackening. [612] Argon is used in incandescent lighting and other applications when diatomic nitrogen is not sufficiently inert. It is even used by museum conservators to protect old materials that are prone to deteriorate due to the gradual oxidation in the air. [613]

Argon is also used for welding, medical instruments, preserving wine, thermal insulation in windows, and microelectronics. Argon is used as an inert gas shield in many forms of welding, including tungsten inert gas welding. [614] Cryosurgery procedures, such as cryoablation, use liquefied argon to destroy cancer cells. When argon is used as a gas laser, it emits a blue-green color. Blue argon lasers are used in surgery to weld arteries, destroy tumors, and correct eye defects. [615]

Argon gas is used in plasma globes and provides a protective atmosphere for growing silicon and germanium crystals. It is also used in technical scuba diving to inflate dry suits because of its low thermal conductivity and inertness. [616]

Argon has no known biological functions. Large amounts of pure argon gas can be considered dangerous in closed areas because it is denser than air and will cause people to suffocate. Argon is difficult to detect because it is colorless, odorless, and tasteless. [617]

18 IN SCRIPTURE

TSADIK (צ) is 18. The number eighteen in Scripture speaks of pillars, rededication, deliverance, and the dwelling place of God.

Speaking of pillars, Solomon set up two bronze (i.e., copper) pillars on the porch of the First Temple. Each pillar was 18 cubits high and was surmounted by a capital of carved lilies, 5 cubits high. The pillar on the right was called *Jachin*, meaning "He establishes." The pillar on the left (north) was called *Boaz*, meaning "swiftness." "*13 Now King Solomon sent and brought Huram from Tyre. 14 He was the son of a widow from the tribe of Naphtali, and his father was a man of Tyre, a bronze worker; he was filled with wisdom and understanding and skill in working with all kinds of bronze work. So he came to King Solomon and did all his work. 15 And he cast two pillars of bronze, each one EIGHTEEN cubits high, and a line of twelve cubits measured the circumference of each*" (1 Kings 7:13-15).

Eighteen is a number connected to God's discipline and deliverance: "*6 Then the children of Israel again did evil in the sight of the Lord, and served the Baals and the Ashtoreths, the gods of Syria, the gods of Sidon, the gods of Moab, the gods of the people of Ammon, and the gods of the Philistines; and they forsook the Lord and did not serve Him. 7 So the anger of the Lord was hot against Israel; and He sold them into the hands of the Philistines and into the hands of the people of Ammon. 8 From that year they harassed and oppressed the children of Israel for EIGHTEEN years—all the children of Israel who were on the other side of the Jordan in the land of the Amorites, in Gilead*" (Judges 10:6-8).

Eighteen is also a number associated with rededication, as in the forever feast of Passover. "*²³ But in the EIGHTEENTH year of King Josiah this Passover was held before the Lord in Jerusalem. ²⁴ Moreover Josiah put away those who consulted mediums and spiritists, the household gods and idols, all the abominations that were seen in the land of Judah and in Jerusalem, that he might perform the words of the law which were written in the book that Hilkiah the priest found in the house of the Lord ²⁵ Now before him there was no king like him, who turned to the Lord with all his heart, with all his soul, and with all his might, according to all the Law of Moses; nor after him did any arise like him*" (2 Kings 23:23-25).

Additionally, 18 is a number associated with God's Dwelling Place according to Ezekiel's vision of the New Jerusalem. "*³⁰ These are the exits of the city. On the north side, measuring four thousand five hundred cubits ³¹ (the gates of the city shall be named after the tribes of Israel), the three gates northward: one gate for Reuben, one gate for Judah, and one gate for Levi; ³² on the east side, four thousand five hundred cubits, three gates: one gate for Joseph, one gate for Benjamin, and one gate for Dan; ³³ on the south side, measuring four thousand five hundred cubits, three gates: one gate for Simeon, one gate for Issachar, and one gate for Zebulun; ³⁴ on the west side, four thousand five hundred cubits with their three gates: one gate for Gad, one gate for Asher, and one gate for Naphtali. ³⁵ All the way around shall be EIGHTEEN thousand cubits; and the name of the city from that day shall be: THE LORD IS THERE*" (Ezekiel 48:30-35).

QUANTUM 90

Not only does TSADIK (צ) have an ordinal value of 18, but a numerical value of 90. Therefore, let's explore the 90th element on the Periodic Table.

Atomic Number 90—Thorium (Th). Quantumly speaking, thorium is a weakly radioactive metallic chemical element that has the atomic number **90**, the atomic weight **232.04**, and the atomic symbol **Th**. A single thorium atom contains 90 protons, 90 electrons, and 142 neutrons.

Thorium is a member of the actinide family. The actinide elements are located in Row 7 of the Periodic Table. They have atomic numbers between 90 and 103. [618] Thorium's name comes from the Norse god of thunder due to its power. It was discovered in the mineral thorite ($ThSiO_4$). Morten Esmark found a black mineral on Løvøya island, Norway, and gave a sample to his father Jens Esmark, a noted mineralogist. The elder Esmark was not able to identify it and sent a sample to Swedish chemist Jöns Jakob Berzelius for examination in 1828; consequently, Berzelius determined it was a new element. [619]

When pure, thorium is a silvery-white metal that is air-stable and retains its luster for several months. When contaminated with the oxide, thorium slowly tarnishes, becoming gray and finally black (forming thorium dioxide). [620] The physical properties of thorium are greatly influenced by the degree of contamination with the oxide. [621] Pure thorium is soft, very ductile, and can be cold-rolled, swaged, and drawn. Thorium is dimorphic, changing at 1400°C from a cubic to a body-centered cubic structure. [622]

Thorium oxide has the highest melting point of all oxides. Only a few elements, such as tungsten, and a few compounds, such as tantalum carbide, have higher melting points. Thorium is slowly attacked by water, but does not dissolve readily in most common acids, except hydrochloric. [623] Thorium is an important alloying agent in magnesium because it imparts greater strength and creep resistance at high temperatures. Thorium oxide is used as an industrial catalyst. Thorium can be used as a source of nuclear power. Thorium is about as abundant as lead and about three times as abundant as uranium. There is probably more energy available from thorium than from both uranium and fossil fuels. [624]

Non-radioactivity-related uses of thorium have been in decline since the 1950s due to environmental concerns largely stemming from the radioactivity of thorium and its decay products. Most thorium applications use its dioxide (sometimes called "thoria" in the industry), rather than the metal. This compound has a melting point of 3300 °C (6000 °F), the highest of all known oxides; only a few substances have higher melting points. [625] This helps the compound remain solid in a flame, and it considerably increases the brightness of the flame; this is the main reason thorium is used in gas lamp mantles. All substances emit energy (glow) at high temperatures, but the light emitted by thorium is nearly all in the visible spectrum, hence the brightness of thorium mantles. [626] Although thorium's use in gas lamp mantles is still common, it is being progressively replaced with yttrium since the late 1990s. [627]

Thorium is odorless and tasteless. The chemical toxicity of thorium is low because thorium and its most common compounds (mostly the dioxide) are poorly soluble in water, which means it precipitates out before entering the body as hydroxide. [628]

Natural thorium decays very slowly compared to many other radioactive materials. The emitted alpha radiation cannot penetrate human skin. As a result, handling small amounts of thorium, such as those in gas mantles, is considered safe, although the use of such items may pose some risks. Exposure to airborne thorium (contaminated dust) can lead to an increased risk of cancers of the lung, pancreas, and blood due to the ability of alpha radiation to penetrate internal organs. [629] As with all radioactive materials, thorium is dangerous to the health of humans and animals. It must be handled with great caution. Living cells that absorb radiation are damaged or killed. [630]

When added to glass, thorium dioxide helps increase its refractive index and decrease dispersion. Such glass finds application in high-quality lenses for cameras and scientific instruments. [631] The radiation from these lenses can darken them and turn them yellow over a period of years and it degrades film, but the health risks are minimal. [632] Yellowed lenses may be restored to their original colorless state by lengthy exposure to intense ultraviolet radiation. Thorium dioxide has since been replaced in this application by rare-earth oxides, such as lanthanum due to providing similar effects but not being radioactive. [633]

90 IN SCRIPTURE

TSADIK (צ) is 90. The New Jerusalem Temple that Ezekiel foresaw had a length of ninety cubits. *Matthew Poole's Commentary* speaks of the proportions of Ezekiel's Temple can easily be laid together to make up the total length of ninety cubits. The Temple and Oracle with their walls are seventy cubits in length. The Temple's Porch is eleven cubits, and its chambers and walls are nine cubits (70+11+9=90). The length of ninety cubits speaks of the reach of righteousness, or the reach of the righteous one(s) while its width of seventy speaks of the radiance of spiritual light AYIN (ע). "*⁸I also saw an elevation all around the temple; it was the foundation of the side chambers, a full rod, that is, six cubits high. ⁹ The thickness of the outer wall of the side chambers was five cubits, and so also the remaining terrace by the place of the side chambers of the temple. ¹⁰ And between it and the wall chambers was a width of twenty cubits all around the temple on every side. ¹¹ The doors of the side chambers opened on the terrace, one door toward the north and another toward the south; and the width of the terrace was five cubits all around. ¹² The building that faced the separating courtyard at its western end was seventy cubits wide; the wall of the building was five cubits thick all around, and its length NINETY cubits"* (Ezekiel 41:8-12).

Sarah was 90 years old when she gave birth to her miraculous son of promise—Isaac. After Hagar gave birth to Ishmael, God told Abraham to change Sarai's name to "Sarah," announcing she would give birth to a son. Whereupon Abraham burst into laughter, which is the meaning of Isaac's name. ⁶³⁴ "*¹⁵ Then God said to Abraham, 'As for Sarai your wife, you shall not call her name Sarai, but Sarah shall be her name. ¹⁶ And I will bless her and also give you a son by her; then I will bless her, and she shall be a mother of nations; kings of peoples shall be from her.' ¹⁷ Then Abraham fell on his face and laughed, and said in his heart, 'Shall a child be born to a man who is one hundred years old? And shall Sarah, who is NINETY years old, bear a child?'"* (Genesis 17:15-17).

The son of promise—Isaac—was a foreshadowing of the Righteous Seed who is Christ. "*¹⁶ Now to Abraham and his Seed were the promises made. He does not say, 'And to seeds,' as of many, but as of one, 'And to your Seed,' who is Christ. . . . ²⁷ For as many of you as were baptized into Christ have put on Christ. ²⁸ There is neither Jew nor Greek, there is neither slave nor free, there is neither male nor female; for you are all one in Christ Jesus. ²⁹ And if you are Christ's, then you are Abraham's seed, and heirs according to the promise"* (Galatians 3:16,27-29).

RIGHTEOUSNESS IN CHRIST

"Melchizedek" is translated as the King of Righteousness. Hebrews 7 speaks of Melchizedek being both the 'king of righteousness' as well as the 'king of peace.' "*¹ For this Melchizedek, king of Salem, priest of the Most High God, who met Abraham returning from the slaughter of the kings and blessed him, ² to whom also Abraham gave a tenth part of all, first being translated 'KING OF RIGHTEOUSNESS,' and then also king of Salem, meaning 'king of peace'*" (Hebrews 7:1-2).

Righteousness is a dove-tail joint where Melchizedek and the Bride of Christ meet. One is the king of righteousness (Heb. 7:2). The other is given bright and clean fine linen to wear equivalent to the righteous deeds of the saints (Rev. 19:7-8). Fundamentally, righteousness is not a State of Doing, but a State of Being. The Hebrew word for "righteousness" is *tsedeq* (Strong's H6664). It's what is right naturally, morally, or legally. Abstractly, *tsedeq* can mean equity; or figuratively, it can mean prosperity. *Tsadeq* can be translated as just, justice, or a cause.

A cause is a reason for action. It's a motive. A cause is also an agent that brings something about, like God's Throne. *"Clouds and thick darkness surround Him; righteousness and justice are the foundation [habitation] of His Throne" (Psalms 97:2).* Therefore, righteousness is literally a place where God is at home. Righteousness is a permanent structure in which He dwells, which is supported by the root meaning behind the word *tsedeq*.

When we dig to the Hebraic root of 'righteousness,' we see that it means TO BE, or to cause to make right in a moral or forensic sense." [635] " 'To be' is the primitive root of 'righteousness,' which also means to cleanse or clear self, to be just, or do justice, or turn to righteousness. 'To be or not to be' really is the right question when it comes to righteousness. My mentor teaches that we have been created as human beings, not human doings. We have to first be, not do; but if we don't do what we say, even the world knows that we're hypocrites. Have you ever noticed that many Christians name and claim "I am the righteousness of God in Christ Jesus;" then shortly thereafter treat people terribly? There's a major disconnect in the Body of Christ when people manifest demonic activity and at the same time claim to be righteous.

Righteousness flows from intimacy. The Bride is clothed in righteousness, but what that really means is that she is enthroned in righteousness. Many of us have been taught that righteousness has nothing to do with what we do. That paradigm needs to shift a quark or two. *" 13 For everyone who partakes only of milk is unskilled in the word of RIGHTEOUSNESS, for he is a babe. 14 But solid food belongs to those who are of full age, that is, those who by reason of use have their senses exercised to discern both good and evil" (Hebrews 5:13-14).* Melchizedek is all about the Word of Righteousness. In fact, the word "order" in the Order of Melchizedek fundamentally means "word."

Don't forget that *"all Scripture is inspired by God and profitable . . . for training in RIGHTEOUSNESS; that the man of God may be adequate, equipped for every good work" (2 Timothy 3:16).* Therefore, we can easily say that the Order of Melchizedek is connected to the Word of the King of Righteousness. Furthermore, we can say that the Bride will resonate at the same righteous frequency as her Beloved Bridegroom. Those things in our lives that are not plumb with God's measuring line of righteousness (according to His Word) will cause us to resonate at a lower frequency than a pure and spotless one.

Righteousness is probably one of the most undervalued and misunderstood commodities of the Kingdom of God. Think about it. God instructs us to seek first His righteousness (Matt. 6:33). He tells us to hunger and thirst for righteousness, and that we are blessed when we are persecuted for righteousness' sake. This means that the righteous place we dwell "in Christ" gets enhanced through persecution. We can also say that the end or purpose behind us being persecuted for His Name's sake is to exhibit the character of Christ, which is founded both in righteousness and justice (Psa. 97:2)."[636]

QUANTUM QUARK (צ)

In summary, TSADIK (צ) is the eighteenth Hebrew Living Letter. In the beginning, this Hebrew letter was called TSADE, which in ancient Hebrew means a fishhook. Today, it is called TSADIK (צ), which means a righteous person. The pictograph for TSADIK (צ) symbolizes pull forward, something inescapable, desire, trouble, and a harvest.

The term TSADDIK also applies to people who emulate God's righteousness by conducting themselves with integrity, truth, and justice. The Divine *Tzaddik* sustains and protects the world. The human *tzaddik* does too. God processes and refines our character, so that we might become the righteousness of God in Christ (2 Cor. 5:21). Matthew 6:33 tells us to seek first the Kingdom of God and His righteousness while *Matthew 5:5* declares *"Blessed are those who hunger and thirst for RIGHTEOUSNESS, for they shall be satisfied."*

Q: Why is righteousness a gift that reigns in life?

Q. How are you seeking to become the righteousness of God in Messiah Yeshua?

KOOF (ק)—19
EMULATION OF HOLINESS

REDEMPTION OF KOOF (ק)

The redemption of the Messiah's Quantum 22 continues with the nineteenth letter of the Hebrew Alphabet. The pronouncement of the sacred letter KOOF (ק) has the sound of *kof* with the sound of "q" as in queen. KOOF (ק) has an ordinal value of 19 and a numerical value of 100. The pictograph of ק is a picture of the back of the head, just like the Holy of Holies is situated in the back of God's Temple.

The redemption of KOOF (ק) reveals the Holy of Holies within mankind and without. In *ALEF-TAV's Hebrew Living™ Letters: 24 Wisdoms Deeper Kingdom Bible Study*, the back of the Messiah's Head is highlighted: "The occipital bone is a bone that covers the back of your head. This area is called the occiput, or posterior noggin. The occipital bone is the only bone in your head that connects with your cervical spine (neck). [637] Or in other words, the occipital bone is the only part of a person's skeletal structure that connects their head to their body. Thus, we understand KOOF (ק) represents the Head of the Messiah's vital connection to His Body.

The occipital bone surrounds a large opening known as the *foramen magnum*. The foramen magnum allows key nerves and vascular structures passage between the brain and spine. Namely, it is what the spinal cord passes through to enter the skull. The brainstem also passes through this opening. [638] The foramen magnum also allows two key blood vessels traversing through the cervical spine, called the vertebral arteries, to enter the inner skull and supply blood to the brain. [639] The main bone of the occiput is trapezoidal in shape and curves into itself like a shallow dish. This occipital bone overlies the occipital lobes of the cerebrum. The occipital lobes of the cerebrum sit at the back of the head and are responsible for visual processing, like perceiving distance and depth, color, object and face recognition, and memory formation. [640] Just like the occipital lobe is in the back of the head and is the smallest lobe of the brain, so is the Holy of Holies, which is in the back of God's Temple, its smallest room." [641]

Never forget that your resurrected triune being—your body, soul, and spirit—in their fullness forms a glorious temple of the Holy Spirit: *"Or do you not know that your body is a temple of the Holy Spirit who is in you, whom you have from God, and that you are not your own?" (1 Corinthians 6:19_{NASB})*. The Greek word for "body" in First Corinthians 6:19 is *soma*, which reveals that the human body being talked about is one that's a sound whole.[642] A body can only be a sound whole when a person's soul and spirit are sound and whole too. *Soma* comes from the Greek word *sozo*,[643] which anchors making the human body whole idea along with healing, deliverance and protection. I would add the ultimate redemption goal for mankind: deliverance from corruption.

In Temple times, if a vessel was found with a KOOF (ק) written on it, it was assumed that its contents were consecrated—sacred and holy (Masser Sheni, 4:11). The Most Holy place in God's Temple is the Holy of Holies. The Holy of Holies is described as the House of the "*Kapporet*" in 1 Chronicles 28:11. Many invite us to believe that the *kapporet* (mercy seat) is an accessory of the Ark of the Covenant; but when one dives into the depths, you will find that everything having to do with the Ark of the Covenant is one. The *kapporet* is the Ark's lid, or its cover. When one studies the word "throne" in Scripture, its primary meaning is covering. Therefore, the *kapporet* symbolizes the Throne of God, or at least a portal to it. It represents the place where the Lord appears (Lev. 16:2).

Additionally, Psalm 80:1 speaks of God being enthroned upon the Cherubim. The Ark of the Covenant has two Cherubim over the Ark, which forms the golden throne in the Holy of Holies.

> Feel the holy reverential fear of the Lord. Hear the song: "Take [us] into the Holy of Holies." The Holy of Holies is a dimensional gateway within His Bride. In humility, the members of the Bride of Christ bow as one. The Oneness Bride walks through the Veil within her bodily temple to behold a fire in between the two Cherubim. At God's command, she takes a step into His mercy seat's Baptism of Fire to be completely purified and purged. She feels the holy fire go through her body, soul, and spirit . . . to their roots. It's a holy consecration. The Corporate Bride of Christ is forged in His Holy of Holies fire, which merges the purified ones together with the righteous and redeemed Quantum 22.[644]

"In the midst of God's Kingdom is a throne and in the midst of the throne are the living creatures that are sometimes called Cherubim. The living creatures have to do with creation. They mark the heavenly sweet spot for our return to our primordial state on earth, as it was in the Garden before the Fall. The following is what Scripture has to say about the Cherubim overall:

[1] Cherubim marks the spot for us to return to our pristine, primeval state (Gen. 3:24).

[2] Cherubim are guardians of the holiness of God "to keep the way of the Tree of Life" (Gen. 3:24), and therefore, they must express His holiness.

[3] Cherubim mark the Dwelling Place of God because He "dwells between the cherubim" (Num. 7:89; 1 Sam. 4:4; 2 Sam. 6:2; 2 Kings 19:15; Psa. 80:1; Psa. 99:1; Isa. 37:16).

[4] Cherubim are carriers of God's throne, which is the seat of His divine glory as well as His royal power (1 Chr. 28:18; Ezek. 1:26-28; Ezek. 10:1)."[645]

Our previous Hebrew Living Letter—TSADIK—centered on righteousness. "The righteousness of God is revealed faith to faith because true righteousness from God Himself is the motive for our faith working through love.

Romans 6 reveals that we are slaves to the one whom we obey, whether to sin unto death or to obedience unto righteousness (Rom. 6:16). Christ's righteousness in us via the Holy Spirit will bring about the process for our becoming the righteousness of God in Christ Jesus here on earth (2 Cor. 5:21), as we partner with Him to become *"a perfect man, unto the measure of the stature of the fullness of Christ" (Ephesians 4:13 $_{KJV}$)*. As we obey the leading of His Spirit within, it will happen just as it did with Yeshua: *"Although He was a Son, He learned obedience from the things which He suffered" (Hebrews 5:8 $_{NASB}$)*. This is the context we need when we read in Romans Chapter 6 that the obedience of His "servants of righteousness" results in righteousness, which then results in their holiness (Rom. 6:19).

Please notice that being holy as He is holy is a command, not a suggestion: *"^{14} As obedient children, not conforming yourselves to the former lusts, as in your ignorance; ^{15} but as He who called you is holy, you also be holy in all your conduct, ^{16} because it is written, 'BE HOLY, FOR I AM HOLY' ^{17} And if you call on the Father, who without partiality judges according to each one's work, conduct yourselves throughout the time of your stay here in fear; ^{18} knowing that you were not redeemed with corruptible things, . . . ^{19} but with the precious blood of Christ, as of a lamb without blemish and without spot" (1 Peter 1:14-19)*. [646]

ANCIENT KOOF (ק)

KOOF (ק) is made up of two letters: RESH (ר) and ZAYIN (ז). ZAYIN descends below the line while RESH hovers above it. This paradoxical union symbolizes the secret of KOOF (ק)—*"There is none holy as God."* In fact, in general, the Hebrew Living Letter KOOF (ק) stands for holiness *(kedushah)*. Of all the letters in the Hebrew Alef-Bet, KOOF is the lowest; therefore, KOOF signifies the emulation of holiness on earth.

"The term 'Holiness of God' conveys the message that He is supremely exalted (Maharah)." Although God's *kedushah* is metaphysical, and therefore [in many ways] inconceivable to man, He still requires of mankind: *"Be holy, as I am holy" (Leviticus 19:2)*. [647] "Man's holiness is a reflection of God's absolute sanctity and purity (R'Bachya). To raise his status in *kedushah*, man must attach himself to God, conforming to His presence and truth ways." [648] "Man's sanctity results from a mastery of all their instincts and inclinations, and placing them at the disposal of God's will (R'Hirsch)." [649] We echo and apply to earthly life the angel's recitation of *"Holy, holy, holy!" (Isaiah 6:3)*. [650]

KOOF (ק) stands for a Temple offering (Magan David). "When the Temple stood, sacrifices brought the worshipers closer to God and brought God's blessing to the entire world." [651]

KOOF (ק) is also the symbol for Growth Cycles. The most obvious manifestation of God's majesty is expressed in nature and its cycles. *Hakafos* (the circular ritual procession on *Hoshanah Rabbah* and Simchas Torah) are manifestations of God's holiness. They are said to possess mystical power, especially circling sevenfold. *Hakafos* have the power to tear down all barriers between God and His people as well as tear down barriers to holiness. Take, for example, the march around Jericho. [652]

"Indeed, at the siege of Jericho in the Canaanite bastion of idolatry, the Israelite warriors circled the city walls while the priests blew the shofar. After one *hakafah* daily for six days, followed by seven *hakatos* on the Sabbath, the walls of Jericho sunk into the ground before Joshua and the Children of Israel (Josh. 6)." [653] According to R'Chiya, the miracle of Jericho was commemorated every *Succos* in the Temple when worshipers circled the Temple Altar. This was done once a day for six days (Yalkut Tehillim 703). On the seventh day of *Succos*—*Hoshanah Rabbah*—the worshipers circled the Temple Altar seven times, paralleling the seven days at Jericho. [654]

COMPOSITE LETTER (ק)

KOOF, KUF, QOPH (ק) is formed by either a combination of the letters FINAL NUN (ן) and KAF (כ), or with a combination of the letters FINAL NUN (ן) and RESH (ר) with the final ן acting as its foot. [655] KOOF (ק) is one of only two Hebrew letters that are composed of two disjointed parts. The other letter is HEI (ה). [656]

The attribute of holiness associated with the sacred letter KOOF (ק) is connected to *"Every tenth one shall be holy to Hashem" (Leviticus 27:32)*; because the numerical value of KOOF is 100 = 10×10 or multiplied holiness. [657]

FULFILLMENT OF KOOF (ק)

KOOF (ק) has a numerical value of 100, which breaks down to a *mispar katan* of 1; thereby, representing the Unity of YHVH. [658] KOOF (ק) fulfills as (קוף) can mean "to round off" or "strength." The numerical value of the letter ק as קוף is 186. [659] The number 186 is also the numerical value of the word (מקום), which is the space and place where the Shekinah rests. *"And the LORD said, Behold, there is a place (מקום) by Me, and thou shalt stand on the rock" (Exodus 33:21 $_{KJV}$)*. [660]

There are a variety of different methods of interpretation, which are classified under the general term *mispar merubah* or "multiplication method." In regard to the number 186, let us simply look at the squared numbers of the Name YHVH. Any number multiplied by itself is called the "number square," which results in creating a two-dimensional square. This square represents an area. It is possible to square the Hebrew letters to come up with the "square value" of each letter, which is the area that the latent power of the Hebrew letter occupies. Summing the squares of the numerical values of the letters YHVH (יהוה) results in the total squared number value for His Tetragrammaton Name. [661]

(י) × (י) = 10 × 10		=	100
(ה) × (ה) = 5 × 5		=	25
(ו) × (ו) = 6 × 6		=	36
(ה) × (ה) = 5 × 5		=	25
	Total	=	186

The *Beis haMikdash* is the place where the *Shekinah* rested, as described in numerous places in the Word of God by the word "place" (מקום). [662]

QUANTUM 19

Atomic Number 19—Potassium (K). Quantumly speaking, potassium is a chemical element that has the atomic number **19**, the atomic weight **39.098**, and the atomic symbol **K**. A single potassium atom contains 19 protons, 19 electrons, and 20 neutrons.

Potassium is the fourth element in the first column of the Periodic Table. It is chemically similar to sodium, which is the alkali metal above it on the Periodic Table. The element potassium was first isolated from potash, which is the ashes of plants. 'Kalium' is potassium's Latin name; hence, its atomic symbol K. It originated from the Arabic *'al qaliy,'* meaning "calcined ashes" (the ashes left over when plant material is burned). A number of modern languages still refer to potassium as kalium. [663]

Potassium is a silvery white metal that is soft enough to easily cut with a knife. [664] When potassium metal is cut, the exposed metal tarnishes quickly, forming a dull oxide coating. When this metal reacts rapidly with atmospheric oxygen, it forms a flaky potassium peroxide within seconds of exposure. Potassium peroxide is an inorganic compound with the molecular formula K_2O_2. [665] Potassium peroxide is highly reactive, oxidizing to a white-to-yellowish solid. While potassium peroxide is not flammable itself, it reacts violently with other flammable materials as well as elemental potassium reacts vigorously on contact with water, producing heat and hydrogen gas. Due to potassium peroxide's unique characteristics, it is used to purify the air as well as a bleach (due to the peroxide). [666]

Elemental potassium does not occur in nature due to its high reactivity of reacting violently with water and oxygen. [667] In nature, potassium occurs only in ionic salts. It is found dissolved in seawater,[668] and occurs in many minerals, such as orthoclase (potassium feldspar), granites, and other igneous rock. [669]

Potassium has such a low melting point that the flame of a candle can cause it to melt. When it burns, it produces a lavender-colored flame. Potassium is the second least dense metal after lithium. It is so light that it can float in water before it violently reacts to it. [670]

Due to potassium being important for plant growth, the largest use of potassium is potassium chloride (KCl). Agricultural fertilizers consume 95% of global potassium chemical production, and about 90% of this potassium is supplied as KCl. [671]

Major potassium chemicals are potassium hydroxide, potassium carbonate, potassium sulfate, and potassium chloride. Megatons of these compounds are produced annually. [672] Industrial applications for potassium include soaps, detergents, gold mining, dyes, glass production, gunpowder, and batteries. [673]

Potassium also plays a vital role in our bodies. Potassium is the eighth or ninth most common element by mass in the human body. The body has about as much potassium as sulfur and chlorine. Only calcium and phosphorus are more abundant (with the exception of the ubiquitous CHON elements). [674] Potassium ions are present in a wide variety of proteins and enzymes. [675]

Potassium levels influence multiple physiological processes in the human body, including:[676] [677]

- hormone secretion and action
- vascular tone
- systemic blood pressure control
- gastrointestinal motility
- acid-base homeostasis
- glucose and insulin metabolism
- mineralocorticoid action
- renal concentrating ability
- fluid and electrolyte balance
- etc.

Potassium ions are vital for the functioning of all living cells. The transfer of potassium ions across nerve cell membranes is necessary for normal nerve transmission. Potassium deficiency and potassium excess can each result in numerous signs and symptoms, including an abnormal heart rhythm and various electrocardiographic abnormalities. Fresh fruits and vegetables are good dietary sources of potassium.

Potassium is present in all fruits, vegetables, meat, and fish. Foods with high potassium concentrations include yams, parsley, dried apricots, milk, chocolate, all nuts (especially almonds and pistachios), potatoes, bamboo shoots, bananas, avocados, coconut water, soybeans, and bran.[678] The United States Department of Agriculture lists tomato paste, orange juice, beet greens, white beans, potatoes, plantains, bananas, apricots, and many other dietary sources of potassium (ranked in descending order according to potassium content). [679]

19 IN SCRIPTURE

KOOF (ק) is 19. The number nineteen in Scripture speaks of the advancement of God's Kingdom, fiery destruction, and inheritance. The Kingdom of God advances from Psalm 19.

"⁷ The law of the Lord is perfect, converting the soul; The testimony of the Lord is sure, making wise the simple; ⁸ The statutes of the Lord are right, rejoicing the heart; The commandment of the Lord is pure, enlightening the eyes; ⁹ The fear of the Lord is clean, enduring forever; The judgments of the Lord are true and righteous altogether. ¹⁰ More to be desired are they than gold, Yea, than much fine gold; Sweeter also than honey and the honeycomb. ¹¹ Moreover by them Your servant is warned, And in keeping them there is great reward. ¹² Who can understand his errors? Cleanse me from secret faults. ¹³ Keep back Your servant also from presumptuous sins; Let them not have dominion over me. Then I shall be blameless, And I shall be innocent of great transgression. ¹⁴ Let the words of my mouth and the meditation of my heart be acceptable in Your sight, O Lord, my strength and my Redeemer" (Psalms 19:7-14).

KOOF (ק) is 19. In the nineteenth year of Nebuchadnezzar, Jerusalem with its Temple was burned. *"¹² Now in the fifth month, on the tenth day of the month (which was the NINETEENTH year of King Nebuchadnezzar king of Babylon), Nebuzaradan, the captain of the guard, who served the king of Babylon, came to Jerusalem. ¹³ He burned the house of the Lord and the king's house; all the houses of Jerusalem, that is, all the houses of the great, he burned with fire" (Jeremiah 52:12-13).*

KOOF (ק) is 19. The Tribe of Naphtali inherited nineteen cities and their villages in the Promise Land. Their land gift from God encompasses the entire western section of the Sea of Galilee. *"³⁸ᵇ NINETEEN cities with their villages. ³⁹ This was the inheritance of the tribe of the children of Naphtali according to their families, the cities and their villages" (Joshua 19:38b-39).*

Isaiah prophesied that the land of Naphtali and Zebulun would see a shining beacon in their lands (Isa. 9:1-2). The Book of Matthew records the fulfillment of this prophecy. *"The land of Zebulun and the land of Naphtali, By the way of the sea, beyond the Jordan, Galilee of the Gentiles: The people who sat in darkness have seen a great light, And upon those who sat in the region and shadow of death Light has dawned" (Matthew 4:15-16).* The fulfillment of this prophecy occurred when Yeshua, who had lived in Nazareth all his life, moved to the Galilean city of Capernaum at the age of 30. Once in the city he started his ministry and began to preach the gospel (Matt. 4:14-17).

KOOF (ק) is 19. Nineteen scrolls of the Book of Isaiah were discovered among the *Dead Sea Scrolls*.

KOOF (ק) is 19. Mary, the mother of Yeshua, is mentioned 19 times in Scripture, connecting her to "holiness."

QUANTUM 100

Not only does KOOF (ק) have an ordinal value of 19, but a numerical value of 100. Therefore, let's explore the 100th element on the Periodic Table.

Atomic Number 100—Fermium (Fm). Quantumly speaking, fermium is a synthetic element that has the atomic number **100**, the mass number **[257]**, and the atomic symbol **Fm**. A single fermium atom contains 100 protons, 100 electrons, and 157 neutrons. Fermium is named after the Italian nuclear physicist Enrico Fermi who made many important scientific discoveries during his lifetime. He was a leader of the U.S. effort to build the world's first fission (atomic) bomb during World War II.[680]

Fermium was first discovered in the fallout from the first successful hydrogen bomb explosion on November 1, 1952 [681] at the Marshall Islands in the Pacific Ocean. For security reasons, this discovery was not announced until 1955. Credit for the discovery of the element fermium goes to a group of University of California scientists under the direction of Albert Ghiorso.

It was called the "Ivy Mike" 10-megaton nuclear test. Initial examination of the debris from the explosion showed a new isotope of plutonium ($^{244}_{94}Pu$). Element 99 (einsteinium) was quickly discovered on filter papers that had been flown through the cloud from the explosion (the same sampling technique that had been used to discover $^{244}_{94}Pu$). Fermium required examining more material—the contaminated coral from the Enewetak atoll—which was shipped to Berkeley, California's University of California Radiation Laboratory for analysis. Approximately two months later, fermium was identified. [682]

Fermium is one of the transuranium elements, which lie beyond uranium on the periodic table. This means that all elements with larger element numbers than uranium—92—are transuranium elements. [683]

"All isotopes of fermium are radioactive. The most stable isotope is fermium-257. Isotopes are two or more forms of an element. Isotopes differ from each other according to their mass number. The number written to the right of the element's name is the mass number. The mass number represents the number of protons plus neutrons in the nucleus of an atom of the element. The number of protons determines the element, but the number of neutrons in the atom of any one element can vary. Each variation is an isotope.

The half-life of fermium-257 is 20.1 hours. The half-life of a radioactive element is the time it takes for half of a sample of the element to break down. A radioactive isotope is one that breaks apart and gives off some form of radiation. For example, suppose 100 grams of fermium-257 is made. Fifty grams of the isotope would be left about one day (20.1 hours) later. After another day (another 20.1 hours), only 25 grams of the isotope would remain."[684]

Fermium is an actinide (encompasses the 15 metallic chemical elements with atomic numbers 89 to 103). Fermium is the heaviest element that can be formed by neutron bombardment of lighter elements; hence, the last element that can be prepared in macroscopic quantities, although pure fermium metal has not yet been prepared. [685]

100 IN SCRIPTURE

KOOF (ק) is 100. Abraham was one hundred years old when his son of promise—Isaac—was born. *"Then Abraham fell on his face and laughed, and said in his heart, 'Shall a child be born to a man who is ONE HUNDRED years old? And shall Sarah, who is ninety years old, bear a child?'" (Genesis 17:17).*

KOOF (ק) is 100. When Jacob moved back to Israel, he bought some land for "100 pieces of money." Remember "Jacob" is shorthand for *"the generation of those who seek Him" (Psalms 24:6).*

"18 Then Jacob came safely to the city of Shechem, which is in the land of Canaan, when he came from Padan Aram; and he pitched his tent before the city. 19 And he bought the parcel of land, where he had pitched his tent, from the children of Hamor, Shechem's father, for ONE HUNDRED pieces of money. 20 Then he erected an altar there and called it El Elohe Israel" (Genesis 33:18-20).

KOOF (ק) is 100. The north and south side of the Court of the Wilderness Tabernacle are both one hundred cubits long. *"9 You shall also make the Court of the Tabernacle. For the south side there shall be hangings for the court made of fine woven linen, ONE HUNDRED cubits long for one side. 10 And its twenty pillars and their twenty sockets shall be bronze. The hooks of the pillars and their bands shall be silver. 11 Likewise along the length of the north side there shall be hangings ONE HUNDRED cubits long, with its twenty pillars and their twenty sockets of bronze, and the hooks of the pillars and their bands of silver"* (Exodus 27:9-11).

KOOF (ק) is 100. One hundred talents of silver were donated and used in the Wilderness Tabernacle for the Sanctuary sockets as well as the bases for the Veil. *"25 And the silver from those who were numbered of the congregation was ONE HUNDRED talents and one thousand seven hundred and seventy-five shekels, according to the shekel of the sanctuary: 26 a bekah for each man (that is, half a shekel, according to the shekel of the sanctuary), for everyone included in the numbering from twenty years old and above, for six hundred and three thousand, five hundred and fifty men. 27 And from the HUNDRED talents of silver were cast the sockets of the sanctuary and the bases of the veil: ONE HUNDRED sockets from the HUNDRED talents, one talent for each socket"* (Exodus 38:25-27).

Approaching the outer court of the Tabernacle of Moses, one would see the bright white linen that represents God's high standard of purity, holiness, and righteousness. The outer fence of the Tabernacle had bronze posts (representing God's judgment due to sin) placed on top of the soil (representing all humanity). The silver on the top of each post represents the "price on our head" for redemption that must be paid (Num. 18:16; Exo. 30:16) to make us righteous (Rom. 3:24-26). The message of sin, righteousness, and judgment is conveyed to anyone approaching the Outer Court via the materials and construction of the Outer Court's fence or wall.

KOOF (ק) is 100. One hundred pomegranates were placed on the Bronze Pillars wreaths of chainwork. *"15 Also he made in front of the temple two pillars thirty-five cubits high, and the capital that was on the top of each of them was five cubits. 16 He made wreaths of chainwork, as in the inner sanctuary, and put them on top of the pillars; and he made ONE HUNDRED pomegranates, and put them on the wreaths of chainwork. 17 Then he set up the pillars before the temple, one on the right hand and the other on the left; he called the name of the one on the right hand Jachin, and the name of the one on the left Boaz"* (2 Chronicles 3:15-17). The pomegranate is a symbol of righteousness, knowledge, and wisdom; because it is said to have 613 seeds, each representing one of the 613 commandments of the Torah.

QUANTUM QUARK (ק)

To summarize, KOOF (ק) is the nineteenth Hebrew Living Letter. KOOF (ק) in ancient Hebrew means the back of the head, which corresponds to the Holy of Holies in God's Temple. The Holy of Holies in the back of a person's head is key to the ultimate redemption goal of mankind, which is one's deliverance from corruption into the fullness of their resurrected state: body, soul, and spirit.

KOOF (ק) is made up of two letters: RESH (ר) and ZAYIN (ז). ZAYIN descends below the line while RESH hovers above it. This paradoxical union symbolizes the secret of KOOF (ק)—*"There is none holy as God."* In fact, in general, the Hebrew Living Letter KOOF (ק) stands for holiness *(kedushah)*. Of all the letters in the Hebrew Alphabet, KOOF is the lowest; therefore, KOOF signifies the emulation of holiness on earth. When the Temple stood, sacrifices brought the worshipers closer to God and brought God's blessing to the entire world. Man's holiness is a reflection of God's absolute sanctity and purity. To raise one's level of holiness, a person must attach himself to God in order to be conformed to His presence and truth-filled ways. Man's sanctity results from a mastery of all their instincts and inclinations, by placing them at the disposal of God's perfect will.

Q: What are your thoughts about the Holy of Holies key to the ultimate redemption goal of mankind?

Q. How does a person raise their level of holiness and sanctity?

ר

RESH (ר)—20
AWAKENED TO THE KING OF KINGS

REDEMPTION OF RESH (ר)

The redemption of the Messiah's Quantum 22 continues with the twentieth letter of the Hebrew Alphabet. The pronouncement of the quantum letter RESH (ר) has the sound of *raysh*. RESH (ר) has an ordinal value of 20 and a numerical value of 200. Its ancient pictograph is a head, which symbolizes Messiah Yeshua, a person, what is the highest, and what is most important. [686]

The redemption of RESH (ר) is all about the brains of the operation of the Body of Christ —Messiah Yeshua—as well as Yeshua's Multimembered Body coming into alignment with their Righteous Head. *"15 But, speaking the truth in love, may grow up in all things into Him who is the Head—Christ— 16 from whom the whole body, joined and knit together by what every joint supplies, according to the effective working by which every part does its share, causes growth of the body for the edifying of itself in love" (Ephesians 4:15-16)*. Messiah Yeshua is the brains behind all of Creation: *"For in Him was created the universe of things, both in the heavenly realm and on the earth, all that is seen and all that is unseen. Every seat of power, realm of government, principality, and authority—it all exists through Him and for His purpose!" (Colossians 1:16 $_{TPT}$)*.

Those who take hold of the True Source receives directly from the brains of the operation of the Body of Christ. Through His life supplying vitality into every part of His Body, the divine righteous Head guides and causes each member to grow into its fullness, even to the utmost where the brains of those vitally connected become just like His brain with the full bandwidth of God's glory. Behold, in this Kingdom Day, the brains of the operation of the Body of Christ will fully hook up with the brains in His wholly redeemed Body (Rom. 8:23). This is the fully redeemed status of God's people, which includes the physical and spiritual righteous realities of intimately knowing the "mind of Christ." The mind of Christ is encoded in humanity's DNA but hidden from man. Behold, the full bandwidth of God's glory in creation where those who are truly and completely in union with the Head Messiah Yeshua will experience their dormant brain pathways and their dormant shadow DNA being released into their full potential. Everybody made in God's own image longs for this!

One of the best ways for mankind to better understand our Creator as the brains of the operation is to study the anatomy and physiology of a human brain. Let's first acknowledge that the two most vital organs for a human body are the heart and the brain, without which there is no life. The human brain is the command-and-control center of a person's nervous system. "All of your emotions, sensations, aspirations, and everything that makes you uniquely individual come from your brain. This complex organ has many functions. It receives, processes, and interprets information. Your brain also stores memories and controls your movements." [687]

Even though the human brain is the command-and-control-center of the central nervous system, it is only one component of it. One of the most important brain connections is to the spinal cord through the brainstem, which is a vital structure that controls many of our automatic functions like breathing, heart rate, and blood pressure. The spinal cord is a long, thin bundle of nervous tissue that extends from the base of the brain down to the lower back. "It's divided into 31 segments, each corresponding to a pair of spinal nerves that branch out to different parts of the body. The spinal cord is our body's information superhighway, relaying messages between the brain and the rest of the body at breakneck speeds. The connection between the brain and spinal cord is a crucial one. The spinal cord information travels along specific pathways called tracts. Ascending tracts carry sensory information up to the brain, while descending tracts bring motor commands down from the brain to the muscles. [688] This constant back-and-forth communication allows us to interact with our environment and respond to stimuli in real-time." [689] This is also true for the two-way constant communication in the Body of Messiah Yeshua. "*4 For just as each of us has one body with many members, and these members do not all have the same function, 5 so in Christ we, though many, form one body and each member belongs to all the others*" (Romans 12:4-5).

"From the brain's enigmatic depths to the spinal cord's intricate network, the central nervous system orchestrates a remarkable dance of perception, action, and everything in between. This dynamic duo, the brain and spinal cord, form the core of our central nervous system (CNS), working in harmony to control every aspect of our existence. From the simplest reflex to the most complex thought, these structures are the maestros of our bodily symphony. Imagine, for a moment, the sheer complexity of this system. Billions of neurons firing in concert, transmitting signals at lightning speed, all while you casually sip your morning coffee. It's mind-boggling, isn't it? Yet, this intricate dance happens effortlessly, every second of every day. The central nervous system is like the body's own internet, a vast network of information highways and processing centers. At its heart lies the brain, our very own supercomputer, nestled safely within the bony fortress of our skull. And stretching from this command center, like a long, delicate cable, is the spinal cord, relaying messages to and from every corner of our body." [690]

Let's do a brief exploration of the wrinkled, walnut shaped wonder sitting at the top position of an upright man. The human brain is a three-pound organ that packs a punch. It's divided into several major regions, each having its own specialized function. First, we have the cerebrum, which is the largest part of the brain. The cerebrum is where thinking, reasoning, emotions, and voluntary movements originate. It is split into two hemispheres, each controls the opposite side of the body. The cerebrum is divided up further into four lobes: frontal, parietal, temporal, and occipital. ⁶⁹¹ Each lobe has its own specialties. The frontal lobe sits right behind the forehead. It is responsible for higher-level thinking, planning, and decision-making. The parietal lobe processes sensory information, which helps a person understand their body's position in space. The temporal lobe deals with auditory processing and memory formation while the occipital lobe at the back of the brain is all about visual processing. ⁶⁹² As we saw with KOOF (ק), the occipital bone covers the back of the head (occiput). Or in other words, the Holy of Holies of the brain. It is one of seven bones that form your skull. ⁶⁹³ At the base of the skull in the occipital bone, there is a large oval opening called the foramen magnum, which allows the passage of the spinal cord.

The next region of the human brain is the cerebellum, which is often called the "little brain." Without the cerebellum, "even the simplest tasks like walking or reaching for a glass of water would become Herculean feats." ⁶⁹⁴ The cerebellum plays a crucial role in coordinating movement, balance, and posture.

Colossians 2:19 illustrates through the Body of Christ holding fast to the Head, it is nourished and woven together. The same is true for the human body. The human brain doesn't work in isolation. When working properly, the spinal cord relays information to and from the rest of the body instantaneously. This is an excellent example of the whole body effectively working by which every part does its share (Eph. 4:16). The spinal cord isn't just a passive conduit, it can also produce its own information, like reflex actions. Have you ever noticed that when you step on a sharp object, your foot jerks away before your brain registers the pain? You can thank the reflex arcs in the spinal cord for that. ⁶⁹⁵

Due to the vital importance of the brain and the spinal cord, the human body has multiple layers of protection. "First up, we have the meninges, three protective layers that envelop both the brain and spinal cord. These membranes—the dura mater, arachnoid mater, and pia mater—provide cushioning and support. They're like the world's most important bubble wrap, protecting our central nervous system from bumps and jolts. Swimming between these layers is cerebrospinal fluid (CSF), a clear, colorless liquid that acts as a shock absorber for the brain and spinal cord. It's like the ultimate waterbed, allowing the brain to float gently within the skull. But CSF does more than just cushion—it also helps remove waste products and distribute nutrients throughout the central nervous system. Of course, we can't forget about the skull and vertebrae. These bony structures form a formidable fortress around our central nervous system. The cranium or skull houses the brain while the vertebrae stack up to form a protective column around the spinal cord. It's nature's version of a high-security vault. Last but certainly not least, we have the blood-brain barrier. This microscopic fortress is a selective semipermeable border of endothelial cells that prevents substances in the bloodstream from haphazardly crossing into the brain tissue. It's like a bouncer at an exclusive club, only letting in the VIPs (essential nutrients) while keeping out the riffraff (potentially harmful substances)." ⁶⁹⁶

At the heart of the central nervous system are neurons, the specialized cells that form the building blocks of our nervous system. These amazing cells have a unique ability to transmit electrical and chemical signals, allowing for rapid communication throughout the body. Neuron connections in the brain form an intricate web, with each neuron potentially connecting to thousands of others. This is synonymous to Christ's electrical-chemical connections to His sparky multimembered body: *"For as the body is one and has many members, but all the members of that one body, being many, are one body, so also is Christ" (1 Corinthians 12:12)*. These neuron connections are called synapses. This is the place where information transfer happens. When a neuron fires, it releases neurotransmitters. Neurotransmitters are chemical messengers that bridge the gap between neurons, allowing the signal to continue its journey. [697]

ANCIENT RESH (ר)

It seems strange that the letter RESH (ר) which stands for "head" and "beginning" appears towards the end of the Hebrew Alphabet. It is significant that RESH (ר) is used to symbolize both a "wicked person" (*Rasa'* רשע) and a "leader, chief, head" (*Rosh* ראש). It teaches that a penitent can attain the extraordinary accomplishment of transforming himself completely from the lowest status to the highest. (Midrash Alpha Beis).

SIMPLE LETTER (ר)

RESH (ר) is one of the seven simple letters. Recall that the simple letters have shapes, which are pictures, that are unique. The seven simple letters whose shapes are distinctive are ZAYIN (ז), KAF (כ), RESH (ר), NUN (נ), YUD (י), VAV (ו), and DALET (ד).

RESH (ר) is very similar to the letter DALET (ד), the only difference between the two shapes is in the upper right-hand corner. The quantum letter DALET (ד) has a sharp corner, which signifies a strong allegiance to the Most High God and His way of life that is articulated in His commandments. In contrast, RESH (ר) curves to accommodate itself into a new direction, which suggests idolatries that easily bend to adjust to the wishes of their followers. [698]

The design of RESH (ר) represents an individual who is bent over—a poor person. The design of DALET (ד) looks very similar, but DALET has a YUD (י) at its upper right-hand corner. The YUD (י) in DALET (ד) represents one who is subservient to God and adheres to every letter of the law (His teaching). [699] The horizontal and vertical lines of RESH (ר) represent intellect and speech. However, since they are not joined with a YUD (י), the speech and intellect of a RESH (ר) individual are for his own gratification. [700]

Additionally, a DALET (ד) is someone who always has in mind that there will be a Day of Judgment due to YUD's connection to the World to come *(Olam HaBa)*. However, a RESH (ר) person does not care what he does. He has no regard for his thoughts or speech because he doesn't believe in the ultimate Day of Judgment. [701] So, the RESH (ר) is the unholy counterpart to the DALET (ד). If a RESH is substituted for the DALET in the word *"echad,"* the Hebrew word becomes *"acher,"* meaning other. "The mere removal of the *dalet's yud* changes the concept of 'one G-d' to "other gods," or idol worship. [702]

FULFILLMENT OF RESH (ר)

RESH (ר) has a numerical value of 200 and fulfills as (רֵישׁ) or (רֹאשׁ) or (רָשׁ)—meaning "beginning," "head," or "poor man," respectively. [703]

QUANTUM 20

RESH (ר) is the 20th letter of the Hebrew Alphabet. ר has an ordinal value of 20 and a numerical value of 200. Always remember that the Hebrew letter KAF (כ) has the numerical value of 20 while RESH (ר)'s ordinal value is 20.

Atomic Number 20—Calcium (Ca). Quantumly speaking, calcium is a chemical element that has the atomic number **20**, the atomic weight **40.078**, and the atomic symbol **Ca**. A single stable calcium atom has 20 protons, 20 electrons, and 20 neutrons. It is in the alkaline earth metal element group.

Calcium's name comes from the Latin term *calx*, meaning lime. This is not referring to the lime fruit; but to calcium oxide (CaO), which is a useful building material derived from heated limestone. [704] Calcium is a reactive, silvery-white, soft metal found in Group 2 of the Periodic Table that tarnishes rapidly in air and reacts with water. [705]

Calcium is an important element for life on Earth. It is the fifth most abundant element in the Earth's crust, the eighth most abundant element found in salt water, and the third most abundant metal, after iron and aluminum. Calcium is rarely found in its elemental form; but is readily found throughout the Earth mostly in various forms of rocks and minerals, such as limestone (calcium carbonate), dolomite (calcium magnesium carbonate), and gypsum (calcium sulfate). [706]

The most common calcium compound found on Earth is calcium carbonate ($CaCO_3$), which is one of the major components of many rocks and minerals. Sedimentary calcium carbonate deposits pervade the Earth's surface, as fossilized remains of past marine life. These sedimentary deposits of calcium occur in two forms: the rhombohedral calcite (more common) and the orthorhombic aragonite (forms in more temperate seas). Minerals of the first type include limestone, dolomite, marble, chalk, and Iceland spar. Aragonite beds make up the Bahamas, the Florida Keys, and the Red Sea basins. Corals, sea shells, and pearls are mostly made up of calcium carbonate. [707]

Calcium compounds are widely used in many industries: in foods and pharmaceuticals for calcium supplementation, in the paper industry as bleaches, as components in cement and electrical insulators, and in the manufacture of soaps.

Calcium and phosphorus are supplemented in foods through the addition of calcium lactate, calcium diphosphate, and tricalcium phosphate. The last is also used as a polishing agent in toothpaste and in antacids. Calcium lactobionate is a white powder that is used as a suspending agent for pharmaceuticals. In baking, calcium phosphate is used as a leavening agent. Calcium sulfite is used as a bleach in papermaking and as a disinfectant. Calcium silicate is used as a reinforcing agent in rubber. Calcium acetate is a component of liming rosin and is used to make metallic soaps and synthetic resins. [708]

Calcium is the fifth most abundant element in the human body. [709] It is an essential element needed in large quantities for good health. [710] The Ca^{2+} ion acts as an electrolyte and is vital to the health of the muscular, circulatory, and digestive systems. The Ca^{2+} ion is indispensable to the building of bone, plus supports the synthesis and function of blood cells. [711] For example, it regulates the contraction of muscles, nerve conduction, and the clotting of blood. As a result, intracellular and extracellular calcium levels are tightly regulated by the body. Calcium can play this role; because the Ca^{2+} ion forms stable coordination complexes with many organic compounds, especially proteins. Calcium also forms compounds enabling the formation of a human skeleton. [712]

Calcium is probably most known to promote the growth of healthy teeth and bones. Calcium phosphate—$Ca_3(PO_4)_2$—is the main component of bones. The average human contains about 1 kilogram of calcium. Children and pregnant women are encouraged to eat foods rich in calcium, such as dairy products, leafy green vegetables, fish, nuts, and seeds. [713]

Everyone knows that calcium is vital for maintaining strong bones and teeth. Scripture relates pleasant words to the health of one's bones: *"Pleasant words are like a honeycomb, sweetness to the soul and health to the bones" (Proverbs 16:24).* Also, recall Ephesians 4's RESH (ר) head-body-bone connection: *"All our direction and ministries will flow from Christ and lead us deeper into Him, the anointed Head of His Body" (Ephesians 4:15 $_{TPT}$).*

20 IN SCRIPTURE

RESH (ר) is 20. In Scripture, twenty is connected to a completed waiting period. Jacob patiently worked and waited twenty years to get his two wives and possessions free from the control of his tricky father-in-law Laban: *"⁴¹ Thus I have been in your house TWENTY years; I served you fourteen years for your two daughters, and six years for your flock, and you have changed my wages ten times. ⁴² Unless the God of my father, the God of Abraham and the Fear of Isaac, had been with me, surely now you would have sent me away empty-handed. God has seen my affliction and the labor of my hands, and rebuked you last night" (Genesis 31:41-42).*

The children of Israel waited twenty years to be free from the oppressive king of Canaan—Jabin. God responded by raising up Deborah and Barak (See Judges 4-5). *"And the children of Israel cried out to the LORD; for Jabin had nine hundred chariots of iron, and for TWENTY years he had harshly oppressed the children of Israel" (Judges 4:3).*

Joseph was sold by his brothers to Midianite traders for twenty shekels of silver who brought him to Egypt: *"Then Midianite traders passed by; so the brothers pulled Joseph up and lifted him out of the pit, and sold him to the Ishmaelites for TWENTY shekels of silver. And they took Joseph to Egypt" (Genesis 37:28).*

RESH (ר) is 20. The framework of the north and south sides of the Wilderness Tabernacle Sanctuary consisted of twenty acacia-wood boards (Exo. 26:18,20). Each board was 10 cubits long and 1-½ cubits broad (Exo. 26:16). The Golden Lampstand was placed in front of the framework on the south side with the Table of Showbread being directly across from it on the Tabernacle's north side. *"¹⁵ And for the tabernacle you shall make the boards of acacia wood, standing upright. ¹⁶ Ten cubits shall be the length of a board, and a cubit and a half shall be the width of each board. ¹⁷ Two tenons shall be in each board for binding one to another. Thus you shall make for all the boards of the tabernacle. ¹⁸ And you shall make the boards for the tabernacle, TWENTY boards for the south side. ¹⁹ You shall make forty sockets of silver under the TWENTY boards: two sockets under each of the boards for its two tenons. ²⁰ And for the second side of the tabernacle, the north side, there shall be TWENTY boards ²¹ and their forty sockets of silver: two sockets under each of the boards" (Exodus 26:15-21)*. The eastern entrance to the Holy Place was 20 cubits long. It was covered with the same costly curtain as the Most Holy Place: *"For the entrance to the tent make a curtain of blue, purple and scarlet yarn and finely twisted linen— the work of an embroiderer" (Exodus 26:36)*.

RESH (ר) is 20. The courtyard for the Wilderness Tabernacle had twenty posts (pillars) and twenty bronze bases on the south side as well as twenty posts and twenty bronze bases on the north side: *"⁹ You shall also make the court of the tabernacle. For the south side there shall be hangings for the court made of fine woven linen, one hundred cubits long for one side. ¹⁰ And its TWENTY pillars and their TWENTY sockets shall be bronze. The hooks of the pillars and their bands shall be silver. ¹¹ Likewise along the length of the north side there shall be hangings one hundred cubits long, with its TWENTY pillars and their TWENTY sockets of bronze, and the hooks of the pillars and their bands of silver" (Exodus 27:9-11)*. Anywhere a person can enter or exit the court that leads to God's Dwelling Place—the Tabernacle—have bronze (i.e., copper) bases supporting it. A base is considered to be a separate architectural feature, which is the bottom of something. It is the structure's support or foundation, like when you pour a concrete foundation for a new home.

RESH (ר) is 20. The embroidered curtain at the entrance of the courtyard for the Wilderness Tabernacle was twenty cubits long. It was made of blue, purple, and scarlet yarn with finely twisted linen. *"For the gate of the court there shall be a screen TWENTY cubits long, woven of blue, purple, and scarlet thread, and fine woven linen, made by a weaver. It shall have four pillars and four sockets" (Exodus 27:16)*.

RESH (ר) is 20. When the LORD instructed Moses to take a census of the whole Israelite community by their clans and families, Moses and Aaron were to list each and every man twenty years old or more who were able to serve in the army (Num. 1:1-3). *"¹ Now the LORD spoke to Moses in the Wilderness of Sinai, in the tabernacle of meeting, on the first day of the second month, in the second year after they had come out of the land of Egypt, saying: ² 'Take a census of all the congregation of the children of Israel, by their families, by their fathers' houses, according to the number of names, every male individually, ³ from TWENTY years old and above—all who are able to go to war in Israel. You and Aaron shall number them by their armies'" (Numbers 1:1-3)*.

RESH (ר) is 20. Samson was from the Tribe of Dan, set apart as a Nazirite. He was a legendary warrior with immense strength who was a judge for Israel for twenty years:

"And he judged Israel TWENTY years in the days of the Philistines" (Judges 15:20).

RESH (ר) is 20. The Ark of the Covenant stayed in Kiriath Jearim for twenty years: *"The Ark remained at Kiriath Jearim a long time—TWENTY years in all"* (1 Samuel 7:2).

RESH (ר) is 20. During the time of King David, the minimum age that a Levite was allowed *"to do the work for the service of the house of the Lord"* (1 Chronicles 23:24) was 20-years-old. After the Babylonian captivity, in the time of Ezra and Nehemiah, Levites 20 years and upwards were assigned *"to oversee the work of the house of the LORD"* (Ezra 3:8).

RESH (ר) is 20. It took Solomon twenty years to finish building the LORD's Temple and the royal palace: *"At the end of TWENTY years, during which Solomon built these two buildings —the temple of the Lord and the royal palace"* (1 Kings 9:10).

200 IN SCRIPTURE

RESH (ר) has a numerical value of 200; therefore, RESH (ר) is 200. The symbolic meaning of the number 200 in Scripture is that of insufficiency on one side of the Messiah's two-edged sword and idolatry on the other.

RESH (ר) is 200. Achar found out the hard way about the insufficiency of money. Having found 200 shekels of silver and other goods among the spoils of Jericho, Achar took them in spite of the Lord's express command not to. After righteous Joshua finds out about Achar's thievery and lies, Joshua orders him to be stoned to death for his trespass against the devoted things. Joshua also ordered that all Achar's possessions be burned (Joshua 7). *"19 And Joshua said to Achar, Give glory to the LORD God of Israel, and give praise to Him; and tell me now what you have done, and do not hide it from me. 20 Achar answered Joshua, and said, Truly I have sinned against the LORD God of Israel, and this is what I did: 21 When I saw among the spoils a beautiful Babylonian tapestry and TWO HUNDRED shekels of silver and a wedge of gold weighing fifty shekels, then I coveted them and took them; and, behold, they are hidden in the earth inside of my tent, and the silver under it. 22 So Joshua sent messengers, and they ran to his tent, and, behold, it was hidden in his tent, and the silver under it. 23 And they took them out of his tent, and brought them to Joshua and to all the people of Israel, and laid them out before the LORD. 24 And Joshua and all Israel with him took Achar the son of Zerah and the silver and the tapestry and the wedge of gold and his sons and his daughters and his oxen and his asses and his sheep and his tent and all he had; and they brought them to the valley of Achar. 25 And Joshua said to him, Why have you troubled us? The LORD shall trouble you this day. And all Israel stoned him with stones, both him and all he had, and burned them with fire. 26 And they raised over him a great heap of stones which remain to this day. So the LORD turned from His fierce anger. Therefore the name of that place is called the Valley of Achar to this day"* (Joshua 7:19-26 $_{Lamsa's\ Aramaic}$).

RESH (ר) is 200. The mother of a man named Micah had two hundred shekels melted down and formed into an idol. Micah started his own house of worship with it. It was a time when everyone did what was right in his own eyes. This idolatrous worship grew and was eventually adopted by the Tribe of Dan, which became the first tribe devoted to idolatry.

Personally, I believe this is the reason that the Tribe of Dan is not found in the Revelation 7 list. This record of idolatry shows the insufficiency of man-made religion to please God. It is insufficient to worship one's own idol at one's own shrine. "*¹ Now there was a man from the mountains of Ephraim, whose name was Micah. ² And he said to his mother, 'The eleven hundred shekels of silver that were taken from you, and on which you put a curse, even saying it in my ears—here is the silver with me; I took it.' And his mother said 'May you be blessed by the Lord, my son!' ³ So when he had returned the eleven hundred shekels of silver to his mother, his mother said, 'I had wholly dedicated the silver from my hand to the Lord for my son, to make a carved image and a molded image; now therefore, I will return it to you.' ⁴ Thus he returned the silver to his mother. Then his mother took TWO HUNDRED shekels of silver and gave them to the silversmith, and he made it into a carved image and a molded image; and they were in the house of Micah. ⁵ The man Micah had a shrine, and made an ephod and household idols; and he consecrated one of his sons, who became his priest. ⁶ In those days there was no king in Israel; everyone did what was right in his own eyes*" (Judges 17:1-6 _{Lamsa's Aramaic}).

RESH (ר) is 200. The other side of the Messiah's two-edged sword in regard to the Scriptural symbology of the number 200 speaks of anointed worshipers, bodyguards, and a generous peacemaking gift. Ezra records that there were two hundred singing men and women that came out of Babylonian captivity that returned to the Promised Land. "*¹ Now these are the people of the province who came back from the captivity, of those who had been carried away, whom Nebuchadnezzar the king of Babylon had carried away to Babylon, and who returned to Jerusalem and Judah, everyone to his own city. ⁶⁴ The whole assembly together was forty-two thousand three hundred and sixty, ⁶⁵ besides their male and female servants, of whom there were seven thousand three hundred and thirty-seven; and they had TWO HUNDRED men and women singers*" (Ezra 2:1,64-65).

After Paul's initial testimony before the Sanhedrin, some Jews vowed to kill him. After the murderous plot was discovered, the Romans used 200 soldiers and 200 spearmen to safely transport Paul from Jerusalem to Caesarea. "*¹¹ But the following night the Lord stood by him and said, 'Be of good cheer, Paul; for as you have testified for Me in Jerusalem, so you must also bear witness at Rome.' ¹² And when it was day, some of the Jews banded together and bound themselves under an oath, saying that they would neither eat nor drink till they had killed Paul. ¹³ Now there were more than forty who had formed this conspiracy. . . . ²³ And he [the commander] called for two centurions, saying, 'Prepare TWO HUNDRED soldiers, seventy horsemen, and TWO HUNDRED spearmen to go to Caesarea at the third hour of the night; ²⁴ and provide mounts to set Paul on, and bring him safely to Felix the governor'*" (Acts 23:11-13,23-24).

Jacob gave his brother Esau 200 ewes and 200 she-goats as a generous peacemaking gift when he returned home to the Promised Land. "*⁹ Then Jacob prayed, 'O God of my father Abraham, God of my father Isaac, Lord, you who said to me, "Go back to your country and your relatives, and I will make you prosper," ¹⁰ I am unworthy of all the kindness and faithfulness you have shown your servant. I had only my staff when I crossed this Jordan, but now I have become two camps. ¹¹ Save me, I pray, from the hand of my brother Esau, for I am afraid he will come and attack me, and also the mothers with their children. ² But you have said, 'I will surely make you prosper and will make your descendants like the sand of the sea, which cannot be counted.'* ¹³ *He spent the night there, and from what he had with him he selected a gift for his brother Esau:* ¹⁴ *TWO HUNDRED female goats and TWENTY male goats, TWO HUNDRED ewes and TWENTY rams,* ¹⁵ *thirty female camels with their young, forty cows and ten bulls, and TWENTY female donkeys and ten male donkeys.* ¹⁶ *He put them in the care of his servants, each herd by itself, and said to his servants, 'Go ahead of me, and keep some space between the herds'" (Genesis 32:9-16).*

QUANTUM QUARK (ר)

To summarize, RESH (ר) is the twentieth Hebrew Living Letter. RESH (ר) in ancient Hebrew means a head. It has an ordinal value of 20 and a numerical value of 200. The pictograph for RESH (ר) symbolizes a person, what is the highest, and what is most important. It can also represent the Head of the Body of Christ—Messiah Yeshua.

The redemption of RESH (ר) is all about the brains of the operation of the Body of Christ. Messiah Yeshua is the brains behind all of Creation. One of the best ways to understand our Creator's brain is through the anatomy and physiology of a person's brain who has been made in God's own image. The human brain is the command-and-control center of the nervous system, which enables thoughts, movements, and emotions. We can say that everything that makes a human a unique individual comes from their brain. All your emotions, sensations, aspirations, and everything that makes you uniquely individual come from your brain.

Q: How does the Head—Messiah Yeshua—guide and direct His Body?

Q. What are your thoughts about the brains of the operation of the Body of Christ?

SHIN (שׁ)—21
COMPLETE PERFECTION OF EL SHADDAI

REDEMPTION OF SHIN (שׁ)

The redemption of the Messiah's Quantum 22 continues with the twenty-first letter of the Hebrew Alphabet. The pronouncement of the Hebrew letter SHIN (שׁ) has the sound of *sheen*. SHIN (שׁ) has an ordinal value of 21 and a numerical value of 300.

Through the first author-led *24 Wisdoms Bible Study* Class,[714] ALEF-TAV (את) shared the facet of the redemption of SHIN (שׁ) that He wants highlighted: SHIN is at the bottom of every person's heart. The lower and larger left ventricle of a human's heart supplies blood to the entire body while the smaller lower right ventricle supplies blood to the lungs. These lower parts of a human heart look like the lines of the Hebrew Living Letter SHIN (שׁ). Therefore, SHIN (שׁ) is written on your heart. SHIN is written on the bottom of your heart.

SHIN is also written on Jerusalem by its surrounding seven hills and three valleys. God has literally marked the landscape of Jerusalem with a SHIN where the lower left ventricle of its SHIN heart is filled with Mount Zion while its lower right ventricle is the City of David with Mount Moriah situated above. King David is the one who reveals that Jerusalem is the city where God chose to put His name (1 Kings 11:36).

SHIN (שׁ) is a name of God—*Shaddai* (שַׁדַּי). More specifically, Almighty God's great name *Shaddai* literally bears the letter SHIN (שׁ). The attributes of *El Shaddai* are: Almighty; All-Powerful; All-Sufficient One; the Judge; the sweet omnipotence of love, sure and all-sufficient resource in time of need; power to fulfill every promise; bountiful; overrides nature to perform miracles and remove mountains; giver and pourer out of self-sacrificial love and blessing; nurturer and comforter that gives strength and quiets the restless, and in Him all fullness dwells. Do not lean on your own understanding, trust the All-Sufficient One from the bottom of your heart.

During the *24 Wisdoms Bible Study* Class for SHIN, a delightful 16-year-old—Liliana Faulkner—elaborated:[715] Ezekiel 11:19-20 speaks of "them" while Ezekiel 36:26 speaks of "you." It's like a promise and a fulfillment. Ezekiel 36 is fifteen chapters after Ezekiel 11. God knew us before the foundations of the earth; therefore, He knows the promises that He's going to give to us directly, saying "you." There is an ALEF-TAV (את) in the personalized verse Ezekiel 36:26:

(וְנָתַתִּי לָכֶם לֵב חָדָשׁ וְרוּחַ חֲדָשָׁה אֶתֵּן בְּקִרְבְּכֶם וַהֲסִרֹתִי אֶת־לֵב הָאֶבֶן מִבְּשַׂרְכֶם וְנָתַתִּי לָכֶם לֵב בָּשָׂר:)

"I will give you a new heart and put a new spirit within you. I will take the heart of stone out of your flesh and give you a heart of flesh" (Ezekiel 36:26). ALEF-TAV (את) is personal. He is the very personal "I am."

Liliana added: "I looked at the biology of the heart that has SHIN written on it—the bottom part. The right side accepts blood that has been de-oxygenated and gives it to the lungs so it can be oxygenated. It reminded me of how the name Yahweh or YHVH is a full breath, and God the Father is oxygen. So, when the right side of the heart is given blood that has been de-oxygenated (i.e., away from the Lord) to the lungs, it's like renewing of one's spirit so it can be useful. That is how Yeshua is the way to the Father—The Way and The Truth and The Life.

The left side of a human heart pumps the oxygenated blood to take oxygen and nutrients to the body. *El Shaddai* is our all-sufficient resource. The blood gets rid of waste and fights infections in the body. Yeshua is cleansing His people with the goal of them being pure and spotless. When you are injured, the blood clots stop you from losing too much blood. So, *El Shaddai* is our comforter in times of injury and prevents injuries from hurting us more than necessary.

We are cleansed by the Blood of Yeshua; therefore, it makes sense that *El Shaddai* has marked every heart made in His image internally with SHIN (ש). Any time I think of blood, I think of Yeshua and how His blood cleanses us, so He really has the ultimate claim on blood. All things come from the heart, so He gives us the opportunity for all our actions to come from Him. It is written on our heart and it's what we are supposed to be doing."

Proverbs 3:3 talks about meditating on *El Shaddai* written on your heart. *The Voice Translation* says: *"Stay focused; do not lose sight of mercy and truth; engrave them on a pendant, and hang it around your neck; meditate on them so they are written upon your heart."* Meditation goes along with things written on your heart. Plus, a pendant is connected to tying God's commandments around your neck in front of your heart. To a Hebrew, a person's neck symbolizes our will and our consciousness.

The Hebrew Living Letters are the fundamental hyperquantum elements of Creation, which we are: *"You are living letters written by Christ, not with ink but by the Spirit of the living God—not carved onto stone tablets but on the tablets of tender hearts" (2 Corinthians 3:3 TPT).* The tender heart and strong heart of *El Shaddai* is at the bottom of everybody's heart.

ANCIENT SHIN (ש)

Divine Power is reflected in the SHIN (ש) that stands high among the Sacred Letters because it represents two Names of God: (שַׁדַּי) the All-Sufficient, Unlimited One, and (שָׁלוֹם) Peace. "Through the establishment of fixed laws in the Universe to protect the integrity of Creation as He intended it to be, God revealed Himself in His Attribute of (שַׁדַּי), the Omnipotent Master of the Universe (R'Hirsch)."[716] "(ש) also stands for the Name (שָׁלוֹם), which denotes *peace* and *perfection*. As the essence of all harmony and perfection, HASHEM is called the *Maker of Peace* (Maharal)."[717] The Hebrew Word Picture for "peace" is *shalom* (שָׁלוֹם). This word picture tells us that *peace* comes when we destroy the authority of chaos (remove the root cause of chaos).[718]

SHIN (שׁ) is also a symbol of Corruption. Within the constellation of the Hebrew Alphabet, SHIN (שׁ) denotes (שֶׁקֶר) falsehood, falsify, lie, lying, and untruth.[719] To counteract the dangerous SHIN (שׁ) with its potential power to ruin mankind, the *Aleph-Beis* ends with the letter TAV (ת), which represents truth. [720]

COMPOUND LETTER (שׁ)

SHIN (שׁ) is a compound letter, which is composed of three VAVs (ו) with a composite value of 18 due to the sacred letter VAV having a numerical value of 6.

VAV (ו) + VAV (ו) + VAV (ו) is 6 + 6 + 6 = 18.

The VAV (ו) chapter in the *ALEF-TAV's Hebrew Living™ Letters: 24 Wisdoms Deeper Kingdom Bible Study* reveals that on one side of the double-edged sword's understanding of "666" we know is the number of the Antichrist Beast. The other side of "666" is the mystery of the number 666, which can also represent the complete light of the righteous Messiah when what the Antichrist has twisted is turned right side up.

FULFILLMENT OF SHIN (שׁ)

SHIN (שׁ) has a numerical value of 300. It fulfills as (שִׁין), which has the numerical value of 360. Therefore, the first fulfillment (שִׁין) has a *mispar katan* of 9 (3+6) that represents absolute truth. שִׁין also translates as either "tooth" or "ivory."[721]

QUANTUM 21

Atomic Number 21—Scandium (Sc). Quantumly speaking, scandium is a chemical element that has the atomic number **21**, the atomic weight **44.956**, and the atomic symbol **Sc**. A single scandium atom has 21 protons, 21 electrons, and 24 neutrons.

Scandium's name is derived from the Latin name *"Scandia,"* which stands for Scandinavia. [722] Scandium is the first element in Group 3 of the Periodic Table. It is a transition metal. Interestingly, scandium is a moderately abundant element. However, it tends to be spread out throughout the earth rather than concentrated in a few places. This makes it difficult to isolate. In fact, scandium is classified as a rare earth element, together with yttrium and lanthanides. Rare earth elements are not really "rare." However, they are difficult to extract from the earth. They are also difficult to separate from each other. [723] Scandium is thought to occur in more than 800 different minerals. [724]

"Under standard conditions, scandium is a silvery-white metal. It is fairly soft and is nearly as light as aluminum. When scandium is first exposed to the air it will tarnish to a slightly pink or yellow color. In its pure form, scandium will react with acids. However, it won't react with oxygen and is very resistant to corrosion. It has a very high melting point, which makes it a good substitute for aluminum in high-temperature applications." [725]

Scandium is sometimes combined with other metals to make alloys. An alloy is made by melting and mixing two or more metals. The mixture has properties different from those of the individual metals. Scandium and aluminum alloys (Al_3Sc) are now somewhat competitive with titanium alloys; however, titanium alloys, which are similar in lightness and strength, are cheaper and much more widely used. [726] "The alloy $Al_{20}Li_{20}Mg_{10}Sc_{20}Ti_{30}$ is as strong as titanium, light as aluminum, and hard as some ceramics." [727]

The main application of scandium by weight is in aluminum-scandium alloys for minor aerospace industry components. They were used in Russian military aircraft, specifically the MiG-21 and MiG-29. [728] Specialized lightweight high-performance sports equipment has also found its niche with scandium-aluminum alloys, including baseball bats,[729] bicycle components and frames,[730] lacrosse sticks, and tent poles. Firearms manufacturing company Smith & Wesson are getting into the scandium alloy action through their production of semi-automatic pistols and revolvers with frames of scandium alloy and cylinders of titanium or carbon steel. [731] Dentists have joined the scandium bandwagon through their use of erbium-chromium-doped yttrium-scandium-gallium garnet (Er,Cr:YSGG) lasers for cavity preparation and in endodontics. [732]

Another commercial use for scandium alloys involves bright lights that come close to replicating natural sunlight. These lights are used for sporting events and movie productions. General Electric patented and made the first scandium-based metal-halide lamps in North America, which are now being produced in all major industrialized countries. These high-intensity discharge lamps are made from scandium triiodide and sodium iodide. They are a powerful white-light source with a high color rendering index. A color rendering index is a quantitative measure of the ability of a light source to reveal the colors of various objects faithfully in comparison with a natural or standard light source. [733] These scandium-based metal-halide lamps allow good color reproduction for motion picture and television cameras. [734]

Elemental scandium is considered non-toxic. However, scandium compounds should be handled as if moderately toxic due to more extensive testing needs to be done. [735]

"Only one naturally occurring isotope of scandium is known, scandium-45. Isotopes are two or more forms of an element. Isotopes differ from each other according to their mass number. The number written to the right of the element's name is the mass number. The mass number represents the number of protons plus neutrons in the nucleus of an atom of the element. The number of protons determines the element, but the number of neutrons in the atom of any one element can vary. Each variation is an isotope.

About 10 radioactive isotopes of scandium are known also. A radioactive isotope is one that breaks apart and gives off some form of radiation. Radioactive isotopes are produced when very small particles are fired at atoms. These particles stick in the atoms and make them radioactive. There are no commercial uses for any radioactive isotope of scandium."[736]

21 IN SCRIPTURE

SHIN (ש) is 21. As the twenty-first Hebrew Living Letter, SHIN (ש) represents the complete perfection reflected in the works of Almighty God (7×3=21). It bears repeating that the Almighty's great name *"Shaddai"* literally bears the letter SHIN (ש).

SHIN (ש) is 21. Twenty-one has two major references in Scripture:

[1] King Zedekiah was crowned at 21-years-old, and
[2] The Prince of Persia delayed an angel who appeared as a glorious man for 21 days.

King Zedekiah was the last king of Judah and the youngest son of Josiah. Zedekiah's real name was "Mattaniah" (2 Kings 24:17). [737] When Nebuchadnezzar enthroned him in place of the rebellious Jehoiachin, his nephew changed his name to "righteousness of Yah" (Zedekiah). Unfortunately, he did not live up to this incredible name.

As a ruler, Zedekiah wavered easily. He was pliant in the hands of any advisor whether prince or prophet. The King of Egypt—Hophra—persuaded Zedekiah to switch loyalty, bringing Babylon's wrath. After a year and a half siege, the walls of Jerusalem yielded. Zedekiah escaped through a hidden gate and fled toward the Jordan. Unfortunately, Zedekiah was overtaken and carried off to meet the King of Babylon. [738]

At Riblah, Zedekiah's sons were slain in his presence, his own eyes were pulled out, and he was bound in fetters. Then, he was taken to Babylon to spend the remainder of his days in a dungeon. As a result of Zedekiah's lack of a firm righteous stance with his vacilating allegiances, Jerusalem was plundered and burned. The nation itself perished while its best population was deported to Babylon as captives. [739] *"Zedekiah was TWENTY-ONE years old when he became king, and he reigned eleven years in Jerusalem" (2 Chronicles 36:11).*

SHIN (ש) is 21. In Daniel 10, a glorious man tells Daniel that he was sent by God immediately when he had prayed, but he was delayed twenty-one days by the Prince of Persia. *"12 Then he said to me, 'Do not fear, Daniel, for from the first day that you set your heart to understand, and to humble yourself before your God, your words were heard; and I have come because of your words. 13 But the prince of the kingdom of Persia withstood me TWENTY-ONE days; and behold, Michael, one of the chief princes, came to help me, for I had been left alone there with the kings of Persia. 14 Now I have come to make you understand what will happen to your people in the latter days, for the vision refers to many days yet to come" (Daniel 10:12-14).*

300 IN SCRIPTURE

SHIN (ש) has a numerical value of 300; therefore, SHIN (ש) is 300. In Scripture, three hundred is connected to Gideon's Army. 300 men were divided into three parties with each carrying a horn and a jar with a torch inside. Each of the 300 did exactly what Gideon did. At the beginning of the middle watch, Gideon and company crept into the Midian camp, they blew their horns as one man and smashed their jars. When they all smashed their jars at the same time, their torches pierced the night, as they cried: *"The sword of the Lord and of Gideon" (Judges 7:15-20).* The Midianites panicked and fled at the sound and light of God's Army.

SHIN (ש) is 300. In Scripture, three hundred is connected to the length of Noah's Ark. "*¹³ And God said to Noah, 'The end of all flesh has come before Me, for the earth is filled with violence through them; and behold, I will destroy them with the earth. ¹⁴ Make yourself an ark of gopherwood; make rooms in the ark, and cover it inside and outside with pitch. ¹⁵ And this is how you shall make it: The length of the ark shall be THREE HUNDRED cubits, its width fifty cubits, and its height thirty cubits*" (Genesis 6:13-15).

The Book of Enoch (1 Enoch) elaborates on the Genesis 6 account:

"In those days, when the children of man multiplied, it happened that there were born unto them handsome daughters, and the angels, the children of heaven, saw them and desired them, and they said to one another, 'Come, let us choose wives for ourselves from among the daughters of man and beget us children.' And Semyaz, being their leader, said unto them, 'I fear that perhaps you will not consent that this deed should be done, and I alone will become (responsible) for this great sin.' But they all responded to him, 'Let us all swear an oath and bind everyone among us by a curse not to abandon this suggestion but to do the deed.' Then they all swore together and bound one another by (the curse). And they were altogether two hundred; and they descended into Ardos, which is the summit of Hermon. And they called the mount, for they swore and bound one another by a curse" (1 Enoch 6:1-7). [740]

Notice that these fallen angels used oaths and curses. Yeshua warns His followers to not swear oaths (Matt. 5:34-37). Additionally, the Book of Enoch expands on the Genesis account of chapter 6 by telling us how these fallen angels revealed corrupt knowledge to humans. "And they took wives unto themselves, and everyone chose one woman for himself, and they began to go unto them. And they taught them magical medicine, incantations, the cutting of roots, and taught them about plants" (1 Enoch 7:1-2). [741]

The Apostle John reveals that the teaching of "magical plants" is *pharmakeia*, which can be translated as witchcraft or sorcery as well as a pharmacist (Rev. 21:8; Rev. 22:15). We are in a time like the days of Noah. We have been warned that this is the time of Satan's last great assault on humanity where corrupt *pharmekeia* drugs are playing a central role. [742] 1 Enoch 8 hints at widespread drug abuse as well as using such fallen angel technology to try to control humanity. [743] "And Azaz'el taught the people (the art of) making swords and knives, and shields, and breastplates; and he showed to their chosen ones bracelets, decorations, (shadowing of the eye) with antimony, ornamentation, the beautifying of the eyelids, all kinds of precious stones, and all coloring tinctures and alchemy.... Amasras' taught incantations and the cutting of roots; and Armaros the resolving of incantations; and Baraqiyal astrology, and Kokarer'el (the knowledge of) the signs, and Tam'el taught the seeing of the stars, and Asder'el taught the course of the moon as well as the deception of man" (1 Enoch 8:1-4). [744]

Notice all the fallen angel knowledge pertains to destruction (or devouring—SHIN): weapons, drugs, witchcraft, the art of deception, makeup leading to lust, and deviant religious practices, like astrology, divining the future through signs, etc. All these appeared to improve the lot of men; however, in actuality, it has enslaved humanity to fallen angel overlords, which "in Christ," we are called to overcome. [745]

Most everyone also knows of the incredible and mysterious story of Enoch pleasing God so much that he was supernaturally taken by God from among the people-group of The Flood. *"So all the days of Enoch were THREE HUNDRED and sixty-five years. And Enoch walked with God; and he was not, for God took him" (Genesis 5:23-24).*

> "He answered, 'Because I am Enoch the son of Jared. When the people-group of the flood sinned and did confuse things, they said to Yahweh, 'Leave us, we don't want to know about you,' then the Sacred One—may He be blessed—removed me from them to be a witness against them in the high heavens to all the inhabitants of earth, so that they wouldn't say, 'The Merciful One is cruel.' " (III Enoch 4 excerpt).[746]

Enoch is not only a forerunner of all those who will arise and shine to be transformed into God's own image, but Messiah Yeshua's transfiguration uniquely portrays the redemption of the 200 fallen angels coming to earth. A large part of the mission of Jesus Christ is that He came to reverse what was started at Mount Hermon which Hebrews believed was the gateway to the underworld. [747]

When Yeshua stood in Caesarea Philippi at the base of Mount Hermon and said, the gates of hell will not prevail against My church, He was literally rebuking the power of the fallen ones—the Watchers. [748] Three days later Yeshua was literally transfigured on Mount Hermon. The Messiah did this on purpose. After His Crucifixion, Yeshua descends into the underworld to preach to the spirits imprisoned there, as if to say: Here I am. Then, He rose from the dead with the keys to the crucial gates—death and Hades. [749]

Never forget the redemption of the destructive side of SHIN by the Almighty, All-Powerful, All-Sufficient One - *El Shaddai*. On the same mountain—Mount Hermon—where the destructive fallen angels descended to earth, Yeshua was transfigured before Peter, James, and John, showing humanity our original form made in God's image. Remember who you are and remember whose you are (Matt. 17:1-8).

ENLARGED SHIN (שׁ)

SHIN (שׁ) holds a notable position in Scripture. From all the Hebrew Living Letters, SHIN (שׁ) was selected as the opening word for King Solomon's holy Song of Songs. This first SHIN (שׁ) in the Song of Songs is written large, just as the BET (ב) is written large in the first word in the Bible—*Bereshit* (בְּרֵאשִׁית). [750]

It is not a coincidence that the letter SHIN (שׁ) and BET (ב) are both compound Hebrew letters, which are made up of three VAVs (ו). Additionally, the enlargement of BET (ב) in Genesis and SHIN (שׁ) in the Song of Songs makes them partners and indicates a combined message.

BET (ב) is the letter of Creation whose purpose is to create a home for mankind made in God's own image where they can mature, so they can come into their rightful place and position.

SHIN (שׁ) is the letter of the mutual love and union between God and Is-real, which fulfills the purpose of God's Creation (Rom. 8). The first Hebrew word in Scripture is *bereshit,* which is translated into English as "in the beginning." One of the ways that the Hebrews find hidden meanings in Scripture is to rearrange the Hebrew letters in a word.

If one rearranges the Hebrew letters of *bereshit* (בְּרֵאשִׁית) in a different order, we get *shir taev*. *Shir taev* means "song of desire." With this, the Hebrew Sages have extrapolated that God created the world in song (harmony) in its crystal-clear pure form. *Shir taev* expresses not only God creating the heavens and earth through song, but the longing of all of creation to glorify God. This is why creation groans for the manifestation of the sons of God because they long to sing their perfect pitch song to their Creator (Rom. 8:22).

King David was an open book before the Lord. He bared his heart to Him. This is one of the reasons why the Sweet Singer of Israel made such pleasing music to the One whom his soul loved—the One who first loved his soul. [751] It is said that *shir taev* (song of desire) corresponds to the Bridal Book—the Song of Solomon. The Song of Songs is the best of all songs. The Hebrew Sages say that the Song of Songs is the song that all other songs sing. Or in other words, it is the underlying song to all rejoicing, high praise, and worship. [752]

QUANTUM QUARK (שׁ)

In summary, SHIN (שׁ) is the twenty-first Hebrew Living Letter. SHIN (שׁ) in ancient Hebrew means a tooth. Its pictograph symbolizes devouring, consuming, destroying, and something sharp. Our SHIN (שׁ) focus is the Messiah being *El Shaddai*. The Almighty's great name *Shaddai* literally bears the letter SHIN (שׁ). The attributes of *El Shaddai* are: Almighty; All-Powerful; All-Sufficient One; the Judge; the sweet omnipotence of love, sure and all-sufficient resource in time of need; power to fulfill every promise; bountiful; overrides nature to perform miracles and remove mountains; giver and pourer out of self-sacrificial love and blessing; nurturer and comforter that gives strength and quiets the restless, and in Him all fullness dwells.

Divine Power is reflected in the SHIN (שׁ) that stands high among the Sacred Letters because it represents two Names of God: (שַׁדַּי) the All-Sufficient, Almighty, Unlimited One and (שָׁלוֹם) Peace. The Hebrew Word Picture for "peace" is *shalom* (שָׁלוֹם). This word picture tells us that peace comes when we destroy the authority of chaos (remove the root cause of chaos).

Q: What attribute of *El Shaddai* stands out most to you and why?

Q. Where would you welcome the peace that destroys the authority of chaos in your life?

ת

TAV (ת)—22
SEAL OF TRUTH

REDEMPTION OF TAV (ת)

The redemption of the Messiah's Quantum 22 continues with the twenty-second letter of the Hebrew Alphabet. TAV (ת) is pronounced just the way it looks—*tav*. It has an ordinal value of 22 and a numerical value of 400. Our redemptive sight zeroes in on the "T" in TAV (ת) and its connection to one of the seven seals of the human body—the thymus gland—that's associated with both the endocrine system and immune system. Behold, the Lion of Judah is speaking about "T Totally Transformed." [753] The "T" in the phrase "T Totally Transformed" stands for "T-cells."

T-cells are born from hematopoietic stem cells found in bone marrow. Bone marrow is key. Hematopoiesis is the process of blood formation. "T-cells" are connected to the thymus gland, because the primary function of the thymus is to train special white blood cells called T-cells or T-lymphocytes. White blood cells (lymphocytes) travel from your bone marrow to your thymus where they mature and become specialized T-cells that defend your body from antibodies and germs. [754]

The thymus gland is the body organ that took scientists the longest time to understand. It is only active when you're born until you reach puberty. [755] Your thymus gland produces most of your T-cells before birth. The rest are made in childhood, and you'll have all the T-cells you need for life by the time you hit puberty. [756] Still, it plays a huge role in training your body to fight infections, and even cancer, for the rest of your life. It is also vital to the body's chemical messaging or endocrine system. [757]

The thymus gland is a secretory gland important to the immune system. One of its main secretions is the hormone thymosin, which stimulates the maturation of T-cells. T-cells help clear the body of damaged cells and pathogens. T-cells are able to bind the T receptor on the target cell's surface that will initiate its eventual death. Despite the thymus gland's essential role in immune health, as we have stated, it is only active [so far] until puberty. However, its actions are instrumental in preventing the body from having an autoimmune response, when the immune system can't distinguish between itself and foreign agents (causing fever, fatigue, and malaise). [758]

Once the white blood cells encounter the thymus gland during childhood, they migrate to the lymph nodes, which stores the immune cells in the body. So, the thymus gland is the recipient of immature T-cells that were created in the bone marrow; but have yet to reach full maturation. Once the thymus receives the T-cells, they are trained to only attack foreign agents. The way this happens is through positive selection [so it is for Transformed Ones]. Only the T-cells that have properly responded to foreign antigens will be selected to survive and eventually migrate to the medulla. Once the surviving T-cells have reached the medulla in a lymph node, they will proceed to mature. The remaining T-cells will go to kill pathogens. [759]

Your thymus gland is in your upper chest behind your breastbone (sternum). It sits between your lungs, just in front of and above your heart. [760] Think of the breath of the Almighty—the sevenfold light of the Holy Spirit—when you contemplate that your thymus gland sits between your lungs. Then, add a vital heart connection because your thymus gland sits just in front of and above your heart.

The Bride of Christ is formed through positive selection. Bridal hearts endure to the end to be found pure (mature) after some arduous and humiliating circumstances. The All-Consuming Fire leads us by His Spirit and His Word to the place where no carnality remains. This is where we become one—bone of His bone and flesh of His divine flesh.

A 2016 study reports that the thymus suppresses the effects of aging, because the hormones released by the thymus inhibit the aging process as well as helping a person's memory. [761] Sounds like characteristics of immortality to me. The Transformation into a resurrected body with a quickening spirit are by the Blood of Yeshua, by the power of the Holy Spirit, and by the enacting [endorsing, decreeing, sanctioning, and performing] of the Father (Rom. 10:9).

Why did Yeshua say, Do this in remembrance of Me (1 Cor. 11:24-25)? The goal of kings and priests made after the Righteous Order of Melchizedek, who are leading their inner fire bride forth, is to bear the record of the testimony of God's DNA in one's body.

Blood is congealed light (1 John 5:6-10). The Word of God was speaking about Adam and Eve as well as His very own beloved in *Genesis 2:23 "bone of my bone, and flesh of my flesh."* First, His Body will be transformed through their bones; and then, their flesh. When a person gets a bone marrow transplant, the DNA in their blood is different than the DNA of their external skin.

Communion facilitates the change of the record of our testimony to become the very DNA of God. Through genuine communion, our bones (marrow) start producing the correct record in our blood first. Then, as we make the right choices, as we walk life out, our body gets conformed into His image. As we rightly examine ourselves, we take on common union in Christ. We become one with Christ and His Body (1 Cor. 11:28). This engagement of the crucifixion and communion process enables an ever-brightening arch of His Presence to form between your blood and Christ's blood as well as between your body and Christ's Body until you are an outward expression of an inward reality—pure white light. The Blood and Body of Yeshua are what enables the kings and priests of the Righteous Order of Melchizedek to return to our original divine design—primordial Light Beings made in God's own image. Or in other words, T Totally Transformed.

ANCIENT TAV (ת)

TAV (ת) is the last Hebrew Living Letter and seal of the word *emet* (אֱמֶת, truth). There is no such thing as just a little falsehood: even a trifle of it turns truth into a lie. Truth is eternal, but when even the smallest numerical unit of (אֱמֶת), namely TAV (ת), is omitted, what remains is (מֵת), death. (Maharal). Maharal states that even an incomplete truth is tantamount to a lie, for falsehood is not the opposite of truth but its absence. [762]

The three letters that spell *emet* are the beginning, middle, and ending letters of the Hebrew Alphabet (ALEF, MEM, TAV). "Why are the letters of falsehood (שֶׁקֶר) consecutive in the Hebrew Alphabet while the letters of truth (אֱמֶת) are spread out?" "To indicate that falsehood is common, but the truth is uncommon." [763] Truth is not only engraved on the Lord of the Universe's signet ring, it's also one of His Names. *"Jesus said to him, 'I am the way, the TRUTH, and the life. No one comes to the Father except through Me'"* (John 14:6).

Truth is found in Christ and His Word. *"¹ In the beginning was the Word, and the Word was with God, and the Word was God. . . . ¹⁴ And the Word became flesh, and dwelt among us; and we saw His glory, glory as of the only Son from the Father, full of grace and TRUTH"* (John 1:1,14 NASB).

Christ's word warriors know the truth, and this is the truth that sets them free. *"³¹ So Jesus was saying to those Jews who had believed Him, 'If you continue in My word, then you are truly My disciples; ³² and you will know the TRUTH, and the TRUTH will set you free'"* (John 8:31-32 NASB).

We are sanctified in the truth and oneness of God's Word. *"¹⁷ Sanctify them in the TRUTH; Your word is TRUTH. ¹⁸ Just as You sent Me into the world, I also sent them into the world. ¹⁹ And for their sakes I sanctify Myself, so that they themselves also may be sanctified in TRUTH. ²⁰ I am not asking on behalf of these alone, but also for those who believe in Me through their word, ²¹ that they may all be one; just as You, Father, are in Me and I in You, that they also may be in Us, so that the world may believe that You sent Me. ²² The glory which You have given Me I also have given to them, so that they may be one, just as We are one; ²³ I in them and You in Me, that they may be perfected in unity, so that the world may know that You sent Me, and You loved them, just as You loved Me. ²⁴ Father, I desire that they also, whom You have given Me, be with Me where I am, so that they may see My glory which You have given Me, for You loved Me before the foundation of the world"* (John 17:17-24 NASB).

The truth of God is exchanged for falsehood when we worship the creature (our own way) rather than the Creator. *"²¹ For even though they knew God, they did not honor Him as God or give thanks, but they became futile in their reasonings, and their senseless hearts were darkened. ²² Claiming to be wise, they became fools, ²³ and they exchanged the glory of the incorruptible God for an image in the form of corruptible mankind, of birds, four-footed animals, and crawling creatures. ²⁴ Therefore God gave them up to vile impurity in the lusts of their hearts, so that their bodies would be dishonored among them. ²⁵ For they exchanged the TRUTH of God for falsehood, and worshiped and served the creature rather than the Creator, who is blessed forever. Amen"* (Romans 1:21-25 NASB).

Redemption comes from the God of truth. *"Into Your hand I entrust my spirit; You have redeemed me, Lord, God of TRUTH."* (Psalms 31:5 NASB).

Glory dwells in the land where graciousness and truth meet. "*⁹ Certainly His salvation is near to those who fear Him, That glory may dwell in our land. ¹⁰ Graciousness and TRUTH have met together; Righteousness and peace have kissed each other. ¹¹ TRUTH sprouts from the earth, And righteousness looks down from heaven. ¹² Indeed, the Lord will give what is good, And our land will yield its produce. ¹³ Righteousness will go before Him And will make His footsteps into a way*" (Psalms 85:9-13 ₙₐₛʙ).

COMPOUND LETTER (ת)

TAV (ת) is a compound letter, which is composed by combining DALET (ד) and NUN (נ) together, whose letters spell out the word DAN (דן)—judgment. Just as the Tribe of DAN (דן) was the rear guard by being the last tribe to set out in their Wilderness travels, so is TAV's position in the Hebrew Alphabet. [764] The letter TAV (ת) translates in Hebrew as "sign" or "mark," and in Aramaic as "continuously."

FULFILLMENT OF TAV (ת)

TAV (ת) fulfills with either the word (תָיו) or (תָו). The numerical value of (תָיו) is 416, which is a multiple of 13 (13 × 32=416). Thirteen is the numerical value of the words "love" (אַהֲבָה) and "one" (אֶחָד). Thirty-two is the numerical value of the word "heart" (לֵב). [765] Being the final letter of the Hebrew Alphabet, TAV represents (לֵב אֶחָד) a loving heart, which has the power to unify everything and everyone in perfect love.

QUANTUM 22

Atomic Number 22—Titanium (Ti). Quantumly speaking, titanium is a chemical element that has the atomic number **22**, the atomic weight **47.867**, and the atomic symbol **Ti**. A single Titanium atom has 22 protons, 22 electrons, and 26 neutrons. Originally, titanium's name was gregorite. The name "titanium" comes from the Titans in Greek mythology. [766]

Titanium is a transition metal that is part of group 4 in the Periodic Table. Titanium is the fourth most abundant metal on Earth as well as the ninth most abundant element on Earth. [767] Titanium is found in almost all living things as well as in bodies of water, rocks, and soils. [768] In nature, titanium is almost always found in igneous rock and its sediments. Its occurrence is widespread. [769] Titanium is not found as a pure element in nature; but it is found in compounds as part of minerals in Earth's crust, such as rutile, ilmenite, and many iron ores. [770]

As a metal, titanium is recognized for its high strength-to-weight ratio. [771] This means it is very light and very strong at the same time. Titanium is twice as strong as aluminum while only weighing 60% more, and it is as strong as steel, yet weighing much less. [772]

Titanium alloys also have high corrosion resistance,[773] fatigue resistance, and high crack resistance. [774] Add to this titanium's ability to withstand moderately high temperatures without creeping, and you understand titanium's use in aircraft, armor plating, naval ships, spacecraft, and missiles. [775] For these industrial applications, titanium is alloyed with aluminum, zirconium, nickel, vanadium, and other elements. Almost 50% of all alloys used in aircraft applications are titanium alloys. In fact, approximately two-thirds of all titanium metal produced is used in aircraft engines and frames. [776]

Titanium is resistant to seawater corrosion; therefore, it is used to make propeller shafts, rigging, and heat exchangers in desalination plants. [777] The former Soviet Union even used titanium to make submarine hulls from titanium alloys. [778]

Due to titanium being biocompatible (non-toxic and not rejected by the body), it has many medical uses, including surgical implements and implants, such as hip balls and sockets (joint replacement) and dental implants that can stay in place for up to 20 years.[779]

Due to titanium's durability and inertness for those with allergies, it has become more popular in jewelry design, particularly titanium rings. When 1% of titanium is alloyed with 24-karat gold, it has the hardness of 14-karat gold and is more durable than pure 24-karat gold while still being able to be marketed as 24-karat gold. [780] Additionally, the lightweight, durable, and dent/corrosion resistance of titanium makes it desirable for watch cases. [781]

Titanium is non-toxic to the human body, even in large doses. An estimated quantity of 0.8 milligrams of titanium is ingested by humans each day, but usually passes through without being absorbed in tissues. [782] Sometimes, titanium bio-accumulates in tissues that contain silica. [783]

When titanium metal is in the form of powder or metal shavings, it can pose a significant fire hazard. When heated in air, titanium metal powder or shavings are an explosion hazard. [784]

22 IN SCRIPTURE

TAV (ת) is 22. Just as there are 22 letters in the Creator's Hebrew Alphabet, God created twenty-two things during the six days of Creation (Gen. 1:1-31).

TAV (ת) is 22. Although most modern translations of the Bible list 39 books for the Old Testament, originally the number of inspired books finalized by Ezra and the Great Assembly is twenty-two. These 22 originally canonized books by Ezra and company correspond to the Quantum 22 Hebrew Living Letters. Yeshua (Jesus) quoted from these original twenty-two canonized books.

TAV (ת) is 22. The word "light" is found 264 times in Scripture. When we divide 264 by 12, we get 22, which represents "light." [785] Take for example, the word "light" appears 22 times in the Gospel of John, seven times in just the first chapter. *"¹ In the beginning was the Word, and the Word was with God, and the Word was God. ² He was in the beginning with God. ³ All things were made through Him, and without Him nothing was made that was made. ⁴ In Him was life, and the life was the LIGHT of men. ⁵ And the LIGHT shines in the darkness, and the darkness did not comprehend it. ⁶ There was a man sent from God, whose name was John. ⁷ This man came for a witness, to bear witness of the LIGHT, that all through him might believe. ⁸ He was not that LIGHT, but was sent to bear witness of that LIGHT. ⁹ That was the true LIGHT which gives LIGHT to every man coming into the world" (John 1:1-9).*

So, TAV (ת) is 22 and represents "light," which is encoded in the Light in the Temple. Recall that the Temple Menorah includes 7 lamps with 22 almonds in various stages of development. Divide 7 into 22 and we get Pi (π), which is a number so close to infinity that it appears to go on forever. Pi is a main feature of a circle because the circumference of a circle is its diameter multiplied by Pi (π). A circle represents eternity or endless life, because it's a never-ending cycle. The fraction 22 divided by 7 never ends, because Pi's random numbers never repeat, even to a hundred trillion calculations as modern computers have shown.

Endless life is the distinguishing hallmark granted to those made after the Righteous Order of Melchizedek: *"¹⁵ And it is yet far more evident if, in the likeness of Melchizedek, there arises another priest ¹⁶ who has come, not according to the law of a fleshly commandment, but according to the power of an endless life. ¹⁷ For He testifies: 'You are a priest forever according to the order of Melchizedek'" (Hebrews 7:15-17).*

Additionally, the 7 into 22 Pi (π) phenomenon is stamped on the last great day of the Feast of Tabernacles *(Sukkot)—Hoshana Rabbah—*which is always held on the 22nd day of the 7th Biblical month. *Hoshana Rabbah* literally means the great hosanna or the great salvation or numerous praise. During the Feast of Tabernacles, Isaiah 12:3 was often quoted, as it is written, *"Therefore with joy you will draw water from the wells of salvation." Sukkot* is known for being "the season of our joy." I'd like to first point out that Jesus had an Aramaic name when He literally walked the earth. This Aramaic name was Yeshua. In Hebrew, Yeshua means "salvation."

Second, let us examine the water-pouring ceremony related to this festival. Even though the water-pouring ceremony is not in the Torah (first five books of the Bible), I bring it up because Yeshua stood on the day of *Hoshana Rabbah* during *Sukkot* and said, *"If anyone is thirsty, let him come to Me and drink. He who believes in Me, as the Scripture said, 'From his innermost being will flow rivers of living water.'" (John 7:37-38 $_{NAS}$).* When Yeshua proclaimed, *"If anyone is thirsty, let him come to Me and drink,"* the water pouring ceremony was different than the previous six days. The previous six days the priests circled God's Altar in a procession singing *Psalm 118:25 "Save now, I pray, O Lord; O Lord, I pray, send now prosperity."* But on the seventh day of the Feast of Tabernacles, it was the people, not the priests, which circled the Altar seven times, as they cried, "Save now!" seven times. This is why the day is called *Hoshanah Rabbah*—the Great Salvation. [786]

TAV (ת) is 22. The last words that Yeshua spoke on the Cross (TAV) were from the first verse of Psalms 22, which was penned by King David: *"My God, My God, why have You forsaken Me? Why are You so far from helping Me, And from the words of My groaning?" (Psalms 22:1). "And about the ninth hour Jesus cried out with a loud voice, saying, 'Eli, Eli, lama sabachthani?' that is, 'My God, My God, why have You forsaken Me?'" (Matthew 27:46).*

TAV (ת) is 22. There are 22 bones in the human skull. 14 of them belong to the facial skeleton and 8 to the neurocranium (braincase).

TAV (ת) is 22. There are 22 amino acids in the human genome, which corresponds to the twenty-two letters in the Hebrew Alphabet. Remember that chromosomes are composed of DNA structured by four bases: A-T and C-G, which are transcribed by the RNA, and are structured in 64 combinations of three letters (codons) to express 22 amino acids.

400 IN SCRIPTURE

TAV (ת) has a numerical value of 400; therefore, TAV (ת) is 400. Once God made a unilateral covenant with Abraham, the Lord reveals to him that his descendants would be afflicted in a foreign land for four hundred years. *"⁴And behold, the word of the Lord came to him, saying, 'This one shall not be your heir, but one who will come from your own body shall be your heir.' ⁵ Then He brought him outside and said, 'Look now toward heaven, and count the stars if you are able to number them.' And He said to him, 'So shall your descendants be.' ⁶ And he believed in the Lord, and He accounted it to him for righteousness. ⁷ Then He said to him, 'I am the Lord, who brought you out of Ur of the Chaldeans, to give you this land to inherit it.' ⁸ And he said, 'Lord God, how shall I know that I will inherit it?' ⁹ So He said to him, 'Bring Me a three-year-old heifer, a three-year-old female goat, a three-year-old ram, a turtledove, and a young pigeon.' ¹⁰ Then he brought all these to Him and cut them in two, down the middle, and placed each piece opposite the other; but he did not cut the birds in two. ¹¹ And when the vultures came down on the carcasses, Abram drove them away. ¹² Now when the sun was going down, a deep sleep fell upon Abram; and behold, horror and great darkness fell upon him. ¹³ Then He said to Abram: 'Know certainly that your descendants will be strangers in a land that is not theirs, and will serve them, and they will afflict them FOUR HUNDRED years. ¹⁴ And also the nation whom they serve I will judge; afterward they shall come out with great possessions. ¹⁵ Now as for you, you shall go to your fathers in peace; you shall be buried at a good old age. ¹⁶ But in the fourth generation they shall return here, for the iniquity of the Amorites is not yet complete'"* (Genesis 15:4-16).

TAV (ת) is 400. In 1 Samuel 22, we are told David flees from King Saul to reside in a cave near Adullam. There approximately four hundred men gather to support him. *"¹ David therefore departed from there and escaped to the cave of Adullam. So when his brothers and all his father's house heard it, they went down there to him. ² And everyone who was in distress, everyone who was in debt, and everyone who was discontented gathered to him. So he became captain over them. And there were about FOUR HUNDRED men with him"* (1 Samuel 22:1-2).

TAV (ת) is 400. According to the command of Cyrus, Darius, and Artaxerxes, the Jewish captives returned to Jerusalem to rebuild their Temple, which was destroyed by the Babylonians. 400 lambs were offered during the rededication ceremony of the Temple. *"¹⁴ So the elders of the Jews built, and they prospered through the prophesying of Haggai the prophet and Zechariah the son of Iddo. And they built and finished it, according to the commandment of the God of Israel, and according to the command of Cyrus, Darius, and Artaxerxes king of Persia. ¹⁵ Now the temple was finished on the third day of the month of Adar, which was in the sixth year of the reign of King Darius. ¹⁶ Then the children of Israel, the priests and the Levites and the rest of the descendants of the* captivity, celebrated the dedication of this house of God with joy. *17 And they offered sacrifices at the dedication of this house of God, one hundred bulls, two hundred rams, FOUR HUNDRED lambs, and as a sin offering for all Israel twelve male goats, according to the number of the tribes of Israel"* (Ezra 6:14-17).

WONDEROUS SIGNS

TAV (ת) is the cross-shaped mark or sign, or simply a mark or a sign. Paleo TAV is an "X" that marks the spot. Behold, the sun, moon, and stars are signs and signals. *"Then God said, 'Let there be lights in the firmament of the heavens to divide the day from the night; and let them be for signs and seasons, and for days and years'" (Genesis 1:14)*. The Hebrew word for "signs" in Genesis 1:14 is *ot* (אוֹת). *Ot* can be translated as a sign, a miraculous sign, a wonder, a banner, a signal, a remembrance, a warning, or a distinguishing mark.

Ot as a signal means that the sun and moon are beacons of light. They are monuments to God's goodness for what He created on the fourth day of creation.

Ot as a miraculous sign or warning correlates to Revelation 6's sixth seal. It is the sixth out of the seven seals for the Earth,[787] which the Lord linked to the seven seals of the human body through the "Redemption of TAV (ת)" featuring the thymus gland and T Totally Transformed: *"12 I looked when He opened the SIXTH SEAL, and behold, there was a great earthquake; and the sun became black as sackcloth of hair, and the moon became like blood. 13 And the stars of heaven fell to the earth, as a fig tree drops its late figs when it is shaken by a mighty wind. 14 Then the sky receded as a scroll when it is rolled up, and every mountain and island was moved out of its place. 15 And the kings of the earth, the great men, the rich men, the commanders, the mighty men, every slave and every free man, hid themselves in the caves and in the rocks of the mountains, 16 and said to the mountains and rocks, 'Fall on us and hide us from the face of Him who sits on the throne and from the wrath of the Lamb! 17 For the great day of His wrath has come, and who is able to stand?'"* (Revelation 6:12-17).

Additionally, the sun and moon are signals of the day of the Lord: *"30 And I will show wonders in the heavens and in the earth: Blood and fire and pillars of smoke. 31 The sun shall be turned into darkness, And the moon into blood, Before the coming of the great and awesome day of the LORD. 32 And it shall come to pass That whoever calls on the name of the LORD shall be saved. For in Mount Zion and in Jerusalem there shall be deliverance, As the LORD has said, Among the remnant whom the LORD calls"* (Joel 2:30-32).

Since TAV (ת) is translated as a "sign" in Hebrew, it is connected to the Biblical Seventh Day Sabbath, which is an eternal sign between YHVH and His people. *"Therefore the children of Israel shall keep the Sabbath, to observe the Sabbath throughout their generations as a perpetual covenant. It is a SIGN between Me and the children of Israel forever; for in six days the Lord made the heavens and the earth, and on the seventh day He rested and was refreshed"* (Exodus 31:16-17).

We are told in Scripture that God worked His signs in Egypt to redeem His people from their enemy. Behold, God is working His signs to free His people from the world. *"40 How often they provoked Him in the wilderness, And grieved Him in the desert!* ⁴¹ *Yes, again and again they tempted God, And limited the Holy One of Israel.* ⁴² *They did not remember His power: The day when He redeemed them from the enemy,* ⁴³ *When He worked His SIGNS in Egypt, And His wonders in the field of Zoan;* ⁴⁴ *Turned their rivers into blood, And their streams, that they could not drink.* ⁴⁵ *He sent swarms of flies among them, which devoured them, And frogs, which destroyed them.* ⁴⁶ *He also gave their crops to the caterpillar, And their labor to the locust.* ⁴⁷ *He destroyed their vines with hail, And their sycamore trees with frost.* ⁴⁸ *He also gave up their cattle to the hail, And their flocks to fiery lightning.* ⁴⁹ *He cast on them the fierceness of His anger, Wrath, indignation, and trouble, By sending angels of destruction among them.* ⁵⁰ *He made a path for His anger; He did not spare their soul from death, But gave their life over to the plague,* ⁵¹ *And destroyed all the firstborn in Egypt, The first of their strength in the tents of Ham.* ⁵² *But He made His own people go forth like sheep, And guided them in the wilderness like a flock;* ⁵³ *And He led them on safely, so that they did not fear; But the sea overwhelmed their enemies.* ⁵⁴ *And He brought them to His holy border, This mountain which His right hand had acquired.* ⁵⁵ *He also drove out the nations before them, Allotted them an inheritance by survey, And made the tribes of Israel dwell in their tents"* (Psalms 78:40-55).

Notice when Moses and Aaron performed signs among them that it was a time of great darkness. We are told that in the dark, God's people did not rebel against His word. *"*²⁶ *He sent Moses His servant, And Aaron whom He had chosen.* ²⁷ *They performed His SIGNS among them, And wonders in the land of Ham.* ²⁸ *He sent darkness, and made it dark; And they did not rebel against His word"* (Psalms 105:26-28).

In this time of the greatest darkness that the world has ever seen, the kings and priests made after the Righteous Order of Melchizedek are a wondrous sign. *"*⁷ *Thus says the Lord of hosts: 'If you will walk in My ways, And if you will keep My command, Then you shall also judge My house, And likewise have charge of My courts; I will give you places to walk Among these who stand here.* ⁸ *Hear, O Joshua, the high priest, You and your companions who sit before you, For they are a wondrous SIGN; For behold, I am bringing forth My Servant the Branch.* ⁹ *For behold, the stone that I have laid before Joshua: Upon the stone are seven eyes. Behold, I will engrave its inscription,' Says the Lord of hosts, 'And I will remove the iniquity of that land in one day.* ¹⁰ *In that day,' says the Lord of hosts, 'Everyone will invite his neighbor under his vine and under his fig tree.'"* (Zechariah 3:7-10).

Remember Yeshua Himself declared I am the Vine and you are the branches. Also, notice the series of three within this five-verse passage: fruit (30), more fruit (60), and much fruit (100). *"*¹ *I am the true vine, and My Father is the vinedresser.* ² *Every branch in Me that does not bear fruit He takes away; and every branch that bears fruit He prunes, that it may bear more fruit.* ³ *You are already clean because of the word which I have spoken to you* ⁴ *Abide in Me, and I in you. As the branch cannot bear fruit of itself, unless it abides in the vine, neither can you, unless you abide in Me.* ⁵ *I am the vine, you are the branches. He who abides in Me, and I in him, bears much fruit; for without Me you can do nothing"* (John 15:1-5).

Thus, the Messianic Title of the Branch is a fullness of Christ's bodily reality of both the Mature Head of Christ and His fully mature Body. *"*¹ *There shall come forth a Rod from the stem* of Jesse, And a Branch shall grow out of his roots. ² The Spirit of the LORD shall rest upon Him, The Spirit of wisdom and understanding, The Spirit of counsel and might, The Spirit of knowledge and of the fear of the LORD" (Isaiah 11:1-2).

The fully mature Body of Christ are those people made in God's image who have grown into the exact same nature as Messiah Yeshua; thus, they are crowned with the Seven Spirits of God and are completely filled to the fullness of the stature of the measure of Christ. The Seven Spirits of God are the seven eyes upon the stone in Zechariah 3:9.

The "stone" with the seven eyes of the Seven Spirits of God represents oneness with our Heavenly Father and Messiah Yeshua due to the Hebrew word "stone" is *even*, which is the conjunction of the Father *(av)* and the Son *(ben)*. Please refer to the BET chapter of the *ALEF-TAV's Hebrew Living™ Letters* book as well as the "Seven Spirits of God" section of the WATER INTO WINE chapter.

QUANTUM QUARK (ת)

In summary, TAV (ת) is the twenty-second Hebrew Living Letter. TAV (ת) in ancient Hebrew means a sign. The picture of a sign is in the shape of a cross, or an "x" that marks the spot, which is seen in its ancient pictograph. The pictograph of the sign embodied in TAV (ת) symbolizes ownership, sealing, making a covenant, joining two things together, and making a sign. The seal of God's seal is the letter TAV (ת). It signifies faith, the conclusion, and the culmination of all 22 quantum forces (letters) active in creation.

TAV (ת) is also the last letter and seal of the word *emet* (אֱמֶת, truth). The three letters that spell *emet* are the beginning, middle, and ending letters of the Hebrew Alphabet (ALEF, MEM, TAV). Truth is found in the Messiah and His Word. Christ's word warriors know the truth, and this is the truth that sets them free (John 8:32).

Q. How do you make sure that you don't miss God's voice that talks through wondrous signs, signals, warnings, etc.?

Q: What is the significance that the three Hebrew letters that spell "truth" cover the entire Hebrew Alphabet?

∞

My prayer is that you never forget that you have been made in God's own image. That you know that your precious image is as gloriously unique as your thumbprint. It's not about fitting in.
It's about discovering the glorious reality of who you truly are.
Your original pattern is simply divine. Your Creator knows you, loves you, and wants to have a relationship with you.
Don't settle for anything less than the righteous Quantum 22 Hebrew Living Letters in your journey back to Eden and beyond.

Additional Resources

"HEBREW LIVING™ LETTERS SPEAK" CLASS YOUTUBE VIDEO RECORDINGS

[1] Hebrew Living Letters Speak—Class #1 – INTRO => https://youtu.be/u96xhyWQcpo

[2] Hebrew Living Letters Speak—Class #2 – ALEF => https://youtu.be/7YnvUuh9ggY

[3] Hebrew Living Letters Speak—Class #3 – BET => https://youtu.be/5KLj7Lb8PPw

[4] Hebrew Living Letters Speak—Class #4 – GIMEL => https://youtu.be/j9P8IMrEODk

[5] Hebrew Living Letters Speak—Class #5A – DALET => https://youtu.be/H9N1zxGT-PG8 and Class #5B - DALET => https://youtu.be/qxYOtI_1UKk

[6] Hebrew Living Letters Speak—Class #6 – HEI => https://youtu.be/TAgciltz9Iw

[7] Hebrew Living Letters Speak—Class #7 – VAV => https://youtu.be/JKiHC15disI

[8] Hebrew Living Letters Speak—Class #8 – ZAYIN => https://youtu.be/eaGzFoc8U8Q

[9] Hebrew Living Letters Speak—Class #9 – CHET => https://youtu.be/a50Bn9Q5_Eo

[10] Hebrew Living Letters Speak—Class #10 – TET => https://youtu.be/jQIUaJSRaPA

[11] Hebrew Living Letters Speak—Class #11 – YUD => https://youtu.be/l0czbvAZA8o

[12] Hebrew Living Letters Speak—Class #12 – KAF => https://youtu.be/fVmc68iv2zA

[13] Hebrew Living Letters Speak—Class #13 – LAMED => https://youtu.be/0QUIm-KIbt9Q

[14] Hebrew Living Letters Speak—Class #14 – MEM => https://youtu.be/SRCQSs4fZog

[15] Hebrew Living Letters Speak—Class #15 – NUN => https://youtu.be/YtD-qqv0TvE

[16] Hebrew Living Letters Speak—Class #16 – SAMECH => https://youtu.be/si3YqX9Is4Y

[17] Hebrew Living Letters Speak—Class #17 – AYIN => https://youtu.be/RRqqIzzW-Eo

[18] Hebrew Living Letters Speak—Class #18 – PEY => https://youtu.be/P5vgytzi_Ec

[19] Hebrew Living Letters Speak—Class #19 – TSADIK => https://youtu.be/GeNXuDhMBHU
[20] Hebrew Living Letters Speak—Class #20 – KOOF => https://youtu.be/PAgA4PVP03I
[21] Hebrew Living Letters Speak—Class #21 – RESH => https://youtu.be/kZLktSzZ1WY
[22] Hebrew Living Letters Speak—Class #22 – SHIN => https://youtu.be/cADIvLGRQ2o
[23] Hebrew Living Letters Speak—Class #23 – TAV => https://youtu.be/dpeCO2s_AjE

GLOSSARY

144000 Firstfruits—The 144,000 Firstfruits cover the broadest spectrum possible within the confines of being from the 12 Tribes of Israel listed in Revelation 7:1-8. They are from full-blooded to one drop, from an elderly adult to a child, from male to female, from black to white, from singles to families to groups. The 144,000 Firstfruits are sometimes described as the Virgin Bride part of the Bride of Christ because they are considered spiritually pure (Rev. 14:4). On Mount Zion stands the Lamb with 144,000 souls with the Lamb's Name and His Father's Name written on their foreheads (Rev. 14:1). The new song that these 144,000 firstfruits sing is the tune coming from their transformed DNA where their mortality has taken on immortality (Rev. 14:3; 1 Cor. 15:53). The Transformation of His Bride/ Remnant/ 144,000 Firstfruits cannot be duplicated by man or Fallen Angels. Yeshua has kept this secret between Himself and His Father. Remember, that the 144,000 are merely the first fruits, which means other Wise Virgins will follow. Watch for the firstfruits transformation of the 144,000 that will crack the glass ceiling and pave the way for others.

Alef-Bet (אָלְפבֵית)—is the origin of the word "Alphabet." ALEF (א) is the first letter of the Hebrew Alphabet while BET (ב) is the second, which starts the succession for all the Hebrew letters. An alphabet is a standardized set of letters that are arranged in a particular order, which represents units of sounds that distinguish words.

Alef-Tav (את)—Not only is ALEF-TAV (את) the first (א) and the last (ת) letters of the Hebrew Alphabet, but את refers to Messiah Yeshua (Jesus Christ) who declares in *Revelation 1:8, "I am the Alpha [ALEF] and the Omega [TAV] . . . who is and who was and who is to come, the Almighty."* The first and the Last (אָלֶף תָו) is also a Messianic reference in Isaiah 41:4, Isaiah 44:6, Isaiah 48:12, and Revelation 22:13. Yeshua is the Word. Yeshua is the ALEF-TAV (את). When we consider Messiah Yeshua is not only the Beginning and the End but also the Creator and Sustainer of all things (Col. 1:16-17), we can understand that *et* (ALEF-TAV) includes all the various objects of creation within heaven and earth. This all began when Father's infinite uncreated light filled the void—empty space—which is our entire created reality.

All-Pervading Light Frequency of the Cosmos—Our DNA is driven and influenced by the all-pervading light frequency of the cosmos. It has a base frequency of 144,000 cycles per second (cps). 12×12 = 144 is a harmonic fractal of the cosmic fractal system that includes all in the living matrix of the interconnected web of life that exists throughout the cosmos. Researchers at first thought that our DNA twinned helices were like a radio transmitter, but we now realize it is more than this. Our DNA has 12 strands, and it is a light antenna.

Almighty God—The Almighty's great name "Shaddai" literally bears the quantum letter SHIN (ש). The attributes of El Shaddai are: Almighty; All Powerful; All Sufficient One; Judge; the sweet omnipotence of love, sure and all-sufficient resource in time of need; power to fulfill every promise; bountiful; overrides nature to perform miracles and remove mountains; giver and pourer out of self-sacrificial love and blessing; nurturer and comforter that gives strength and quiets the restless, and in Him all fullness dwells. Ezekiel's Living Creature, which portrays a wing of God's Righteous Melchizedek Army flying by the Spirit with one another, has the voice of the Almighty (Ezek. 1:24; Ezek. 10:5).

Amplituhedron—Two scientists—Nima Arkani-Hmed and Juroslav Tranka from Cal Tech—discovered that all space and time are the emanation of a single geometric form. They just don't know what it is yet. They call this shape the "amplituhedron" or the "positive grassmanian." It looks like four tetrahedrons stuck together. This Amplituhedron appears to be part of the *Merkabah*—technically it is two *Mer-ka-bahs*. All photons (energy, light) are all tetrahedral *Merkabahs*. That means that all light is an emanation of a single geometric form—the *Merkabah* (3-D Star of David)—which consists of the two tetrahedrons (one pointed up and the other pointed down), which illustrates ascending and descending.

Ascension in Christ—The pattern for Ascension in Christ comes from Ephesians 4:9-10. The plain sense *(peshat)* understanding of Ephesians 4:9-10 is that it is speaking about Messiah Yeshua's death, burial, and resurrection. On the *derash* level, the Bible also speaks of God's people who say that they abide in Christ should walk in the same manner as our Pattern Son—Messiah Yeshua (1 John 2:6). We need to follow Him in His death, burial, and resurrection (Rom. 6:1-6). Or in other words, follow Christ by first descending to the Kingdom of God within a believer (i.e., the current level a person operates in); and then, ascending "in Christ" by the Blood of the Lamb to focus on the perfect will of our Heavenly Father (Rev. 4:1-2; Heb. 10:19-20; John 5:19). This is on the *remez* (hidden) level while the various ascensions in Christ are mainly on the *sod* (mysterious) level. For more on this subject, please refer to the *Ascension Manual*.

Atom—No one has actually seen an atom, but we know they exist based on repeatable physics measurements. *Proverbs 8:26* calls atoms *"the primal dust of the world"* or *"the beginning of the dust"* that cannot be seen. All physical matter is made up of molecules and molecules are made up of atoms. In general, we can say that atoms are electrically neutral because they have a net electrical charge of zero. For instance, the simplest and lightest hydrogen atom has only one proton and one electron while deuterium is a hydrogen atom that has had a neutron added, which forms a heavier nucleus. Each atomic substance has a different configuration of protons and neutrons in its nucleus, plus a unique mass. In an electrically neutral atom, the number of negatively charged electrons always match the number of positively charged protons. Atoms with more neutrons than protons retain the same chemical properties, but some of these isotopes aren't stable. Unstable atoms are radioactive, which means that they emit one or more particles from the nucleus in a random fashion by a process called quantum tunneling.

Atomic Number—is the number of protons in the nucleus of an atom, which determines the chemical properties of an element as well as its place in the Periodic Table. The elements on the Periodic Table are arranged in order of increasing number of protons in the nucleus of the atom.

Atomic Structure—is the structure of an atom. An atom is a complex arrangement of negatively charged electrons arranged in shells about a positively charged nucleus. The nucleus contains most of the atom's mass, and is composed of protons and neutrons, except for hydrogen which only has one proton.

Atomic Weight—The total weight of an atom is called the atomic weight. It is approximately equal to the number of protons and neutrons, with a little extra added for electrons. Atomic weight is sometimes called atomic mass.

Bride of Christ—is the Lamb of God's wife (Rev. 21:9). Messiah Yeshua (Jesus Christ) is called *"the Lamb of God who takes away the sins of the world"* by John the Baptist in *John 1:29*. The Bride of Christ is a multi-membered, corporate organism joined in holy matrimony to the King of Kings. The Bride of the Messiah (Christ) is made up of two parts: the 144,000 Virgin Bride and the Wise Virgins (Matt. 25:1-13). This Bridal Company will be one with the Messiah, just as the Heavenly Father and Yeshua are one (John 17:23). There is a great mystery concerning Christ and His people who bear His Name being one flesh, and of Christ loving the *Ekklesia* so much that He gave Himself for it to sanctify and cleanse it with the washing of the water of the Word (Eph. 5:26). The Messiah's Bride makes herself ready for the marriage of the Lamb through the people's righteous deeds (Rev. 19:7-8). The Bride of Christ is the Dwelling Place of God whose code name is the "New Jerusalem" that accompanies the new heavens and new earth. Revelation 21 pictures the Bride, the Lamb's wife as a golden cube whose wall is 144 cubits, whose foundation has 12 layers with the names of the apostles of the Lamb on each, and whose gates are 12 with the names of the twelve tribes of Israel on each.

Collapsing the Wave Function—A measurement performed by an observer in a quantum experiment causes all the possibilities of the wave function to "collapse" to a single outcome. When we hear, we are observing. When we see, we are observing. When we perceive, we are observing. When we observe, we are collapsing the wave function.

Chemical Elements—These are chemical substances that cannot be broken down into other substances. The basic particle that constitutes a chemical element is the atom. Chemical elements are distinguished from each other by the number of protons in the nuclei of each atom. This is known as its atomic number.

Crystalline (Crystal) Water—Liquid crystals are a unique phase of matter that's gel-like. The vast majority of water in the human body is in a liquid crystalline state. Liquid crystalline water is also called structured water, organized water, or hexagonal water. Molecules remain mobile in the liquid crystalline form, and they tend to move together like a school of fish. Liquid crystals store and transmit information, like solid crystals, but are flexible and much more responsive. The repeating pattern of liquid crystalline provides efficient pathways for the smooth flow of energetic information. The presence of liquid crystal water in the human body is vital for cellular integrity and DNA stability.

Cymatics—Everything that we see, touch, and experience in the natural realm is a result of cymatics, which is the study of sound and vibration made visible. The multiple forms of universal geometry, symmetry, and beauty that emerge through resonance in various mediums allow us to ponder the wonder of the nature of sound, vibration, and form itself. Johannes Kepler puts it this way: "Where there is matter, there is geometry." Geometry is vibrations seen. Geometric shapes, some call sacred geometry, are simply sounds made visible.

David Van Koevering—was a quantum physicist, inventor, musicologist, and visionary. David Van Koevering pioneered electronic musical instruments, helping to create the market for synthesizers. He owned over 600 patents—most of them relating to music. One of his last inventions is the table of elements synthesizer. It tunes the frequencies and frequency partials to the atomic weight vibration of the elements that make up the human body. He discovered the frequencies of your body include 31 fundamental frequencies. Eleven of these frequencies are the dominant frequencies—the active frequencies—and 20 of these frequencies are trace elements in your body.

Dead Sea Scrolls—are the ancient, mostly Hebrew, manuscripts of leather, papyrus, and copper that were first found in 1947 on the northwest shore of the Dead Sea. The discovery of the *Dead Sea Scrolls* is among the most important finds in modern archaeology. The *Dead Sea Scrolls* come from various sites and date from the 3rd century BCE to the 2nd century CE. The term usually refers more specifically to manuscripts found in 11 caves near the ruins of Qumran, which most scholars think was the home of the community that owned the scrolls. The 15,000 fragments (most of which are tiny) represent the remains of 800 to 900 original manuscripts.

Electromagnetic Spectrum—is the range of frequencies of electromagnetic radiation and their respective wavelengths and photon energies. The entire electromagnetic spectrum from lowest to highest frequencies (longest to shortest wavelengths) includes radio waves (radio, television, microwaves, and radar), infrared radiation, visible light, ultraviolet radiation, x-rays, and gamma rays. The electromagnetic spectrum describes all the different kinds of light, including those that the human eye cannot see. Most of the universe is invisible to the human eye. The light that we can see is made up of the individual colors of the rainbow, which represents a very small portion of the electromagnetic spectrum. The brilliant David Van Koevering proclaims: "The entire electromagnetic spectrum was sung into existence by Jesus Christ who sustains all things." We must understand that everything had a higher bandwidth before The Fall.

Electron—is one of the three main particles that make up an atom. Electrons are the negatively charged particles of an atom. Electrons act as the primary carrier of electricity in solids. Together, all the electrons of an atom create a negative charge that balances the positive charge of the protons in the atomic nucleus. For example, a hydrogen atom has just one electron and one proton while a uranium atom has 92 protons and 92 electrons.

Endocrine System—Three different kinds of tissues make up the endocrine system in the human body: endocrine glands, endocrine organs, and endocrine-related tissues. Your endocrine system is in charge of creating and releasing hormones to maintain countless bodily functions. Hormones are chemicals that coordinate different functions in your body by carrying messages through your blood to your organs, skin, muscles, and other tissues. These signals tell your body what to do and when to do it.

Fibonacci Sequence—In mathematics, the Fibonacci sequence is a sequence in which each number is the sum of the two preceding numbers. Individual numbers in the Fibonacci sequence are known as Fibonacci numbers. The fabulous Fibonacci numbers of 0-1-1-2-3-5-8-13-21-34-etc. reveal the prized golden ratio of 1:1.618033988, which surprisingly is not a number but a universal phenomenon of successive Fibonacci wavelengths.

Fractals—are a natural phenomenon and a mathematical set that exhibits a repeating pattern on every scale. The term "fractal" was first used by mathematician Benoit Mandelbrot in 1975. He based it on the Latin word *fractus*, which means fragmented or broken or fractured. Mandelbrot used fractals to extend the concept of theoretical fractional dimensions to geometric patterns in nature. Theoretical fractals are infinitely self-similar [replication is exactly the same at every scale], iterated [act of repeating a process to approach a desired goal], and detailed mathematical constructs having fractal dimensions.

Geometry—Geometry is the visual manifestation of vibration, which is seen through cymatics. Many top scientists in today's world believe that space is "nothing." They call it a vacuum. They say it is structureless, yet it is laden with hyper-dimensional geometries.

Hanukkah—is a Biblical Feast that Messiah Yeshua is shown to have celebrated in John 10:22-23. Hanukkah goes by several names: the Feast of Dedication, the Feast of Lights, and the Feast of Miracles. The word "Hanukkah" means dedication. Its primary focus should be the dedication or re-dedication of one's heart, which is meant to cause your light to become purer, as you become more like the Light of the World—Messiah Yeshua.

Hebrew Living Letters—are the twenty-two letters of the Hebrew Alphabet, which are the *Quantum 22*. They are also called quantum letters as well as sacred letters. These joyous Hebrew letters are the hyperquantum intersecting elements of Creation between the spiritual and natural realms as well as part of the protoplasm of Creation.

Hyperquantum—The Initial Singularity is the place where God intersected the physical universe causing the universe to exist. Our *Echad* God (plurality in one) is the outside agent who was the Creator that started Creation, and who used their joyous sacred letters to be the intersecting elements of Creation between the spiritual and material realms. The Hebrew Living Letters are the hyperquantum substance that flowed out of the Lamb slain before the foundation of the world in the form of the Quantum 22 letters of the Hebrew Alphabet. In physics, hyperquantum pertains to values in which a quantum-mechanical equation approaches infinity. Not only do the Quantum 22 Hebrew Living Letters approach infinity, but they are also infinite because they are one with the Eternal, Self-Existent One.

Infant Universe—There was no material thing at first, just energy in the form of pure uncreated light. There was also the information encoded in the quantum letters of the Word of God. When you read about the Word in the beginning, always keep in mind that the Word is made up of Hebrew letters. *"In the beginning was the Word, and the Word was with God, and the Word was God" (John 1:1)*. *"In the beginning God created [את] the heavens and the earth." (Genesis 1:1)*. Additionally, the infant universe had magnificent and unending possibilities. All possible states of elementary particles in our universe are 10^{90}.

Inflationary Universe—The fresh new universe grew enormously during a period of rapid expansion called "inflation." The inflation expansion of space was driven by energy released in a kind of state transition as the universe cooled, similar to the way energy is released when ice condenses out of chilled water. When the state transition was complete, the universe slowed to a smaller rate of expansion. Keep in mind that scientists know that the universe is still expanding. As the inflationary universe expanded 1000 times, the quark-gluon plasma cooled, which caused protons and neutrons to form. When the universe cooled another 1000 times, light atomic nuclei formed.

Laminin—is the "glue" that holds the human body together. Laminin is in the shape of a cross that's formed from the alpha, beta, and gamma trimeric proteins, which reveals a mysterious picture of the Heavenly Father, the Son, and the Holy Spirit uniting through the laminin cross at the molecular level to sustain and maintain the integrity of the human body.

Living Creature—One dimension of the Living Creature classically depicted in Ezekiel 1 and Ezekiel 10 depicts the One New Man in Christ on a personal level where an individual is joined to the Father, the Son, and the Holy Spirit in a four-fold way so that Christ's Being can be lived out through a person by means of the Seven Spirits of God according to the perfect will of the Father. To become rightly connected to the Father, the Son, and the Holy Spirit, one needs to accept Messiah Yeshua as their Lord and Savior (Rom. 10:9-10; Phil. 2:9-11). This is a personal and loving relationship with the Creator Himself, which is not connected to any religious institution. Each person should be firing to the best of their ability on all cylinders before they connect corporately. Ezekiel 1 and Ezekiel 10 portray a righteous wing of God's Melchizedek Army who flies by His Spirit with one another. Notice that the Book of Ezekiel's Living Creature is made up of four or more living creatures. We know that the plurality of the living creatures become one singular Living Creature through the Spirit that is in each of their wheels within wheels beside each one of them (Ezek. 1:15-16, 20-21). Note that the Cherubim in Ezekiel 10 are equivalent to the Living Creature in Ezekiel 1 (Ezek. 10:20). For more on this subject, please refer to the *Understanding the Order of Melchizedek: Complete Series* book.

Manna—began to fall on the fifteenth day of the second month. Manna in the Wilderness was a foreshadowing of the true Bread of Life that rains down from heaven—Messiah Yeshua. The redeemed mature Body (Bread) of the Messiah attaches to this reality. The central tenet of manna is *"they shall be taught of God" (John 6:45)*. Manna was like coriander seed, which is a strong-smelling seed-like plant in the carrot family. It's native to south-eastern Europe and grows all over Europe, the Middle East, China, India, and Turkey. In the West, coriander is recognized as cilantro. Coriander seed is white (Exo. 16:31). One *omer* of manna was a day's portion of manna for one person. One *omer* is approximately 43.2 average-sized eggs in volume. A day's portion of manna was meant to test God's people to see whether they will walk in His instruction, or not. The Lord disciplines us, so we know what's in our hearts, whether we will keep His commandments or not. He humbles us by feeding us manna, which we do not know, that we may understand that man does not live by bread alone, but by everything that proceeds out of the mouth of the Lord (Exo. 16:31).

Mel Gel—Mel Gel is about gelling as one with the High Priest of the Righteous Order of Melchizedek where God's people slow down, move in Christ and rest in Him.

Melchizedek (מַלְכִּי־צֶדֶק**, also transliterated Melchisedech or Malki Tzedek)**—"Melchizedek" is translated as the King of Righteousness. Hebrews 7:1-2 speaks of Melchizedek being both the "king of righteousness" as well as the "king of peace." Righteousness is a dove-tail joint where Melchizedek and the Bride of Christ meet. One is the king of righteousness (Heb. 7:2). The other is given bright and clean fine linen to wear equivalent to the righteous deeds of the saints (Rev. 19:7-8). Fundamentally, righteousness is not a State of Doing, but a State of Being. "To be" is the primitive root of 'righteousness,' which also means to cleanse or clear self, to be just, or do justice, or turn to righteousness. 'To be or not to be' really is the right question when it comes to righteousness. There are at least three levels of Melchizedek that I know of. First, we have the High Priest made after the Order of Melchizedek— Messiah Yeshua (Christ Jesus). Yeshua is the ONLY high priest of the Righteous Order of Melchizedek; and therefore, its head (Hebrews 2:17; 4:14-15; 5:1,5,10; 6:20; 7:26; 8:1; 9:11; 10:21). Having only one High Priest at a time is the Biblical pattern. Therefore, we can be one with the High Priest of the Order of Melchizedek; and thus, High Priests too, but only in our oneness with Yeshua. Second, there is the angel called Melchizedek who holds the same rank as Lucifer did before he fell. Just like Lucifer is called a Sheltering Cherub in Ezekiel 28:16, so is this cherubic Melchizedek. Some call the cherubic Melchizedek the Chief Chancellor of God's Treasury Room. Thirdly, there is the Genesis 14 Melchizedek whom the Hebrews record was Shem—the righteous son of Noah. Shem earned the honorary title of "Melchizedek." Note: Shem currently is the highest five-star general of the Righteous Order of Melchizedek in the heavenly realms.

Merkabah—*Merkabah* is a Hebrew word where *"Mer"* means light, *"Ka"* means spirit, and *"Bah"* means body. All photons (energy, light) are all tetrahedral *Merkabahs*. That means that all light is an emanation of a single geometric form—the *Merkabah*—which consists of the two tetrahedrons (one pointed up and the other pointed down), which illustrates the ascending and descending process done through Ascension in Christ. The Biblical *Merkabah* Process is the key to the complete transformation of our bodies into light. The *Merkabah* is all about changing your Kingdom of God within through the work of Messiah Yeshua; and then, manifesting it outward. The *Merkabah* is foundational to all life and light. Everyone has been designed by our Creator to come back to this single point of origin in Christ.

Metatron—The righteous Metatron image of Messiah Yeshua (Jesus Christ) is a multidimensional, multi-membered eternal reality. The righteous Metatron can do the miraculous works of God; because everyone in the righteous Metatron Matrix is one in Christ, tapping into the Righteous Head. Metatron is Biblical because it can be found in the Pseudo-Jonathan translation of Genesis 5:24. The righteous Metatron is the undifferentiated state of the Messiah. "Undifferentiated" means no difference; therefore, Metatron consists of the fully mature Head Messiah Yeshua and His 100-percent fully mature members of His Body who have pressed into being made into His exact image. His fully mature Body of Christ is an eternal reality of the fullness of the Heavenly One New Man in Christ (Messiah), and it can be classified as the Metatron Oneness Matrix with the Heavenly Father the Son, and the Holy Spirit. There is an unrighteous or false Metatron Mithra who is connected to Christmas and the demonic control of spacetime.

Metatron's Cube (c=13)—The thirteen circles of Metatron's Cube are the fundamental blueprint for all atomic structures. From this matrix—the womb of creation—of 13 spheres, all the five platonic solids of the physical realm can be created.

Molecule—is the simplest unit of a chemical substance that has all the properties of that substance. A molecule is a group of two or more atoms held together by attractive forces known as chemical bonds. For instance, a water molecule is the smallest unit that is still water. A water molecule can be divided into two hydrogen atoms and one oxygen atom.

Moon Model—Dr. Robert Moon gave us the Moon Model. He was a physicist, chemist, and engineer who applied the five Platonic solids as a configuration of the atomic nucleus. Basically, Dr. Moon was saying space can be quantized, which means space is not empty. It has structure. He solved quantum physics by the concept of a geometric nucleus where the nucleus isn't just some big ball of protons, but the protons are shells. He taught that protons aren't particles, but waves and waves are geometric.

Mystic—The word "mystic" comes from the Latin *mysticus*, which means "of mysteries." The Word of God has at least 21 references to the term "mystery" and "mysteries." Know that a mystery is a religious truth that one can know only by revelation and may not be fully understood. A mystic relates to the mysterious, has a feeling of awe or wonder, and is a follower of the mystical way of life. A mystical way of life is having a spiritual meaning of reality. It involves having direct communion with God or ultimate reality.

Mysticism—is known as becoming one with God. The *Merriam-Webster Dictionary (10th Edition)* describes the noun "mysticism" as the experience of mystical union or direct communion with ultimate reality reported by mystics. Mysticism is the belief that direct knowledge of God, spiritual truth, or ultimate reality can be attained through subjective experience (such as intuition or insight). According to Dr. Elizabeth Alvilda Petroff, mysticism has been called "the science of the love of God," and "the life which aims at union with God." Its emphasis is on the spiritual life as a progressive climb — sometimes a steep and arduous one.

Neutron—is a subatomic particle whose symbol is n or n^0, which has a mass slightly greater than a proton. Neutrons and protons constitute the nuclei of atoms.

New Jerusalem—The holy city that comes out of the spiritual realm (heaven) is constructed with God's white living stones, who are the pure and spotless individuals that make up the wife of the Lamb (1 Pet. 2:5; Rev. 2:17; Rev. 21:1-9). The New Jerusalem is God's Dwelling Place that is being built in this Kingdom Day, which includes new heavens and a new earth in which righteousness dwells (2 Pet. 3:13; Rev 19:7-8). Messiah Yeshua will eternally reside with His Bride; thus, the New Jerusalem is a metaphor for the Tabernacle of God among men where the Lord permanently dwells with and in His people.

Nucleus—The quantum model of the atom consists of protons, neutrons, and electrons. Protons and neutrons are bound together by nuclear forces to form a nucleus while tiny electrons orbit the core of the atom (i.e., the nucleus).

Occipital Bone—is the only part of a person's skeletal structure that connects a human body's head to their body. The occipital bone surrounds a large opening known as the foramen magnum, which allows passage of the spinal cord, brainstem, and two key vertebral arteries that supply blood to the brain.

One New Man—This term is taken from Ephesians 2:15-16 for Jews and Gentiles being united in Christ (Messiah).

Order of Melchizedek—The Righteous Order of Melchizedek is the highest spiritual order on earth that manifests heaven—making the kingdoms of this earth, the kingdoms of our God (Rev. 11:15). As a king and priest of the Order of Melchizedek, you are called to bring heaven to earth. This happens when we follow the Pattern Son Yeshua by only doing what we see the Father doing (John 5:19). Scripturally, when we are born-again, we merely SEE the Kingdom of God (John 3:3) and are immature sons at this point. When we are born of water and the Spirit, we ENTER the kingdom of God (John 3:5). When anyone enters the Kingdom of God, they are commissioned into the Righteous Order of Melchizedek by the High Priest Himself. It is during this step that Manifested Sons, or the Maturing Sons, are produced. This is where the New Living Creature of the One New Man in the Messiah who is a wing of His Righteous Melchizedek Army takes shape. To INHERIT the Kingdom of God, we must get rid of all the works of the flesh (Gal. 5:19-21; 1 Cor. 6:9-10). This is where Mature or Maturing Sons become His Pure and Spotless Bride. It is also where a person gets grafted into the righteous Oneness Metatron Matrix.

Periodic Table of Elements—is widely used in chemistry, physics, and other sciences. Every kind of atom that mankind has so far discovered appears on the Periodic Table of Elements where each atom has its own unique properties. For example, a lead atom has 82 protons, 82 electrons, and 125 neutrons, which is a heavy element. Versus helium, which causes balloons to float in the air. The periodic table organizes all discovered chemical elements in rows (called periods) and columns (called groups) according to increasing atomic numbers. Scientists use the periodic table to quickly refer to information about an element, like atomic mass and atomic symbol.

Phi (ϕ)—is not just a linear number sequence determined from the Fibonacci numbers as in 3-5-8-13-etc where 5+8=13 and 8+13=21. The Laws of Phi are defined as a system of coherent consciousness that links the galactic, human, and quantum realities. Its secret is how it oscillates. How one term like 8 is divided into the next term 13 to give a proportion close to 1.618. As we continually divide previous terms into the next larger term, the division is always a number that lies above then below this mythical unreachable plateau of 1.618; thus, Phi is the universally governing oscillation. It's a set of cascading proportions that approach the ideal—the omniscient Phi Wave. PHI (ϕ) is the golden ratio that we see in flowers, in crystals, and in all of creation, including galaxies.

Pi (π)—is the ratio of a circle's circumference to its diameter, which is said to be approximately equal to 3.14159... Pi's random numbers never repeat, even to a hundred trillion calculations as modern computers have shown; therefore, Pi is considered to be a number so close to infinity that it appears to go on forever. A circle represents eternity or endless life because of its never-ending cycle. The total number of Hebrew Living Letters (22) divided by the number of simple letters (7) is equivalent to Pi (π) which means that the quantum 22 letters are connected to God's limitless, uncreated light and the Father of Light's voice as well as the complete cycle of creation. This is emphasized in the Temple Menorah, which includes 7 lamps and 22 almonds in various stages of development.

Pineal Gland—is the seat of the soul that produces what is called the mind's eye, which are mental/spiritual images. The pineal gland is reddish gray in color, resembles a pinecone shape, and is about the size of a grain of rice (5-8 mm). It is located in the center of a human brain between the two hemispheres, tucked in a groove where the two halves of the thalamus join. Unlike most of the brain, the pineal gland is not isolated from the body by the blood-brain barrier.

Planck Time—According to quantum mechanics theory, the smallest division of time is Planck time, an unimaginably short duration of 10^{-43} seconds. For all practical purposes, events separated by a duration of a Planck time are instantaneous. Scientists believe they worked out how the universe unfolded from one Planck unit of time forward. Yet, no one knows what happened before this because our scientific equations and observations break down at the singularity, the beginning.

Platonic Solids—The Platonic Solids are a group of three-dimensional shapes that have been known since antiquity. There are only five geometric solids whose faces are composed of identical polygons: [1] Tetrahedron (4-faces), [2] Cube (6-faces), [3] Octahedron (8-faces), [4] Icosahedron (12-faces), and [5] Octahedron (20-faces). These five Platonic Solids are the basic geometries of life at all levels of reality. All Platonic Solids can be created from Metatron's Cube.

Proton—is one of the three main particles that make up an atom. Protons are found in the nucleus of an atom. This is the tiny dense region in the center of an atom. One proton has a positive electrical charge of one (+1) and a mass of 1 atomic mass unit (1 amu), which is 1.67×10^{-27} kilograms.

Protoplasm of Creation—The Quantum 22 Hebrew Living Letters are part of the protoplasm of Creation. They are the individual spiritual forces that originally belonged, and still belong, to the nature of Christ. The alphanumeric Hebrew letters were birthed from Yeshua's essential, intrinsic essence, which was originally included wholly in the Word. The firstborn of all creation is the very thing that constitutes every living substance. Behold, the Hebrew Living Letters are the elemental material of the Word of God in creation. Behold, the Hebrew letters spiral into the DNA of Creation, which is the protoplasm of the Universe.

Quanta—Quantum physics studies elements of nature that are divided into discrete units or energy packets called "quanta." This is where we get the word "quantum." An invisible quanta of energy is the smallest unit into which something in our physical realm can be partitioned. In 1918, Max Plank was awarded the Nobel Prize for his discovery that all of nature is made up of these invisible quanta of energy.

Quantum—means the smallest possible discrete unit of any physical property, such as energy or matter.

Quantum 22—are the twenty-two letters of the Hebrew Alphabet. Through the Lamb slain before the foundation of the world, the Hebrew Living Letters became flesh; thus, establishing the Lord's hyperquantum link between the spiritual and natural realms.

Quantum Entanglement—a phenomenon that occurs when two or more objects are connected in such a way that they can be thought of as a single system, even if they are very far apart. The state of one object in that system can't be fully described without information on the state of the other object. Likewise, learning information about one object automatically tells you something about the other and vice versa.

Quantum Leap—The *Merriam-Webster Dictionary (10th Edition)* defines "quantum leap" as an abrupt change, sudden increase, or dramatic advance. The term "quantum leap" is rarely used in a scientific context; but it originates as a synonym of quantum jump, which describes an abrupt transition (as of an electron, an atom, or molecule) from one discrete energy state to another.

Quantum Mechanics—is a fundamental theory of physics, which is the foundation of all quantum physics. Quantum mechanics is a science dealing with the behavior of matter and light on the atomic and subatomic scale. Quantum mechanics is complex and extremely mathematical, yet its overarching concepts can be grasped when they are taken one simple step at a time. Quantum mechanics is an insult to the strictly material view of the world. More than any other scientific theory, quantum mechanics reveals the mind of God.

Quantum Model of the Atom—is the basis of every material thing as well as the first success of quantum mechanics. The quantum model of the atom consists of protons, neutrons, and electrons. Protons and neutrons are bound together by nuclear forces to form a nucleus while tiny electrons orbit the core of the atom (i.e., the nucleus).

Quantum Particles—Everything in the universe can be reduced to particles. Particle theorist Mary Gaillard says: "We basically think of particles as a point-like object, and yet particles have distinct traits, such as charge and mass." Quantum particles straddle the worlds of mathematics and reality with an uncertain footing. There are different facets to quantum particles. We could say that a particle is a collapsed wave function, a quantum excitation of a field, an irreducible representation of a group, might be vibrating strings, a deformation of the qubit ocean, or what we measure in detectors. (*What is a Particle?* Natalie Wolchover (senior editor). https://www.quantamagazine.org/what-is-a-particle-20201112/). It is basically impossible to predict what a quantum particle will do at any given moment.

Quantum Physics—is the study of matter and energy at the most fundamental level. It aims to uncover the properties and behaviors of the very building blocks of nature. Quantum Physics is the basis for everyone's reality of their own experiences. However, it is erroneous to say that things don't exist until they are measured, or that reality is whatever we want it to be because those concepts leave out the ultimate observer—the Creator Himself.

Quantum Tunneling—is a quantum-mechanical phenomenon when a particle is able to penetrate through a potential energy barrier that is higher in energy than the particle's kinetic energy. Atoms with more neutrons than protons retain the same chemical properties, but some of these isotopes aren't stable. Unstable atoms are radioactive, which means that they emit one or more particles from the nucleus in a random fashion by a process called quantum tunneling. During fission in nuclear reactors, tunneling takes place within the fuel of the radioactive atoms. Electrical components making modern electronics possible also rely on the phenomena of quantum tunneling of electrons to function.

Quantum Uncertainty Principle—Formulated by the German physicist and Nobel laureate Werner Heisenberg in 1927, the uncertainty principle states that we cannot know both the position and speed of an atomic particle (such as a proton or electron) with perfect accuracy. The more we nail down the particle's position, the less we know about its speed and vice versa. Any object with wave-like properties will be affected by this principle. Quantum objects are special because they all exhibit wave-like properties by the very nature of quantum theory. (*What is the Uncertainty Principle and Why Is It Important?* https://scienceexchange.caltech.edu/topics/quantum-science-explained/uncertainty-principle)

Quarks—are the fundamental particles (elements) of matter that combine to make protons and neutrons.

Quark-Gluon Plasma—Scientists know that the earliest matter to appear in the primordial universe was a super-hot and super-dense liquid, identified as quark-gluon plasma. Quark-gluon plasma has been observed in high-energy physics laboratories.

Sacred Geometry—Sacred geometric patterns exist all around us, creating the fundamental structure of life in the universe. These patterns can be broken down into a language of mathematics. Many consider Platonic Solids sacred geometry's "building blocks" (tetrahedron, hexahedron, octahedron, dodecahedron, and icosahedron). Sacred geometry reveals the idea that everything is connected, and that God created and sustains the universe through sound made visible (a geometric shape). Sacred geometry is considered an ancient science that explores and explains the energy patterns that create and unify all things and reveals the precise way that the energy of creation organizes itself.

Seven Seals of the Human Body—are the seven endocrine glands and organs that only Messiah Yeshua can open and shift into their fullness to facilitate the transformation of a human body into a resurrected body and a quickening spirit. *"Behold, the Lion of the tribe of Judah, the Root of David, has prevailed to open the scroll and its seven seals" (Revelation 5:5).* The seven seals of the human body are the pituitary gland, the pineal gland, the thyroid and parathyroid glands, the thymus gland, the adrenal gland, the hypothalamus, and the pancreas.

Seven Spirits of God—Not only do the Seven Spirits of God make up the nature of the Messiah and the original nature of man, but they're also the chief princes of highest praise that are inextricably connected to God's Dwelling Presence Glory—*Shekinah*. Isaiah 11:2 describes the Seven Spirits of God as the Spirit of the Lord, the Spirit of Wisdom, the Spirit of Understanding, the Spirit of Counsel, the Spirit of Might (Power), the Spirit of Knowledge, and the Spirit of the Fear of the Lord. The precious blood of Yeshua (Jesus), the Hebrew Living Letters, and the Seven Spirits of God were components of creating our physical world, which is all essences of the exact re-presentation in the Son (Heb. 1:3). Messiah Yeshua is pure white light while the Seven Spirits of God are the rainbow colors that make up that spotless white light. These seven make up the Nature of Christ. The Seven Spirits of God are unique to man. When God formed man of the dust into a clay vessel of the ground, He breathed into his nostrils the breath of life (Gen. 2:7). When God breathed into man's nostrils the breath of life, man became a living being made in God's own image through the impartation of the Seven Spirits of God (Job 33:4). When God breathed the breath of life into Adam's nostrils, the Seven Spirits of God were woven throughout mankind's body, soul, and spirit, thus, making each person a temple of the Holy Spirit.

Seventh Day of Creation—is the Biblical day of rest. The fourth commandment of the Big Ten tells us to remember the Sabbath and keep it holy. *"For in six days the LORD made the heavens and the earth, the sea and everything that is in them, and He rested on the seventh day; for that reason the LORD blessed the Sabbath day and made it holy" (Exodus 20:11).* Behold, He is my Sabbath Rest, and I am His.

Singularity—In 1989, Astrophysicist George Smoot and other scientists of the Cosmic Background Explorer (COBE) satellite program basically discovered that the universe started as a single point in the distant past. Their discoveries essentially forced cosmology into agreement with the Book of Genesis. George Smoot remarked, "If you're religious, it's like looking at God." The Initial Singularity was not just the beginning of matter and energy, but it was also the beginning of time and space.

Spacetime—is also called the spacetime continuum. The fabric of spacetime is a conceptual model that combines the three dimensions of space with the fourth dimension of time. Albert Einstein helped develop the idea of spacetime through his theory of relativity. According to the best of current physics theories, spacetime explains the motion of massive objects in space as well as the unusual relativistic effects that arise when traveling near the speed of light.

Subatomic Particles—are also called elementary particles, which are various self-contained units of matter and energy. More than 200 subatomic particles have been detected so far, and most appear to have a corresponding antiparticle (antimatter).

T-Cells—are connected to the thymus gland, because the primary function of the thymus is to train special white blood cells called T-cells or T-lymphocytes. Lymphocytes travel from bone marrow to the thymus where they mature and become specialized T-cells that defend a human body from infections and diseases. The body's thymus gland produces most T-cells before birth. The rest are made during childhood until a person hits puberty.

Tohu Bohu—The Hebrew phrase for "void" in Genesis 1:2 is *tohu bohu*, which means empty and void. *Tohu bohu* describes the condition of the earth immediately before the "let there be light" declaration in Genesis 1:3.

Transformation—Messiah Yeshua will transform the physical bodies of those who press on toward the goal to win the supreme and heavenly prize to which God in Christ Jesus is calling us upward (Phil. 3:14). Transformation is a heavenly inheritance (1 Pet. 1:4). Transformation happens to those who inherit of the Kingdom of God (Gal. 5:19-21; 1 Cor. 6:9-10). These are the ones who have incorruption and immortality put on them. The bodies of the

Transformed Ones are raised in incorruption, raised in glory, raised in power, raised a spiritual body, and made a quickening spirit (1 Cor. 15:42-54). Transformation is an instantaneous change, but also part of the process that comes from loving our Beloved. It is a natural flow, which the enemy cannot duplicate because it only flows from being connected in oneness to the True Head of Messiah Yeshua.

Uncreated Light in Genesis—In the infant universe, there was no material thing at first, just energy in the form of pure uncreated light. Scripture reveals that this uncreated light is the Light of the World (John 8:12)—Messiah Yeshua—who is also called the Word of God (Rev. 19:13); and don't forget about the information encoded in the letters of the Word of God. When you read about the Word in the beginning, always keep in mind that the Word is made up of Hebrew letters from ALEF to TAV. *"In the beginning was the Word, and the Word was with God, and the Word was God" (John 1:1). "In the beginning God created* [את] *the heavens and the earth." (Genesis 1:1).*

Wave-Particle Duality—is when physical entities, such as light and electrons, possess both wave-like and particle-like characteristics. For example, the behavior of light cannot be solely explained by models of particles or waves. If light is only a particle, then why does it refract when traveling from one medium to another? If light is only a wave, why does it dislodge electrons? All behavior of light can be explained by combining the two models.

Word of God—The sword of the Spirit is the Word of God (Eph. 6:17). After God's true and righteous judgments on Babylon, several things are revealed in Revelation 19: it's time for the marriage supper of the Lamb, the Bride of Christ has made herself ready, and heaven opens to the One sitting on a white horse whose robe is dipped in blood and His name is called The Word of God. The Bible is called the Word of God, which is living, active, and sharper than any two-edged sword (Heb. 4:12). Scriptures in the Holy Bible are simply revealing the testimony of Jesus Christ and His-story, which is our story too (Rev. 1:2). That's why some people will say that the Word of God is a person, not words on a piece of paper. However, these words are important; because faith comes by hearing, and hearing by the Word of God (Rom. 10:17). And it is *"by faith we understand that the worlds were framed by the word of God, so that the things which are seen were not made of things which are visible"* (Heb. 11:3).

Zero Point Energy—is also called Divine Point Energy. Zero Point Energy is a cubic centimeter of nothing—no temperature, no magnetic energy, no electromagnetic energy, and no radiation. Zero Point Energy is the God Particle that the Father of Lights used to speak the physical realm into existence. Zero Point Energy is where Messiah Yeshua is pulling energy to cause you and everything else to exist. It is part of the glory of God above the speed of light.

RECOMMENDED READING

- *ALEF-TAV's Hebrew Living™ Letters: 24 Wisdoms Deeper Kingdom Bible Study* by Robin Main.
- *Ascension Manual: Mystic Mentoring (in Christ)* by Robin Main.
- *BOOK of LETTERS, The: A Mystical Alef-bet (Sefer Otiyot)* by Lawrence Kushner.
- *Chumash, The: The Stone Edition* by Rabbi Nosson Scherman.
- *Complete World of the Dead Sea Scrolls, The* by Philip R. Davies, George J. Callaway, R. Phillip
- *Dead Sea Scrolls, The—A New Translation* by Michael Wise, Martin Abegg Jr, Edward Cook
- *Finding God in Science: Extraordinary Evidence for the Soul and Christianity, A Rocket Scientist's Gripping Odyssey* by Michael O'Connell.
- *Gate of Heaven: The History and Symbolism of the Temple in Jerusalem, The* by Margaret Barker.
- *Hebrew Word Pictures* by Frank T. Seekins
- *Hidden Messages in Water, The* by Masaru Emoto
- *Hidden Tradition of the Kingdom of God, The* by Margaret Barker.
- *Letters of Light: A Mystical Journey Through the Hebrew Alphabet* by Rabbi Aaron L Raskin.
- *Made in Heaven: A Jewish Wedding Guide* by Rabbi Aryeh Kaplan.
- *Messiah and His Hebrew Alphabet* by Dick Mills and David Michael.
- *Quantum Glory: The Science of Heaven Invading Earth* by Phil Mason.
- *Sabbath, The* by Abraham Joshua Heschel.
- *SANTA-TIZING: What's wrong with Christmas and how to clean it up* by Robin Main.
- *Songs of Ascents, The: Psalms 120 to 134 in the Worship of Jerusalem's Temples* by David C. Mitchell.
- *Temple Theology: An Introduction* by Margaret Barker.
- *Understanding the ALEF-BEIS: Insights into the Hebrew Letters and Methods for Interpreting Them* by Dovid Leitner.
- *Understanding the Order of Melchizedek: Complete Series* by Robin Main.
- *Wisdom in the Hebrew Alphabet, The.* by Rabbi Michael L. Munk
- *WORD, THE: The Dictionary that Reveals the Hebrew Source of Englis* by Isaac E. Mozeson.

BIBLIOGRAPHY

Barker, Margaret. *Temple Theology: An Introduction.* London, England, Society for Promoting Christian Knowledge, 2004.

Barker, Margaret. *The Gate of Heaven: The History and Symbolism of the Temple in Jerusalem.* Sheffield, England, Sheffield Phoenix Press, 2008.

Barker, Margaret. *The Hidden Tradition of the Kingdom of God.* London, England, Society for Promoting Christian Knowledge, 2007.

Bentorah, Chaim. *Learning God's Love Language.* Travelers Rest, South Carolina, True Potential, Inc., 2018.

Bullinger, E.W. *Numbers in Scripture.* London, England: Eyre & Spottiswoode (Bible Warehouse) Ltd., 1921.

Charlesworth, James H. and Newsom, Carol A. with Rietz, H.W.L. *The Dead Sea Scrolls. Hebrew, Aramaic, and Greek Texts with English Translations*; [the Princeton Theological Seminary Dead Sea Scrolls project]. Vol. 4B. *Angelic Liturgy: Songs of the Sabbath Sacrifice.* Louisville: Westminster John Know Press, 1999.

Chumney, Edward. *The Seven Festivals of the Messiah.* Shippensburg, Pennsylvania: Treasure House (Destiny Image Publishers, Inc.), 2001.

Davies, Philip R. Brooke, George J. Callaway, Phillip R. *The Complete World of the Dead Sea Scrolls.* New York, New York. 2011

Emoto, Masaru. *The Hidden Messages in Water.* New York, New York: Atria Books, 2001.

Ford, Kenneth. *101 Quantum Questions: What You Need to Know about the World You Can't See.* Cambridge, Massachusetts: Harvard University Press, 2011.

Ginsburgh, Rabbi Yitzchak. *The Hebrew Letters: Channels of Creative Consciousness.* Jerusalem, Israel: Linda Pinsky Publications, 1990.

Heschel, Abraham Joshua. *The Sabbath.* Boston, Massachusetts, 2003.

Ifrah, Georges , *The Universal History of Numbers: From prehistory to the invention of the computer.*, John Wiley and Sons, ISBN 0-471-39340-1. Translated from the French by David Bellos, E.F. Harding, Sophie Wood and Ian Monk, 2000.

James, Tim. *Elemental: How the Periodic Table Can Now Explain (Nearly) Everything.* New York, New York: Abrams Press, 2020.

Kaplan, Rabbi Aryeh. *Made in Heaven: A Jewish Wedding Guide.* Brooklyn, New York: Moznaim Publishing Corporation, 1983.

Kumar, Manjit. *Quantum, Einstein, Bohr, and the Great Debate About the Nature of Reality.* New York: W. W. Norton & Company, 2008.

Kushner, Lawrence. *The BOOK of LETTERS: A Mystical Alef-bet (Sefer Otiyot).* Woodstock, Vermont: Jewish Lights Publishing, 15th Anniversary 2nd Edition, 1990.

Leitner, Dovid. *Understanding the ALEF-BEIS: Insights into the Hebrew Letters and Methods for Interpreting Them.* Nanuet, New York: Feldheim Publishers, 2007.

Livio, Mario. *The Accelerating Universe.* New York, New York: John Wiley & Sons, Inc., 2000.

Main, Robin. *ALEF-TAV's Hebrew Living™ Letters: 24 Wisdoms Deeper Kingdom Bible Study.* Loveland, Colorado: Sapphire Throne Ministries Publishing, 2023.

Main, Robin. *Ascension Manual: Mystic Mentoring (in Christ).* Loveland, Colorado: Sapphire Throne Ministries Publishing, 2016.

Main, Robin. *SANTA-TIZING: What's wrong with Christmas and how to clean it up.* Xulon Press, 2008.

Main, Robin. *Understanding the Order of Melchizedek: Complete Series.* Sapphire Throne Ministries Publishing, Loveland, Colorado, 2017.

Mason, Phil. *Quantum Glory: The Science of Heaven Invading Earth.* Maricopa, Arizona: XP Publishing, 2010.

Merriam-Webster. *Merriam-Webster's Colligate Dictionary (10th Edition).* Springfield, Massachusetts: Merriam-Webster Incorporated, 1993.

Mills, Dick and Michael, David. *Messiah and His Hebrew Alphabet.* Orange, California: Dick Mills Ministries, 1994.

Mitchell, David C. *The Songs of Ascents: Psalms 120 to 134 in the Worship of Jerusalem's Temples.* Newton-Mearns, Scotland UK: Campbell Publications, 2015.

Mozeson, Isaac E. *THE WORD: The Dictionary that Reveals the Hebrew Source of English.* New York, New York: SPI Books, 2000.

Munk, Rabbi Michael L. *The Wisdom in the Hebrew Alphabet.* Brooklyn, New York: Mesorah Publishing, Ltd., 1998.

O'Connell, Michael. *Finding God in Science: Extraordinary Evidence for the Soul and Christianity, A Rocket Scientist's Gripping Odyssey.* Orange County, California: Eigen Publishing, 2022.

Raskin, Rabbi Aaron L. *Letters of Light: A Mystical Journey Through the Hebrew Alphabet.* Brooklyn, New York: Rabbi Aaron L. Raskin and Sichos in English.

Rosenblum, Bruce & Kuttner, Fred. *Quantum Enigma: Physics Encounters Consciousness.* New York, New York: Oxford University Press.

Seekins, Frank T. *Hebrew Word Pictures.* Phoenix, Arizona: Living Word Pictures, Inc., 2001.

Scherman, Rabbi Nosson. *The Chumash: The Stone Edition.* Rahway, New Jersey: Mesorah Publications, Ltd., 2015.

Strong, James LL.D, S.T.D. *The New Strong's Exhaustive Concordance of the Bible.* Nashville, Tennessee: Thomas Nelson Publishers, 1995, 1996.

Van Koevering, David. *Sound Doctrine - Quantum Physics, Time, Space, and Matter* (6 Audio CDs). Albany, Oregon: The Elijah List, 2008.

Walker, Evan Harris. *The Physics of Consciousness.* Cambridge, Massachusetts: Perseus Publishing, 2000.

Weeks, Mary Elvira. *The Discovery of the Elements* (6th ed). Easton, PA: Journal of Chemical Education, 1956.

Wells, D. *The Penguin Dictionary of Curious and Interesting Numbers.* London, England: Penguin Group, 1987.

Wise, Michael. Abegg Jr., Martin. Cook, Edward. *The Dead Sea Scrolls—A New Translation.* San Francisco, California : Harper San Francisco: A Division of Harper Collins Publishers, 2005.

Wolf, Fred Alan. *Star Wave, Mind, Consciousness, and Quantum Physics.* New York, New York: Macmillian Publishing Company, 1984.

Wolf, Fred Alan, *Taking the Quantum Leap.* New York, New York: Harpercollins Publishers, 1981, 1989.

ENDNOTES

[1] *"Sound Doctrine: Quantum Physics, Time, Space, and Matter. The Physics of Worship"* by David Van Koevering, Audio-CD-6, Session 5.

[2] *Scientists Just Captured The Flash of Light That Sparks When a Sperm Meets an Egg* by Bec Crew. https://www.sciencealert.com/scientists-just-captured-the-actual-flash-of-light-that-sparks-when-sperm-meets-an-egg, 27 April 2016.

[3] *The Wisdom of the Hebrew Alphabet* by Rabbi Michael L. Munk, p. 26

[4] Ibid. p. 25

[5] *Dead Sea Scrolls. Hebrew, Aramaic, and Greek Texts with English Translations.* Vol. 4B. *Angelic Liturgy: Songs of the Sabbath Sacrifice.* by James H. Charlesworth & Carol A. Newsom with H.W.L. Rietz, p. 3

[6] *The Word: The Dictionary That Reveals The Hebrew Source of English* by Isaac E. Mozeson, p. 3

[7] Ibid.

[8] Hebrew for Christians, https://www.hebrew4christians.com/

[9] *Mystic Mentoring Group Ascension in Christ* on April 19, 2019 – Flaming Swords #15.

[10] *Mystic Mentoring Group Ascension in Christ* on October 11, 2015 – Deeper Kingdom Discoveries #6.

[11] "Introduction to Torah" – Part 1 CD by Monte Judah. *Lion and Lamb Ministries.* www.lionlamb.net

[12] *Dead Sea Scrolls: A New Translation* by Michael Wise, Martin Abegg Jr. and Edward Cook, 130. The Coming of Melchizedek section, p. 455-457, 11Q13

[13] *Ascension Manual* by Robin Main. https://www.amazon.com/Ascension-Manual-Mystic-Mentoring-Christ/dp/0578188511/

[14] *Bereshit: In the Beginning of What?* By Reb Jeff. 10/18/2011. https://www.rebjeff.com/blog/bereshit-in-the-beginning-of-what

[15] Tosef. Sotah, vi.2

[16] Sifre, Num. 132, p. 49a

[17] *Holy Spirit* by Joseph Jacobs, Ludwig Blau. https://www.jewishencyclopedia.com/articles/7833-holy-spirit

[18] Ibid.

[19] *The Chumash, The Stone Edition* by Rabbi Nosson Scherman. Parashas Bereishis (Genesis) 1:2, p. 3

[20] *Molecular Properties of Red Wine Compounds and Cardiometabolic Benefits* by Melissa M. Markoski, Juliano Garavaglia, Aline Oliveira, Jessica Olivaes & Aline Marcadenti. https://www.ncbi.nlm.nih.gov/pmc/articles/PMC4973766/

[21] *Water into Wine, Gummy Bear Testing* by Steve K. Ritter. *Chemical & Engineering News*, January 26, 2015, Volume 93, Issue 4. https://cen.acs.org/articles/93/i4/Water-Wine-Gummy-Bear-Testing.html

[22] *The Secret of Light* by Walter Russell, Back cover.

[23] *The Passion Translation: The New Testaments with Psalms, Proverbs, and Song of Songs.* 2020 Edition. John 2. Footnote *d* 2:10. Translated by Dr. Brian Simmons

[24] Rethinking reality: Is the entire universe a single quantum object? by Heinrich Päs. 5 July 2023. https://www.newscientist.com/article/mg25834460-800-rethinking-reality-is-the-entire-universe-a-single-quantum-object/

[25] Lexicon: Strong's G2889 – *kosmos.* https://www.blueletterbible.org/lexicon/g2889/kjv/tr/0-1/

[26] *What is quantum entanglement?* By Jesse Emspak. May 16, 2023. https://www.space.com/31933-quantum-entanglement-action-at-a-distance.html

[27] *Fields Are White #1. Mystic Mentoring* Group Ascension in Christ.

[28] *Mystic Mentoring* Group Ascension on April 11, 2024

[29] *Glorious Things Unfolding* email by Christine Beadsworth on April 3, 2024, Fresh Oil Releases. christine@freshoilreleases.co.za

[30] *Understanding the Order of Melchizedek: Complete Series* by Robin Main. https://amazon.com/Understanding-Order-Melchizedek-Robin-Main/dp/0998598240/

[31] For more about the Seven Seals of the Human Body, please refer to the "Redemption of ZAYIN" section in Part II of this book (ZAYIN Chapter).

[32] *Mystic Mentoring* Group Ascension on 15 August 2022.

[33] *Merriam-Webster's Collegiate Dictionary* (10th Ed.), corruption, p. 261
[34] Ibid., incorrupt, p. 589
[35] Ibid., incorruptible, p. 589
[36] Ibid., incorruption, p. 589
[37] Ibid., p. 580
[38] Ibid., immortality, p. 580
[39] Ibid., mortal, p. 758
[40] Ibid., transfiguration, p. 1253
[41] Ibid., transfigure, p. 1253
[42] Ibid., transform, p. 1253
[43] Ibid., transformation, p. 1254
[44] *Mystic Mentoring Group Ascension* on 8 August 2022
[45] Majestic Lions #35 – *Mystic Mentoring* Group Ascension on May 6, 2019
[46] *12 Strands of DNA – Our Spiritual Heritage* by Kate A. Spreckley. 12 May 2008. https://spiritlibrary.com/spirit-pathways/12-strands-of-dna-our-spiritual-heritage
[47] *Golden Ratio in DNA*. http://www.goldenratioanddarwin.com/dns.html
[48] *DNA & The Word* by Ken Cagle. https://www.scribd.com/doc/84822037/20609128-dna-the-word
[49] *Shadows of Messiah – DNA* by Levi Madison. 27 June 2013. https://shadowsofmessiah.blogspot.com/2013/06/shadows-of-messiah-dna.html
[50] National Human Genome Research Institute. https://www.genome.gov/genetics-glossary/Epigenomev
[51] *Shadows of Messiah – DNA* by Levi Madison. 27 June 2013. https://shadowsofmessiah.blogspot.com/2013/06/shadows-of-messiah-dna.html
[52] *DNA (Divine Nature Activated)*. http://www.atam.org/DNA.html
[53] *How Many Cells Are in the Human Body? Fast Facts*. https://www.healthline.com/health/number-of-cells-in-body#daily-cell-death
[54] *DNA (Divine Nature Activated)*. http://www.atam.org/DNA.html
[55] Cell. *The Nature and Function of Cells* by Bruce M. Alberts. *Encyclopedia Britannica*. https://www.britannica.com/science/cell-biology
[56] Cell. https://kids.britannica.com/kids/article/cell/352933
[57] Cell. The Structure of Biological Molecules. *Encyclopedia Britannica*. https://www.britannica.com/science/cell-biology
[58] *The God Code* by Gregg Braden, p. 84.
[59] DNA Revelation by Order of Melchizedek. Feb. 26, 2014. https://atam.org/dna-revelation
[60] *How Many Cells Are in Your Body*? By Carl Zimmer. https://www.nationalgeographic.com/science/article/how-many-cells-are-in-your-body
[61] DNA Revelation by Order of Melchizedek. Feb. 26, 2014. https://atam.org/dna-revelation
[62] *Burnt Offering* by Morris Jastrow, Jr., J. Frederic McCurdy, Kaufmann Kohler, Louis Ginzberg. *Jewish Encyclopedia*. https://www.jewishencyclopedia.com/articles/3847-burnt-offering
[63] Ibid.
[64] *Antiquities of the Jews*, 8:3:6 by Josephus Flavius
[65] Sea, Brazen. *Encyclopaedia Biblica: A Critical Dictionary of the Literary, Political and Religion History, the Archeology, Geography and Natural History of the Bible* (1899), edited by Thomas Kelly Cheyne and J. Sutherland Black.
[66] *Golden Glory #35. Mystic Mentoring* Group Ascension on 28 November 2023. www.mysticmentoring.com
[67] Ibid.
[68] *DNA Can Be Influenced And Reprogrammed By Words And Frequencies* - Russian DNA Discoveries. 21 October 2005. Sandpoint, ID.
[69] *DNA: Pirates of the Sacred Spiral* by Dr. Leonard G. Horowitz. January 1, 2004. https://www.amazon.com/DNA-Pirates-Leonard-G-Horowitz/dp/0923550453
[70] DNA Revelation by Order of Melchizedek. Feb. 26, 2014. https://atam.org/dna-revelation
[71] *22 Amino Acids in Human Genome and 22 Letters in the Hebrew Alphabet*. June 9, 2024, 8:49 am https://godssecret.wordpress.com/category/dna/
[72] *What is a Chromosome?* https://www.yourgenome.org/theme/what-is-a-chromosome/
[73] *DNA Proves the Bible True* by Max Jividen. https://www.facebook.com/reel/8412421428822535
[74] *Researchers Discover that DNA Naturally Fluoresces* by Northwestern University. August 15, 2016. https://phys.org/news/2016-08-dna-naturally-fluoresces.html
[75] *Making Sense of Medicine: Biophotons: The Light of Life* by Bob Keller. Oct. 17, 2019. https://www.newburyportnews.com/news/lifestyles/making-sense-of-medicine-biophotons-the-light-of-life/article_3e22dadb-8d40-50de-a226-d91a829444e0.html
[76] *Biophotons: The Light in Our Cells*. https://www.esalq.usp.br/lepse/imgs/conteudo_thumb/Biophotons.pdf
[77] Ibid.

[78] *Biophotons: The Human Body Emits, Communicates with, and is made from Light* by Sayer Ji founder of GreenMedInfo.com. June 25, 2013. https://warrentonwellness.com/biophotons-the-human-body-emits-communicates-with-and-is-made-from-light/

[79] *Quantum Glory: The Science of Heaven Invading Earth* by Phil Mason, p. 55

[80] Ibid., p. 57

[81] *Planck's Constant* by Gavin Wright. https://techtarget.com/whatis/definition/Plancks-constant

[82] "quantum mechanics" definition. https://www.britannica.com/science/quantum-mechanics-physics

[83] *Finding God in Science* by Michael O'Connell, p. 105

[84] *101 Quantum Questions: What You Need to Know about the World You Can't See* by Kenneth Ford. p. 226-227

[85] *Finding God in Science* by Michael O'Connell, p. 104

[86] Ibid., p. 113

[87] Ibid., p. 121

[88] *How Does God Redeem Evil or Demonic Things?* Oct. 19, 2017. https://sapphirethroneministries.wordpress.com/2017/10/19/how-does-god-redeem-evil-or-demonic-things/

[89] *SANTA-TIZING: What's wrong with Christmas and how to clean it up* by Robin Main. https://www.amazon.com/SANTA-TIZING-Whats-wrong-Christmas-clean/dp/1607911159/

[90] *My Last Christmas.* Nov. 28, 2012. https://santatizing.wordpress.com/2012/11/28/my-last-christmas/

[91] *North Pole Going South?* Dec. 21, 2012. https://santatizing.wordpress.com/2012/12/21/north-pole-going-south/

[92] *Golden Scribe Angel* by Robin Main. Dec. 1, 2014. https://santatizing.wordpress.com/2014/12/01/golden-scribe-angel/

[93] *Melchizedek Enoch in Merkabah Form (Metatron)* by Robin Main. Dec. 1, 2014. https://santatizing.wordpress.com/2014/12/01/melchizedek-enoch-in-merkabah-form-metatron/

[94] *Who is Metatron?* by Robin Main. Dec. 28, 2019. https://sapphirethroneministries.wordpress.com/2023/02/17/who-is-metatron-2/

[95] *Metatron is Biblical* by Robin Main. Oct. 31, 2018. https://sapphirethroneministries.wordpress.com/2018/10/31/metatron-is-biblical/

[96] *Oneness Metatron Matrix – Fully Mature Sons* by Robin Main. Nov. 19, 2017. https://sapphirethroneministries.wordpress.com/2017/11/30/oneness-metatron-matrix-fully-mature-sons/

[97] *Are We Called To Be Metatron?* by Robin Main. Nov. 1, 2018. https://sapphirethroneministries.wordpress.com/2018/11/01/are-we-called-to-be-metatron/

[98] *Metatron – Oneness of the Messiah Yeshua* by Robin Main. July 1, 2017. https://sapphirethroneministries.wordpress.com/2017/07/01/metatron-oneness-of-the-messiah-yeshua/

[99] *Festive Tethers to Space-Time* by Robin Main. Dec. 16, 2016. https://sapphirethroneministries.wordpress.com/2017/11/07/festive-tethers-to-space-time/

[100] *Purpose of Group Ascensions* by Robin Main. Mar. 8, 2015. https://sapphirethroneministries.wordpress.com/2017/02/07/purpose-of-group-ascensions/

[101] *What is Ascension in Christ?* by Robin Main. July 2, 2015. https://sapphirethroneministries.wordpress.com/2017/02/07/what-is-ascension-in-christ/

[102] *Ascension Manual* by Robin Main. https://www.amazon.com/dp/0578188511

[103] *Understanding the Order of Melchizedek: Complete Series* by Robin Main. https://www.amazon.com/Understanding-Order-Melchizedek-Robin-Main/dp/0998598240/

[104] *Quantum, Einstein, Bohr, and the Great Debate About the Nature of Reality* by Manjit Kumar, p. 242-243

[105] *Quantum Enigma: Physics Encounters Consciousness* by Bruce Rosenblum & Fred Kuttner, p. 67

[106] *Progress in Religion: his acceptance speech for the Templeton Prize* by Freeman Dyson at Washington National Cathedral on 9 May 2000. Also, Edge 68–May 16, 2000, p. 6 http://www.edge.org/documents/archive/edge68.html

[107] *Star Wave, Mind, Consciousness, and Quantum Physics* by Fred Alan Wolf, p. 202

[108] *The Physics of Consciousness* by Evan Harris Walker, p. 97

[109] *Dark Matter and the Origin of Cosmic Structure* by David Schramm, Sky & Telescope, Oct. 1994, p.29

[110] *Finding God in Science* by Michael O'Connell, p. 88

[111] Ibid., p. 132

[112] Ibid., p. 135

[113] Ibid., p. 91

[114] Mark Peplow, NATURE, Published 19 April 2005. See also *Quark-gluon plasma flows like water, calculations suggest"* by Sam Jarman. Physics World, 9 June 2021. http://physicsworld.com/a/quark-gluon-plasma-flows-like-water-calculations-suggest/

[115] *Finding God in Science* by Michael O'Connell, p. 92

[116] *The Accelerating Universe* by Mario Livio, p. 54

[117] Ibid., p. 54

[118] Ibid., p. 113

[119] *Finding God in Science* by Michael O'Connell, p. 93

[120] Ibid.

[121] *Recipe for a HUMAN revealed: Biologist calculates the chemical formula for a person* by Richard Gray for MailOnline: https://www.dailymail.co.uk/sciencetech/article-3028890/Recipe-HUMAN-revealed-Biologist-calculates-chemical-formula-person.html

[122] *Hearing the Frequencies of the Periodic Table* audio clip by David Van Koevering embedded in *These are the Sounds that David Van Koevering Heard in Heaven."* Sounds of Healing - John Tussey Music. https://www.youtube.com/watch?v=5X8jFPhC5v0&t=852s

[123] *My Story – Introduction to and Recording with Frequencies of the Periodic Table of Elements and Testimonies from Listeners* by John Tussey. https://johntussey.com/john-s-story-periodic-table-of-elements-music

[124] *"Sound Doctrine: Quantum Physics, Time, Space, and Matter. The Physics of Worship"* by David Van Koevering, Audio-CD-6, Session 5.

[125] *What is Cymatics?* https://journeyofcuriosity.net/pages/what-is-cymatics-how-to-explained

[126] *Golden Ratio: Living Mathematics of Nature* by Jain108. https://jain108academy.com/course/golden-ratio/

[127] *ROBERT MOON: Space is Quantized. Atomic Structure Based on the Nesting of Paired Platonic Solids* by Jain108 Academy. https://www.youtube.com/watch?v=O9w-ny6Xxn0

[128] *The Geometric Basis for the Periodicity of the Elements* by Laurence Hecht, https://21sci-tech.com/Articles%202004/Spring2004/Periodicity.pdf

[129] *A Tale of Three Cubes* by Robin Main. https://youtu.be/U4PD9_F2xvI

[130] John 19:17

[131] From the noun גלגלת (*gulgoleth*), skull or head, which in turn derives from the verb גלל (*galal*), to roll. https://www.abarim-publications.com/Meaning/Golgotha.html

[132] *Understanding the Alef-Beis: Insights into the Hebrew Letters and Methods for Interpreting Them* by Dovid Leitner, p. 7

[133] Ibid., p. 8

[134] Ibid.

[135] Ibid., p. 8-9

[136] Ibid. p. 9

[137] Ibid.

[138] Ibid., p. 138

[139] Ibid., p. 13

[140] Ibid.

[141] Ibid.

[142] Ibid., p. 14-15

[143] Ibid., p. 15-16

[144] Ibid., p. 176

[145] Ibid., p. 33-34

[146] Ibid., p. 16

[147] Ibid.

[148] Ibid., p. 18

[149] Ibid., p. 16

[150] Ibid., p. 141

[151] Ibid. p. 140-141

[152] Ibid., p. 142-144

[153] Ibid., p. 157-158

[154] Ibid., p. 176

[155] Ibid.

[156] Ibid., p. 152

[157] *Artists Ascension #85*—AA Meeting. *Mystic Mentoring* Group Ascension in Christ on March 18, 2024.

[158] *A Book of Letters: A Mystical Alef-bait* by Lawrence Kushner. א. p. 21-22

[159] *Letters of Light* by Rabbi Aaron L. Raskin, p. 11

[160] *The Book of Letters* by Lawrence Kushner. א. p.23

[161] *The Wisdom in the Hebrew Alphabet* by Rabbi Michael L. Munk, p. 46

[162] Ibid., p. 47

[163] Additions in brackets are mine

[164] *Understanding the Alef-Beis: Insights into the Hebrew Letters and Methods for Interpreting Them* by Dovid Leitner, p. 13

[165] Ibid., p. 188-189

[166] Ibid., p. 188

[167] "Hydrogen". https://www.rsc.org/periodic-table/element/1/hydrogen

[168] "Hydrogen." *Encyclopædia Britannica.*

[169] This Month in Physics History. January 1, 1925. *Cecelia Payne-Gaposchkin and the Day the Universe Changed* by Richard Williams, APS News, January 2015 (Volume 24, Number 1).

[170] *Importance of Hydrogen* by Mikhail Polenin. https://sciencing.com/importance-hydrogen-5434321.html

[171] *Hydrogen Halo Lifts the Veil of our Galactic Home* by University of Arizona. https://phys.org/news/2017-04-hydrogen-halo-veil-galactic-home.html

[172] *Importance of Hydrogen by Mikhail Polenin*. https://sciencing.com/importance-hydrogen-5434321.html

[173] *The Wisdom of the Hebrew Alphabet* by Rabbi Michael L. Munk, p. 52-53

[174] Ibid., p. 53

[175] Ibid.

[176] Ibid., p. 54

[177] See the "quark" definition in the Glossary in the back of this book about quarks being an elementary particle.

[178] *Mystic Mentoring* Artists Group Ascension #49 – Joseph's Seven Spirit of God Rainbow Tent on 02-15-2021

[179] *Mystic Mentoring* Artists Group Ascension #51 – Fire Water Fingers on 04-12-2021

[180] *The Book of Letters: A Mystical Alef-bait* by Lawrence Kushner. ⌧. p. 24

[181] Ibid., p. 26

[182] *Letters of Light* by Rabbi Aaron L. Raskin, p. 30-31

[183] Ibid., p. 31

[184] *Understanding the Alef-Beis: Insights into the Hebrew Letters and Methods for Interpreting Them* by Dovid Leitner. ב. p. 201

[185] "Helium". https://www.ducksters.com/science/chemistry/helium.php

[186] "Helium". https://www.rsc.org/periodic-table/element/2/helium

[187] "Helium Chemical Element". https://www.britannica.com/science/helium-chemical-element

[188] *Guide to the Elements: Revised Edition* by Albert Stwertka (1998). p. 24.

[189] "Helium". https://www.rsc.org/periodic-table/element/2/helium

[190] "LHC: Facts and Figures" CERN. Archived from the original (PDF) on 2011-07-06.

[191] *The Wisdom in the Hebrew Alphabet* by Rabbi Michael L. Munk. p. 55.

[192] Ibid.

[193] *Understanding the Order of Melchizedek: Complete Series* by Robin Main. https://www.amazon.com/Understanding-Order-Melchizedek-Robin-Main/dp/0998598240/

[194] *The Wisdom of the Hebrew Alphabet* by Rabbi Michael L. Munk, p. 62-63

[195] Proteins and Polypeptides: "Proteins are organic compounds that contain four elements: nitrogen, carbon, hydrogen, and oxygen. To comprehend the full scope of proteins, it is crucial to understand various properties, including the basic biological molecule, peptides, polypeptide chains, amino acids, protein structures, and the processes of protein denaturation. According to IUPAC, polypeptides with a molecular mass of 10,000 Da or more are classified as proteins. At times, the term 'proteins' refers to molecules with 50-100 combined amino acids. Each protein contains one or more polypeptide chain. The chemical properties and order of the amino acids determines the structure and function of the polypeptide. Each of these polypeptide chains is made up of amino acids, which are joined together in a detailed, deliberate and specific order." https://peptidesguide.com/proteins.html

[196] "Cell Adhesion & Migration" by Monique Aumailley. (Jan. 1, 2013) PMID 23263632. https://www.ncbi.nlm.nih.gov/pmc/articles/PMC3544786/

[197] "A Simplified Laminin Nomenclature" by M. Aumailley, L. Bruckner-Tuderman, WG Carter, R. Deutzmann, D. Edgar, P. Ekblom, et al. (August 2005). *Matrix Biology*. **24** (5): 326-332. doi: 10.1016/j.matbio.2005.05.006

[198] Trimeric: a compound, complex, or structure made up of three components. *Farlex Partner Medical Dictionary* © Farlex 2012. https://medical-dictionary.thefreedictionary.com/trimeric

[199] "Form and Function: The Laminin Family of Heterotrimers" by H. Colognato, PD Yurchenco. (June 2000). *Developmental Dynamics*. **218** (2) : 213-234. doi: 10.1002/(SICI)1097-0177 (200006) 218:2 < AID-DVDY1>3.0CO; 2-R. PMID 10842354.

[200] *Extracellular Matrix: A Practical Approach* by MA Haralson, JR Hassell (1995). IRL Press.

[201] "Cell Adhesion & Migration" by Monique Aumailley. (Jan. 1, 2013) PMID 23263632. https://www.ncbi.nlm.nih.gov/pmc/articles/PMC3544786/

[202] Ibid.

203"The Nature and Biology of Basement Membranes by A. Pozzi, PD Yurchenco, RV Iozzo. (January 2017). *Matrix Biology*. 57-58: 1-11. doi: 10.1016/j.matbio.2016.12.009. PMC 5387862.

204The Good Rev. https://www.facebook.com/photo/?fbid=738905768244108&set=a.123501043117920

205"Metastatic Potential Correlates with Enzymatic Degradation of Basement Membrane Collagen" by LA Liotta, K. Tryggvason, S. Garbisa, I. Hart, CM Foltz, S. Shafie. (March 1980). *Nature*. **284** (5751): 67-68. doi: 10.1038/284067a0. PMID 6243750.

206"The Nature and Biology of Basement Membranes" by A. Pozzi, PD Yurchenco, RV Iozzo. (January 2017). *Matrix Biology*. 57-58: 1-11. doi: 10.1016/j.matbio.2016.12.009. PMC 5387862.

207*Letters of Light* by Rabbi Aaron L. Raskin, p. 35

208*The Wisdom of the Hebrew Alphabet* by Rabbi Michael L. Munk, p. 71

209Ibid., p. 74

210*The Book of Letters* by Lawrence Kushner. ג. p. 28

211*The Wisdom of the Hebrew Alphabet* by Michael L. Munk, p.71

212*Understanding the Alef-Beis: Insights into the Hebrew Letters and Methods for Interpreting Them* by Dovid Leitner. The Letter ג. p. 205

213Ibid., p. 206

214Ibid.

215Ibid.

216Ibid.

217"Lithium". https://www.rsc.org/periodic-table/element/3/lithium

218"Lithium". https://www.ducksters.com/science/chemistry/lithium.php

219Ibid.

220Nuclear Weapon Design. Federation of American Scientists (21 October 1998). fas.org

221"Lithium Chemical Element". https://www.rsc.org/periodic-table/element/3/lithium

222"Lithium Chemical Element". https://www.britannica.com/science/lithium-chemical-element

223*Wisdom 24 Class – Wisdom 6 – DALET* (ד) on March 30, 2024

224Refer to "The Seven Spirits of God" and "Transformed Ones" sections of the *Water into Wine* chapter

225*Understanding the Alef-Beis: Insights into the Hebrew Letters and Methods for Interpreting Them* by Dovid Leitner. The Letter ד. p. 210

226Ibid.

227"Beryllium". https://www.osha.gov/beryllium

228"Beryllium Chemical Element". https://www.britannica.com/science/beryllium

229*Understanding the Order of Melchizedek: Complete Series* by Robin Main.

230*The Book of Letters: A Mystical Alef-bait* by Lawrence Kushner. ד. p. 31

231*Hebrew Word Pictures* by Frank T. Seekins, p. 26,138

232*Mystic Mentoring* Group Ascension on 24 January 2015.

233*The Book of Letters: A Mystical Alef-bait* by Lawrence Kushner. ה. p. 33

234*The Wisdom of the Hebrew Alphabet* by Rabbi Michael L. Munk, p. 85

235Ibid., p. 88

236Ibid.

237Ibid.

238Ibid., p. 87

239*Understanding the Alef-Beis: Insights into the Hebrew Letters and Methods for Interpreting Them* by Dovid Leitner. The Letter ה. p. 213-214

240Ibid., p. 215

241*The Wisdom of the Hebrew Alphabet* by Michael L. Muck, p. 86

242*Understanding the Alef-Beis: Insights into the Hebrew Letters and Methods for Interpreting Them* by Dovid Leitner. The Letter ה. p. 214

243"Don't Call Element Boron Boring". https://chemistrytalk.org/boron-element/

244"Boron". https://www.ducksters.com/science/chemistry/boron.php

245"Boron Chemical Element". https://www.britannica.com/science/boron-chemical-element

246Ibid.

247"Boron". https://kids.britannica.com/students/article/boron/310311

248www.britannica.com

249"What is Boron?" https://borates.today/boron/

250"Understanding the Effect of Boron in Steels" by D. Scott MacKenzie, Ph.D., FASM. https://thermalprocessing.com/understanding-the-effect-of-boron-in-steels/

251 Ibid.

252 Ibid.

253 Ibid.

254 "Positive Amplitudes in the Amplituhedron" by Nima Arkani-Hamed, Andrew Hodges, Jaroslav Trnka, https://arxiv.org/abs/1412.8478

255 *The Dead Sea Scrolls: Hebrew, Aramaic, and Greek Texts with English Translations*. Edited by James H. Charlesworth and Carol A. Newsom

256 www.nature.com

257 Ibid.

258 *Merriam-Webster's Collegiate Dictionary* (10th ed.), honeycomb, p. 556

259 *The Wisdom of the Hebrew Alphabet* by Rabbi Michael L. Munk, p. 95

260 *Understanding the Alef-Beis: Insights into the Hebrew Letters and Methods for Interpreting Them* by Dovid Leitner. The Letter ו. p. 226-227

261 Ibid., p. 226

262 Ibid., p. 228

263 "9 Essential Facts About Carbon" by Eva von Schaper. https://www.mentalfloss.com/article/504856/9-diamond-facts-about-carbon

264 "Carbon Chemical Element". https://www.britannica.com/science/carbon-chemical-element

265 "9 Essential Facts About Carbon" by Eva von Schaper. https://www.mentalfloss.com/article/504856/9-diamond-facts-about-carbon

266 Ibid.

267 Ibid.

268 "What is a Diamond?". https://www.diamondsourceva.com/education/diamonds-what-is-a-diamond.asp

269 "The rise of graphene" by A. K. Geim & K. S. Novoselov (26 February 2007). *Nature Materials*. 6 (3): p. 183–191.

270 "Graphene Oxide". https://nanografi.com/graphene/graphene-oxide/

271 See the VAV chapter in *ALEF-TAV's Hebrew Living™ Letters: 24 Wisdoms Deeper Kingdom Bible Study* by Robin Main, p. 86-97

272 *The Wisdom of the Hebrew Alphabet* by Rabbi Michael L. Munk, p. 95

273 Ibid., p. 96

274 The Sign of Jonah that Yeshua Himself referred to in Luke 11:29 was the Bar-Sagale Eclipse on 15 June 763 BC. Remember, there was a total solar eclipse at Christ's Crucifixion too. https://sapphirethroneministries.wordpress.com/2024/03/27/only-sign-given-will-be-the-sign-of-jonah/

275 21 August 2017 to 8 April 2024 is 6 years, 6 months, and 6 days. Refer to VAV for more on "6".

276 *Endocrine System*. https://my.clevelandclinic.org/health/body/21201-endocrine-system

277 Ibid.

278 *What does the lymphatic system do?* Written by Markus MacGill. Medically reviewed by Angelica Balingit, MD. Updated on January 22, 2024. https://www.medicalnewstoday.com/articles/303087#anatomy

279 *Proof that life begins at conception*. Spearhead Missions. https://www.facebook.com/reel/519827314052139

280 Dr. Zach Bush MD, https://www.facebook.com/reel/422253477530306

281 Endocrine System. https://my.clevelandclinic.org/health/body/21201-endocrine-system

282 Messiah Yeshua (Jesus Christ) confirmed revelatory information shared by Dr. Sharon Montes, MD to Robin Main (i.e., Her guess that the seven seals of the human body are the endocrine system and people's focus are wrong when they focus on chakras rather than Christ) as well as Yeshua personally revealing the exact seven seals of the human body that He opens and shifts into their fullness to facilitate the transformation of a human being into a resurrected body and a quickening spirit: [1] the pituitary gland, [2] the pineal gland, [3] the thyroid and parathyroid glands, [4] the thymus gland, [5] the adrenal gland, [6] the hypothalamus, and [7] the pancreas. A soul can only carry the DNA of the One who created and birthed it. Robin Main on September 12, 2024

283 *The Wisdom of the Hebrew Alphabet* by Rabbi Michael L. Munk, p. 104

284 Ibid., p. 105

285 Ibid., p. 106

286 Ibid.

287 Ibid., p. 108

288 *Hebrew Word Pictures* by Dr. Frank T. Seekins, p. 36

289 *Letters of Light* by Rabbi Aaron L. Raskin, p. 71-72

290 *Understanding the Alef-Beis: Insights into the Hebrew Letters and Methods for Interpreting Them* by Dovid Leitner. The Letter ו. p. 246

291 "Nitrogen". https://www.rsc.org/periodic-table/element/7/nitrogen

292 "Abundances of the Elements" by Hans Suess & Harold Urey (1956). *Reviews of Modern Physics*. 28 (1): p. 53.

293 "Nitrogen Summary". https://www.britannica.com/summary/nitrogen

294 "Amino acid metabolism" by W. Sakami & H. Harrington (1963). *Annual Review of Biochemistry*. 32 (1): p. 355–398.

295 news-medical.net/life-sciences/What-Chemical-Elements-are-Found-in-the-Human-Body.aspx

[296] "Understanding Nitrogen" http://cawood.co.uk/blog/understanding-nitrogen/

[297] *nal.usda.gov*

[298] https://modernfarmer.com/2016/11/pilgrims-no-idea-farm-luckily-native-americans/

[299] "Biological Nitrogen Fixation" by Stephen C. Wagner. https://www.nature.com/scitable/knowledge/library/biological-nitrogen-fixation-23570419/

[300] www.hebrew4christians.com

[301] *Ascend to Ever Rest* by Robin Main. http://sapphirethroneministries.wordpress.com/2018/02/21/ascend-to-ever-rest/

[302] *The Sabbath* by Abraham Joshua Heschel, p. 5

[303] Ibid., p. 7

[304] Ibid., p. 8

[305] Ibid., p. 11

[306] *The Sabbath* by Abraham Joshua Heschel, p. 10

[307] *Letters of Light* by Rabbi Aaron L. Raskin, p. 81-82

[308] *Made in Heaven: A Jewish Wedding Guide* by Rabbi Aryeh Kaplan, p. 133-148

[309] Ibid.

[310] Ibid.

[311] Ibid., p. 186-190

[312] Group Ascension with Robin Main and Danielle Chambers on 20 November 2014

[313] *The Wisdom of the Hebrew Alphabet* by Rabbi Michael L. Munk, p. 112

[314] Ibid., p. 113-114

[315] *Ask Dr. Brown* by Michael L. Brown

[316] *The Wisdom of the Hebrew Alphabet* by Rabbi Michael L. Munk, p. 114

[317] *Understanding the Alef-Beis: Insights into the Hebrew Letters and Methods for Interpreting Them* by Dovid Leitner. The Letter ח. p. 250

[318] "Oxygen Summary". https://www.britannica.com/summary/oxygen

[319] "Oxygen". https://pubchem.ncbi.nlm.nih.gov/element/Oxygen

[320] www.coachup.com

[321] Ibid.

[322] Ibid.

[323] *Letters of Light* by Rabbi Aaron L. Raskin, p. 83

[324] Ibid.

[325] "Aggregate Frequencies of Body Organs" by Awadhesh Kumar Maurya & Amit Sharma. Proceedings of IEEEFORUM International Conference, 20th August 2017. New Delhi, India. https://digitalxplore.org/up_proc/pdf/317-150634077720-24.pdf

[326] Ibid.

[327] Ibid.

[328] Ibid.

[329] *iCare – PEMF*. An exciting new inexpensive way of optimizing your body for health and longevity. https://icare-pemf.com/

[330] *Journal of Ecological Engineering*. http://www.jeeng.net/Effects-of-432-Hz-and-440-Hz-Sound-Frequencies-on-the-Heart-Rate-Egg-Number-and-Survival,134038,0,2.html

[331] *The Battle of Frequencies: Unveiling the Secrets of 432Hz vs 440Hz Tuning*. https://www.sensoryland.com/blog1/440hz-vs-432hz

[332] *Bridal Veil #13 – Mystic Mentoring* Group Ascension on 26 July 2017. https://www.youtube.com/watch?v=1N89cLhEU3Q

[333] Cody Cameron Main shared on 11-23-2024

[334] *The Wisdom of the Hebrew Alphabet* by Rabbi Michael L. Munk, p. 119

[335] Ibid.

[336] Ibid., p. 257

[337] *Understanding the Alef-Beis: Insights into the Hebrew Letters and Methods for Interpreting Them* by Dovid Leitner. The Letter ט. p. 256

[338] Ibid., p. 256

[339] "Fluorine Chemical Element". https://www.britannica.com/science/fluorine

[340] https://www.creative-chemistry.org.uk/alevel/core-inorganic/periodicity/trends7

[341] www.britannica.com

[342] "Fluorine". http://www.chemistryexplained.com/elements/C-K/Fluorine.html

[343] *Fluorine-Rich Planetary Environments are Possible Habitats for Life* by Nedjiko Budisa, Vladimir Kubyshkin, Dirk Schulze-Makuch. https://ncbi.nlm.nih.gov/pmc/articles/PMC4206852/#:~:text=Fluorine

344 "What is Fluorine?" https://www.vedantu.com/chemistry/fluorine

345 *New Strong's Concise Dictionary of the Words in the Hebrew Bible*, 446. p. 8

346 Ibid., 2497. p. 44

347 *Hebrew Word Pictures* by Frank T. Seekins, p. 48

348 *Mystic Mentoring Group Ascension: Joyful Trumpets #31* on 2 June 2020

349 Likutei Maharan

350 *Sound Doctrine: Quantum Physics, Time, Space, and Matter. The Physics of the Supernatural Realm* (CD-2) by David Van Koevering

351 *Zero Point Where Time And Space Are One* by Dr. Stephen E. Jones

352 *The Wisdom of the Hebrew Alphabet* by Rabbi Michael L. Munk, p. 119

353 Ibid.

354 Ibid.

355 *Understanding the Alef-Beis: Insights into the Hebrew Letters and Methods for Interpreting Them* by Dovid Leitner. The Letter י. p. 259

356 Ibid.

357 "Neon Chemical Element". https://www.britannica.com/science/neon-chemical-element

358 "Facts About Neon" by Rachel Ross. https://www.livescience.com/28811-neon.html

359 "Neon Chemical Element". https://www.britannica.com/science/neon-chemical-element

360 *Chemistry for Higher Tier*. R. Gallagher & P. Ingram (2001-07-19). University Press. p. 282.

361 https://www.scientificamerican.com/article/how-do-neon-lights-work/

362 "Gases Used in Neon Signs" by John Papiewski. https://sciencing.com/gases-used-neon-signs-5581339.html

363 "Neon". https://www.vedantu.com/chemistry/neon

364 "About Neon". https://www.americanelements.com/ne.html

365 *Mishnah 5:5,* Ramban

366 Please refer to RESH for the details about quantum 20 as well as the number 20.

367 *Merriam-Webster's Collegiate Dictionary* (10th ed.), wing, p. 1356

368 *Understanding the Order of Melchizedek: Complete Series* by Robin Main, p. 153-156

369 *The Passion Translation* [2020 Edition], p. 293

370 *What does Kallah mean in Hebrew?* https://thebridaltip.com/what-does-kallah-mean-in-hebrew

371 *Mystic Mentoring Group Ascension – Intimate Embrace #8* held on May 23, 2022

372 Ascension in Christ vision received by Susanna Wilson on Aug. 8, 2024

373 *The Wisdom of the Hebrew Alphabet* by Rabbi Michael L. Munk, p. 135

374 Ibid.

375 Ibid.

376 Ibid.

377 *Understanding the Alef-Beis: Insights into the Hebrew Letters and Methods for Interpreting Them* by Dovid Leitner. The Letter כ. p. 268

378 "Sodium". https://www.rsc.org/periodic-table/element/11/sodium

379 "Sodium Chemical Element". https://www.britannica.com/science/sodium

380 "Ibid.

381 "Salt and Sodium." https://www.hsph.harvard.edu/nutritionsource/salt-and-sodium/

382 "Sodium Chloride". https://www.encyclopedia.com/science-and-technology/chemistry/compounds-and-elements/sodium-chloride

383 "Sodium Chemical Element". https://www.britannica.com/science/sodium

384 Ibid.

385 "Sodium Bicarbonate – Uses, Side Effects, and More". https://www.webmd.com/drugs/2/drug-11325/sodium-bicarbonate-oral/details

386 dougaddison.com/2017/08/a-deeper-look-at-the-number-11

387 Quiet Time Vision of Robin Main on 6 September 2000.

388 *The Wisdom of the Hebrew Alphabet* by Rabbi Michael L. Munk, p. 139-140

389 *Understanding the Alef-Beis: Insights into the Hebrew Letters and Methods for Interpreting Them* by Dovid Leitner. The Letter ל. p. 275

390 Ibid.

391 *Magnesium, From the Sea to the Stars* by Patrick H. Shea. https://www.sciencehistory.org/distillations/magnesium-from-the-sea-to-the-stars

392 "Magnesium Chemical Element". https://www.britannica.com/science/magnesium

393 "Magnesium Summary". https://www.britannica.com/summary/magnesium

394 "Magnesium Chemical Element". https://www.britannica.com/science/magnesium

[395] "Magnificent Magnesium". https://chemistrytalk.org/magnesium-element/
[396] *The Stone Edition of the Chumash*, p. 469
[397] Ibid.
[398] Ibid.
[399] *Amethyst in Water* by Ehud Abrahamson. 24 Feb. 2024. https://www.facebook.com/100089647486554/videos/926034102529760
[400] https://mentalhealthdaily.com/2014/04/15/5-types-of-brain-waves-frequencies-gamma-beta-alpha-theta-delta/
[401] *The Stone Edition of the Chumash*, p. 449
[402] Ibid., p. 691
[403] *DNA: 12-Stranded* by Jain108 Academy
[404] "Zinc Chemical Element". https://www.britannica.com/science/zinc
[405] Abundance of elements in the earth's crust and in the sea, *CRC Handbook of Chemistry and Physics*, 97th edition (2016–2017), p. 14-17
[406] "Zinc Chemical Element". https://www.britannica.com/science/zinc
[407] *Zinc, the brain and behavior* by C.C. Pfeiffer, E.R. Braverman. https://pubmed.ncbi.nlm.nih.gov/7082716/
[408] *The Physiological, Biochemical, and Molecular Roles of Zinc Transporters in Zinc Homeostasis and Metabolism* by Taiho Kambe, Tokuji Tsuji, Ayako Hashimoto & Naoya Itsumura. https://journals.physiology.org/doi/full/10.1152/physrev.00035.2014
[409] *British Medical Journal*. 326 (7386): p. 409–410
[410] *American Journal of Clinical Nutrition*. 51 (2): p. 225–227
[411] "Zinc Summary". https://www.britannica.com/summary/zinc
[412] www.tekhelet.com
[413] *The Hidden Messages in Water* by Masaru Emoto, p. xv
[414] *Where Does the Body's Vitality Come From?* By A Midwestern Doctor, Mar. 10, 2023. https://midwesterndoctore.com/p/what-actually-happens-with-water
[415] What is Structured Water? https://www.dancingwithwater.com/the-new-science-of-water
[416] Ibid.
[417] Ibid.
[418] Ibid.
[419] Ibid.
[420] *Where Does the Body's Vitality Come From?* By A Midwestern Doctor, Mar. 10, 2023. https://midwesterndoctore.com/p/what-actually-happens-with-water
[421] Much more on this topic is in: *Understanding the Order of Melchizedek: Complete Series* by Robin Main. https://www.amazon.com/Understanding-Order-Melchizedek-Robin-Main/dp/0998598240/
[422] *New state of water molecule discovered* by Ron Walli, Oak Ridge National Laboratory. April 22, 2016. https://phys.org/news/2016-04-state-molecule.html
[423] *The Wisdom of the Hebrew Alphabet* by Rabbi Michael L. Munk, p. 146
[424] Ibid.
[425] Ibid.
[426] *Understanding the Alef-Beis: Insights into the Hebrew Letters and Methods for Interpreting Them* by Dovid Leitner. The Letter מ. p. 277
[427] "Aluminium Guide". https://aluminium-guide.com/
[428] "The World of Pure Aluminum". https://www.alumiplate.com/solutions/high-purity-aluminum-guide/
[429] "Aluminum (Al)". https://thechemicalelements.com/aluminum/
[430] "Aluminum". *Ullmann's Encyclopedia of Industrial Chemistry* by W.B. Frank (2009).
[431] *The Astrophysical Journal*. 591 (2): p. 1220–1247
[432] *Handbook of Isotopes in the Cosmos : Hydrogen to Gallium* by Leiden, p. 129–137
[433] *Metallic Materials Specification Handbook*. Springer Science & Business Media 11 June 2021
[434] *The Dead Sea Scrolls: Hebrew, Aramaic, and Greek Texts with English Translations*. Edited by James H. Charlesworth and Carol A. Newsom
[435] Ibid.
[436] Ibid.
[437] *The Book of Letters: A Mystical Alef-bait* by Lawrence Kushner. מ. p. 53
[438] *Nature's Building Blocks* by John Emsley, p. 506-510
[439] "The Element Zirconium." https://education.jlab.org/itselemental/ele040.html
[440] *A Guide to the Elements* by Albert Stwertka, p. 117-119

441 "The Evolution of Thermal Barrier Coatings in Gas Turbine Engine Applications" by Meier, SM, Gupta DK, *Journal of Engineering for Gas Turbines and Power*, p. 250-257

442 *Zirconium and Its Compounds* by Francis P. Venable www.biomedrarebooks.com/product/899/Zirconium-and-Its-Compounds

443 "Abnormal trace metals in man: zirconium" by Henry A. Schroeder. *Journal of Chronic Diseases*. 19 (5): p. 573–586

444 *Zirconium Biomedical and nephrological applications* by Lee DBN, Roberts M, Bluchel CG, Odell RA, p. 550-556

445 "zircon". *Online Etymology Dictionary* by Douglas Harper.

446 study.com/academy/lesson/what-is-the-element-zirconium-used-for-lesson-for-kids.html

447 b. Sukkah 53b

448 *Mystic Mentoring Group Ascension, Artists Ascension* on April 19, 2017

449 *Learning God's Love Language* by Chaim Bentorah, p.86

450 *The Wisdom of the Hebrew Alphabet* by Rabbi Michael L. Munk, p. 151

451 Ibid., p. 152

452 Ibid., p. 153

453 *Understanding the Alef-Beis: Insights into the Hebrew Letters and Methods for Interpreting Them* by Dovid Leitner. The Letter ב. p. 281

454 "Silicon". https://www.rsc.org/periodic-table/element/14/silicon

455 "Metalloids". https://www.ducksters.com/science/chemistry/metalloids.php

456 "Silicon Element Facts". https://www.chemicool.com/elements/silicon.html

457 "Silicon Summary". https://www.britannica.com/summary/silicon

458 "Silicon". https://www.ducksters.com/science/chemistry/silicon.php

459 "Silicon Summary". https://www.britannica.com/summary/silicon

460 "SILICON". *Annual Review of Plant Physiology and Plant Molecular Biology* by Emanuel Epstein, (1999). 50: p. 641–664.

461 *Interrelations between Essential Metal Ions and Human Diseases*. Metal Ions in Life Sciences. Vol. 13. Springer. Chapter 14. Silicon: The Health Benefits of a Metalloid by Keith R. Martin, p. 451-473

462 *The Physiological Role of the Silicon and its AntiAtheromatous Action. Biochemistry of Silicon and Related Problems* by Fragny M. Loeper, (1978). p. 281–296.

463 *How Semiconductors Work* by Marshall Brain. https://electronics.howstuffworks.com/diode.htm

464 "Silicon". https://www.rsc.org/periodic-table/element/14/silicon

465 Ibid.

466 *Reflections of Redemption in Nisan, Part 5* by Hannah Weiss. https://news.kehila.org/reflections-of-redemption-in-nisan-part-5/

467 Ibid.

468 Ibid.

469 Ibid.

470 Ibid.

471 Ibid.

472 Ibid.

473 "Tin". https://www.ducksters.com/science/chemistry/tin.php

474 "Tin Summary". https://www.britannica.com/summary/tin

475 "Tin Chemical Element". https://www.britannica.com/science/tin

476 Ibid.

477 Ibid.

478 *Tin demand to decline – International Tin Association*. Mining.com. 18 October 2019.

479 "Tin". https://www.ducksters.com/science/chemistry/tin.php

480 *The Dead Sea Scrolls: A New Translation* by Michael Wise, Martin Abegg Jr., & Edward Cook, p.592-593

481 *The Midrash Says on Shemot* by Rabbi Moshe Weissman, p. 82

482 *Mystic Mentoring* Group Ascension on 3 October 2024 (The Feast of Trumpets).

483 "Water-Drawing, Feast of" by Executive Committee of the Editorial Board, Judah David Eisenstein. *The Jewish Encyclopedia*. https://jewishencyclopedia.com/articles/14794-water-drawing-feast-of

484 Ibid.

485 *Tabernacles, Feast of* by Joseph Jacobs & H.G. Friedmann. *The Jewish Encyclopedia*. https://jewishencyclopedia.com/articles/8113-ingathering-feast-of

486 *Crux for God's Dwelling Presence* by Robin Main. https://santatizing.wordpress.com/2013/12/12/crux-for-gods-dwelling-presence/

487 *The Jewish Encyclopedia*

488 *The Songs of Ascents* by David C. Mitchell, p. 33

489 Ibid.

490 "Radak on Psalms Commentary." https://www.sefaria.org/Radak_on_Psalms

491 *The Wisdom of the Hebrew Alphabet* by Rabbi Michael L. Munk, p. 159

492 *Shabbos 104a*, p. 236

493 *The Wisdom of the Hebrew Alphabet* by Rabbi Michael L. Munk, p. 160

494 Ibid., p. 161

495 *Understanding the Alef-Beis: Insights into the Hebrew Letters and Methods for Interpreting Them* by Dovid Leitner. The Letter ס. p. 287

496 Ibid.

497 "Phosphorus". https://www.rsc.org/periodic-table/element/15/phosphorus

498 "Phosphorus Chemical Element". https://www.britannica.com/science/phosphorus-chemical-element

499 Ibid.

500 Ibid.

501 "Phosphorus". https://www.ducksters.com/science/chemistry/phosphorus.php

502 Ibid.

503 "Phosphorus Chemical Element". https://www.britannica.com/summary/phosphorus-chemical-element

504 "Neodymium". https://www.sciencelearn.org.nz/resources/2800-neodymium

505 Abundance of elements in the earth's crust and in the sea, *CRC Handbook of Chemistry and Physics,* 97th edition (2016–2017), p. 14-17

506 "Neodymium Chemical Element". https://www.britannica.com/science/neodymium

507 Ibid.

508 "Neodymium". https://www.sciencelearn.org.nz/resources/2800-neodymium

509 *OG* by Emil G. Hersh, E. Schreiber, Wilhelm Bacher, Jacob Zallel Lauterbach. https://jewishencyclopedia.com/articles/11672-og

510 *Sweet Mysteries #23 – Mystic Mentoring* Group Ascension on 23 September 2024.

511 Refer to the ZAYIN (ז) chapter where seven of the endocrine glands and organs are revealed to be the seven seals of the human body, which only Messiah Yeshua can open and shift into their fullness to facilitate the transformation of a human being into a resurrected body and a quickening spirit. One of the seven seals for the human body is the pineal gland.

512 "Human pineal physiology and functional significance of melatonin." *Frontiers in Neuroendocrinology.* **25** (3-4); 177-195. doi:10.1016/j.yfrne.2004.08.001

513 "Development of the Pineal Gland: Measurement with MR" by M. Sumida, AJ Barkovich, TH Newton. (February 1996). *American Journal of Neuroradiology*. 17 (2): 233-236. PMC 8338352. PMID 8938291.

514 "The historical appearance of the human pineal gland from puberty to old age" by E. Tapp, M. Huxley. (October 1972). *The Journal of Pathology*. 108 (2) : 137-144. doi:10.1002/path.1711080207. PMID 4647506.

515 "An Algorithmic Approach to the Brain Biopsy – Part I" by BK Kleinschmidt-DeMasters, RA Prayson. (November 2006). *Archives of Pathology & Laboratory Medicine*. 103 (11) : 1630-1638. doi:10.5858/2006-130-1630-AAATTB.

516 *Medical Neuroscience* by TC Pritchard, KD Alloway. Hayes Barton Press. (1999).

517 *Melatonin and the Mammalian Pineal Gland* by J. Arendt. Ed. 1. London. Chapman & Hall, 1995, p. 17

518 *How the pineal gland became an obsession for both spiritualists and sci-fi writers* by Matthew Roza. *Salon* Magazine. https://www.salon.com/2021/05/15/how-the-pineal-gland-became-an-obsession-for-both-spiritualists-and-sci-fi-writers/

519 "Vividness of Visual Imagery Depends on the Neural Overlap with Perception in Visual Areas" by N. Dijkstra, S.E. Bosch, M.A.J. van Gerven. *The Journal of Neuroscience*, 37 (5), 1367 LP-1373. (2017).

520 "The Role of Area 17 in Visual Imagery: Convergent Evidence from PET and RTMS" by S.M. Kosslyn, A. Pascual-Leone, O. Felician, S. Camposano, J.P.I. Keenan, W. Ganis, K.E. Sukel, N.M. Alpert. (2 April 1999). *Science*. 284 (5411) : 167-170. doi:10.1126/science.284.5411.167.

521 *Descartes and the Pineal Gland*. Updated Sep. 18, 2013. https://plato.stanford.edu/entries/pineal-gland/

522 *How the pineal gland became an obsession for both spiritualists and sci-fi writers* by Matthew Roza. *Salon* Magazine. https://www.salon.com/2021/05/15/how-the-pineal-gland-became-an-obsession-for-both-spiritualists-and-sci-fi-writers/

523 "Synchronizing Effects of Melatonin on Diurnal and Circadian Rhythms" by M. Pfeffer, HW Korf, H. Wicht. *Gen Comp Endocrinal* 2018; 258: 215-221. https://pubmed.ncbi.nlm.nih.gov/28533170/

524 "A Survey of Molecular Details in the Human Pineal Gland in the Light of Phylogeny, Structure, Function, and Chronobiological Diseases" by JH Stehle, A. Saade, O. Rawashdeh, K. Ackermann, A. Jilg, T. Sebesteny, et al. *Pineal Res*. 2011; 51: 17-43. https://pubmed.ncbi.nlm.nih.gov/21517957/

525 "N,N-dimethyltryptamine and the pineal gland: Separating fact from myth" by DE Nichols. *J Psychopharmacol*. 2018;32:30–36. https://pubmed.ncbi.nlm.nih.gov/29095071/

526 *The morphological and functional characteristics of the pineal gland by* Bogdan Alexandru Gheban, Ioana Andreea Rosca, and Maria Crisan. https://www.ncbi.nlm.nih.gov/pmc/articles/PMC6709953/

527 "N,N-dimethyltryptamine and the pineal gland: Separating fact from myth" by DE Nichols. *J Psychopharmacol.* 2018;32:30–36. https://pubmed.ncbi.nlm.nih.gov/29095071/

528 *How the pineal gland became an obsession for both spiritualists and sci-fi writers* by Matthew Roza. *Salon* Magazine. https://www.salon.com/2021/05/15/how-the-pineal-gland-became-an-obsession-for-both-spiritualists-and-sci-fi-writers/

529 *The Wisdom of the Hebrew Alphabet* by Rabbi Michael L. Munk, p. 159

530 Ibid., p. 172

531 Ibid., p. 172-173

532 Ibid., p. 173

533 Ibid.

534 Ibid.

535 *Understanding the Alef-Beis: Insights into the Hebrew Letters and Methods for Interpreting Them* by Dovid Leitner. The Letter ע. p. 292

536 Ibid.

537 "Sulfur." https://en.mimi.hu/chemistry/sulphur.html

538 "Sulfur Chemical Element". https://www.britannica.com/science/sulfur

539 "Sulfur." https://www.ducksters.com/science/chemistry/sulfur.php

540 "Sulfur." *Encyclopedia of Analytical Science (Second Ed.)* by J. Räisänen, 2005. https://www.sciencedirect.com/topics/medicine-and-dentistry/sulfur

541 *Chemistry of the Elements* (2nd ed.), by N.N. Greenwood & A. Earnshaw.

542 "Sulfur Chemical Element". https://www.britannica.com/science/sulfur

543 *Sulphur surplus: Up to our necks in a diabolical element* by Laurence Knight. (19 July 2014). BBC.

544 *Sulphur and the Human Body*. The Sulfur Institute. Retrieved 3 April 2021.

545 "What is the body made of?" *New Scientist*. 3 November 2021.

546 "Nutritional essentiality of sulfur in health and disease" by Yves Ingenbleek & Hideo Kimura, *Nutrition Reviews*. 71 (7). p. 413–432. (July 2013).

547 "Sulfur." *Encyclopedia of Analytical Science (Second Ed.)* by J. Räisänen, 2005. https://www.sciencedirect.com/topics/medicine-and-dentistry/sulfur

548 *Radical Prayer #4: The Anointing of the Sons of Issachar* by Jamie Rohbaugh. https://www.fromhispresence.com/anointing-of-the-sons-of-issachar/

549 Ibid.

550 Ibid.

551 Ibid.

552 "The Element Ytterbium". https://education.jlab.org/itselemental/ele070.html

553 "Ytterbium Chemical Element". https://www.britannica.com/science/ytterbium

554 *The Elements, in Handbook of Chemistry and Physics* (81st ed.) by C. R. Hammond, (2000)

555 *Magnetism of Rare Earth* by M. Jackson, (2000). The IRM quarterly 10(3): 1

556 *Inorganic Chemistry* 3rd ed. by G.L. Miessler and D.A. Tarr. (2010) Pearson/Prentice Hall publisher

557 *The Elements, in Handbook of Chemistry and Physics* (81st ed.) by C. R. Hammond, (2000)

558 "Facts About Ytterbium." https://www.livescience.com/38423-ytterbium.html

559 "Ytterbium". https://www.rsc.org/periodic-table/element/70/ytterbium

560 *NATIONS AND LANGUAGES, THE SEVENTY* by Kaukmann Kohler, Isaac Broyde. https://jewishencyclopedia.com/articles/11382-nations-and-languages-the-seventy

561 *JACOB, called also Israel* by Emil G. Hirsch, M. Seligsohn, Solomon Schechter, Julius H. Greenstone. https://jewishencyclopedia.com/articles/8381-jacob

562 *The Wisdom of the Hebrew Alphabet* by Rabbi Michael L. Munk, p. 175

563 *Understanding the Order of Melchizedek: Complete Series* by Robin Main, p. 262. https://www.amazon.com/Understanding-Order-Melchizedek-Robin-Main/dp/0998598240/

564 *Understanding the Order of Melchizedek: Complete Series* by Robin Main, p, p. 262-263

565 Ibid., p. 263

566 Ibid., p. 355

567 Ibid., p. 356

568 https://drjoedispenza.net/blog/scientist-proves-dna-can-be-reprogrammed-by-words-and-frequencies/

[569] *Researchers Reveal How Our Words Can Physically Change our DNA* by Power of Positivity. Last modified May 21, 2023. https://www.powerofpositivity.com/our-dna-impacted-by-words/

[570] Apostle Natasha GrBich, Facebook post on 6 April 2024. 10:35 AM.

[571] *The Wisdom of the Hebrew Alphabet* by Rabbi Michael L. Munk, p. 180

[572] Ibid.

[573] Ibid.

[574] Ibid., p. 181

[575] *Understanding the Alef-Beis: Insights into the Hebrew Letters and Methods for Interpreting Them* by Dovid Leitner. The Letter פ. p. 302

[576] Ibid., p. 301

[577] Ibid.

[578] "Chlorine Chemical Element". https://www.britannica.com/science/chlorine

[579] "Chlorine Summary". https://www.britannica.com/summary/chlorine

[580] "Weaponry: Use of Chlorine Gas Cylinders in World War I". *historynet.com*. 2006-06-12

[581] *Facts About Chlorine* by Agata Blaszczak. https://www.livescience.com/28988-chlorine.html

[582] *Chlorine*. noaa.gov. 15 October 2015

[583] "Ohio Derailment More Deadly Than Chernobyl, Experts Warn" by Mike Adams (naturalnew.com) and Eric Coppolino (@PlanetWaves) on 02-22-2023. https://banned.video/watch?id=63f5590a0fcabd3bc3d197cc

[584] "What is dioxin?" https://science.howstuffworks.com/question220.htm

[585] "Halogen: Chemical Element Group". https://www.britannica.com/science/halogen

[586] "The discovery of the elements. XVII. The halogen family" by Mary Elvira Weeks (1932). *Journal of Chemical Education*. 9 (11): 1915.

[587] *Facts About Chlorine* by Agata Blaszczak. https://www.livescience.com/28988-chlorine.html

[588] *Introduction to Torah* CD by Monte Judah. Lion and Lamb Ministries. www.lionlamb.net

[589] Ibid.

[590] "Mercury Chemical Element." https://www.britannica.com/science/mercury-chemical-element

[591] "The Elements" by C.R. Hammond, CRC Handbook of Chemistry and Physics (86th ed).

[592] *AIEEE Chemistry* by E. Ramanathan. Sura Books. p. 251.

[593] "Mercury as an Antisyphilitic Chemotherapeutic Agent" by G J O'Shea. *Journal of the Royal Society of Medicine*. 83 (June 1990): p. 392–395.

[594] "Autism: a form of lead and mercury toxicity" by Heba A. Yassa (November 2014). *Environmental Toxicology and Pharmacology*. 38 (3): p. 1016–1024.

[595] "Mercury". https://kids.britannica.com/kids/article/mercury/353455

[596] *Hebrew Word Pictures* by Frank T. Seekins. Tsadik, p. 80

[597] *Blue Letter Bible*. Strong's G1343. δικαιοσύνη dikaiosýnē, dik-ah-yos-oo'-nay https://www.blueletterbible.org/lexicon/g1343/kjv/tr/0-1/

[598] The crucifixion process is the Baptism of Fire—Matthew 3:11; Matthew 16:24; Galatians 2:20.

[599] *The Wisdom in the Hebrew Alphabet* by Michael L. Munk, p. 190

[600] Ibid.

[601] Ibid.

[602] *Understanding the Alef-Beis: Insights into the Hebrew Letters and Methods for Interpreting Them* by Dovid Leitner. The Letter צ. p. 304

[603] Ibid.

[604] Ibid., p. 304-305

[605] "Argon." https://www.rsc.org/periodic-table/element/18/argon

[606] *Argon Element Facts* by Dr. Doug Stewart https://www.chemicool.com/elements/argon.html

[607] "Argon." https://www.rsc.org/periodic-table/element/18/argon

[608] *Argon Element Facts* by Dr. Doug Stewart https://www.chemicool.com/elements/argon.html

[609] "Argon Chemical Element." https://www.britannica.com/science/argon-chemical-element

[610] "Argon." https://www.ducksters.com/science/chemistry/argon.php

[611] Ibid.

[612] Ibid.

[613] "Argon". https://www.newworldencyclopedia.org/entry/Argon

[614] Ibid.

[615] *Nature's Building Blocks* by J. Emsley (2001). Oxford University Press. pp. 44–45.

[616] "Argon". https://www.newworldencyclopedia.org/entry/Argon

[617] "Argon". https://pubchem.ncbi.nlm.nih.gov/compound/Argon

[618] "Thorium". http://www.chemistryexplained.com/elements/T-Z/Thorium.html
[619] "Thorium". https://pubchem.ncbi.nlm.nih.gov/element/Thorium
[620] "Thorium". Periodic Table of Elements: LANL. https://periodic.lanl.gov/90.shtml
[621] "Thorium". https://pubchem.ncbi.nlm.nih.gov/element/Thorium
[622] Ibid.
[623] Ibid.
[624] "Thorium". https://www.rsc.org/periodic-table/element/90/thorium
[625] *Nature's building blocks: an A–Z guide to the elements* by J. Emsley, J. (2011). pp. 544–548.
[626] "Candoluminescence and radical-excited luminescence" by H.F. Ivey (1974). *Journal of Luminescence*. 8 (4): p. 271–307.
[627] *The Book of Non-electric Lighting: The classic guide to the safe use of candles, fuel lamps, lanterns, gaslights, & fire-view stoves* by Tim Matson. p. 60.
[628] "Toxicological Profile for Thorium". Agency for Toxic Substances and Disease Registry U.S. Public Health Service. 1990. p. 4.
[629] "Thorium: Radiation Protection". United States Environmental Protection Agency. 1 October 2006.
[630] "Thorium". http://www.chemistryexplained.com/elements/T-Z/Thorium.html
[631] *The Elements, in Handbook of Chemistry and Physics* (81st ed.) by C.R. Hammond (2004).
[632] "Thoriated Camera Lens (ca. 1970s)". Oak Ridge Associated Universities. 2021.
[633] "Thorium and Thorium Compounds" by W. Stoll (2005). *Ullmann's Encyclopedia of Industrial Chemistry*.
[634] *SARAH (SARAI)* by Emil G. Hirsch, Wilhelm Bacher, Jacob Zallel Lauterbach, Joseph Jacobs, Mary W. Montgomery. https://jewishencyclopedia.com/articles/13194-sarah-sarai
[635] *Understanding the Order of Melchizedek: Complete Series* by Robin Main, p. 11-13. https://www.amazon.com/Understanding-Order-Melchizedek-Robin-Main/dp/0998598240/
[636] *Understanding the Order of Melchizedek: Complete Series* by Robin Main, p. 11-13. https://www.amazon.com/Understanding-Order-Melchizedek-Robin-Main/dp/0998598240/
[637] www.spineuniverse.com/
[638] Ibid.
[639] Ibid.
[640] Ibid.
[641] *ALEF-TAV's Hebrew Living™ Letters: 24 Wisdoms Deeper Kingdom Bible Study* by Robin Main. KOOF. p. 243
[642] *The New Strong's Exhaustive Concordance of the Bible* by James Strong, LL.D., S.T.D., Greek Dictionary of the New Testament, G:4983 *soma*, p. 88
[643] Ibid., G:4982 *sozo*, p. 88
[644] *Mystic Mentoring* Group Ascension (in Christ) on October 10, 2024.
[645] *Understanding the Order of Melchizedek: Complete Series* by Robin Main, p. 157-158
[646] Ibid., p. 16-17
[647] *The Wisdom in the Hebrew Alphabet* by Michael L. Munk p. 195
[648] Ibid.
[649] Ibid.
[650] Ibid., p. 196
[651] Ibid., p. 197
[652] Ibid., p. 196
[653] *The Wisdom in the Hebrew Alphabet* by Michael L. Munk, p. 196
[654] Ibid.
[655] *Understanding the Alef-Beis: Insights into the Hebrew Letters and Methods for Interpreting Them* by Dovid Leitner. The Letter ק. p. 306
[656] Ibid.
[657] Ibid.
[658] Ibid., p. 305
[659] Ibid., p. 305-306
[660] Ibid., p. 153
[661] Ibid., p. 152-153
[662] Ibid., p. 153
[663] *Element Oddities: 11 Confusing Chemical Symbols Explained* by Compound Interest. https://www.compoundchem.com/2016/02/02/confusing-elements/

[664] "Potassium/ Chemical element" by Adam Augustyn. *Encyclopedia Britannica*.

[665] *Handbook of Chemistry and Physics* (87 ed.) by David R. Lide (1998i). p. 477

[666] *Chemical Principles 6th Ed* by Steven S. Zumdahl (2009). p. A22

[667] "Potassium" by Arnold F. Holleman, Egon Wiberg, Nils Wiberg (1985).

[668] "The Sodium and Potassium Content of Sea Water" by D.A. Webb (April 1939). *The Journal of Experimental Biology* (2): p.183

[669] "Potassium" by M. L. Halperin, K.S. Kamel (1998-07-11). *The Lancet* 352 (9122). p. 135–140.

[670] "Potassium". https://www.ducksters.com/science/chemistry/potassium.php

[671] *Chemistry of the Elements (2nd ed.)* by Norman N. Greenwood; Alan Earnshaw (1997). p.73

[672] Potassium compounds". *Ullmann's Encyclopedia of Industrial Chemistry* by H Schultz; et al. (2006). Vol. A22. p. 39–103

[673] "Potassium". https://www.ducksters.com/science/chemistry/potassium.php

[674] *Chemistry* by Raymond Chang (2007). McGraw-Hill Higher Education. p. 52.

[675] "Chapter 8. Sodium and Potassium Ions in Proteins and Enzyme Catalysis" by Milan Vašák & Joachim Schnabl (2016). In Astrid, Sigel; Helmut, Sigel; Roland K.O., Sigel (eds.). *The Alkali Metal Ions: Their Role in Life*. Metal Ions in Life Sciences. Vol. 16. Springer. p. 259–290.

[676] "Disorders of potassium metabolism" by ID Weiner, S. Linus, CS Wingo (2014).. In J. Freehally, RJ Johnson, J. Floege (eds.). *Comprehensive clinical nephrology* (5th ed.). St. Louis: Saunders. p. 118.

[677] "Disorders of potassium balance" by DB Mount & K Zandi-Nejad (2011). In MW Taal, GM Chertow, PA Marsden, KL Skorecki, AS Yu, BM Brenner (eds.). *The kidney* (9th ed.). pp. 640–688.

[678] "Potassium Food Charts". Asia Pacific Journal of Clinical Nutrition. 2021-04-29.

[679] "Potassium Content of Selected Foods per Common Measure, sorted by nutrient content" (PDF). USDA National Nutrient Database for Standard Reference.

[680] "Fermium". http://www.chemistryexplained.com/elements/C-K/Fermium.html

[681] "Einsteinium and Fermium" by Albert Ghiorso (2003). *Chemical and Engineering News*. 81 (36): p. 174–175

[682] Ibid.

[683] "Fermium". http://www.chemistryexplained.com/elements/C-K/Fermium.html

[684] Ibid.

[685] "Fermium, Mendelevium, Nobelium, and Lawrencium" by Robert J. Silva (2006). In Lester R. Morss, Norman M. Edelstein, Jean Fuger (eds.). *The Chemistry of the Actinide and Transactinide Elements*. Vol. 3 (3rd ed.). Dordrecht: Springer. pp. 1621–1651.

[686] *Hebrew Word Pictures* by Frank T. Seekins, Reysh, p. 88

[687] *What is the brain*? https://my.clevelandclinic.org/health/body/22638-brain

[688] Ibid.

[689] *Brain and Spinal Cord: The Central Nervous System's Dynamic Duo* by NeuroLaunch editorial team. September 30, 2024. https://neurolaunch.com/brain-and-spinal-cord/

[690] Ibid.

[691] Ibid.

[692] Ibid.

[693] The Anatomy of the Occipital Bone by Laura Barhum. Updated June 26, 2023.Medically reviewed by Jenny Sweigard, MD. https://www.verywellhealth.com/occipital-bone-anatomy-4692834

[694] *Brain and Spinal Cord: The Central Nervous System's Dynamic Duo* by NeuroLaunch editorial team. September 30, 2024. https://neurolaunch.com/brain-and-spinal-cord/

[695] Ibid.

[696] Ibid.

[697] Ibid.

[698] *Understanding the Alef-Beis: Insights into the Hebrew Letters and Methods for Interpreting Them* by Dovid Leitner. The Letter ר. p. 308

[699] *Letters of Light* by Rabbi Aaron L. Raskin, p. 196

[700] Ibid.

[701] Ibid.

[702] Ibid., p. 197

[703] *Understanding the Alef-Beis: Insights into the Hebrew Letters and Methods for Interpreting Them* by Dovid Leitner. The Letter ר. p. 307

[704] *Facts About Calcium* by Traci Pedersen. https://www.livescience.com/29070-calcium.html

[705] "Calcium". https://www.rsc.org/periodic-table/element/20/calcium

[706] "Calcium". https://www.ducksters.com/science/chemistry/calcium.php

[707] *Chemistry of the Elements* (2nd ed.). Norman N. Greenwood and Alan Earnshaw (1997), p. 108

[708] *Calcium and Calcium Alloys* by Stephen E. Hluchan and Kenneth Pomerantz, p. 485–487

[709] "Calcium". Linus Pauling Institute, Oregon State University, Corvallis, Oregon. 1 September 2017.

[710] "Calcium: Fact Sheet for Health Professionals". Office of Dietary Supplements, US National Institutes of Health. 9 July 2019.

[711] "Introduction: From Rocks to Living Cells" by Martha Sosa Torres & Peter M.H. Kroneck. pp. 1–32

[712] Ibid.

[713] "Calcium". https://www.ducksters.com/science/chemistry/calcium.php

[714] For our Spirit-led curriculum, we used *ALEF-TAV's Hebrew Living™ Letters: 24 Wisdoms Deeper Kingdom Bible Study* by Robin Main. (https://amzn.to/3NCDXy9).

[715] *Wisdom 23 – SHIN* Class on 2024-05-29.

[716] *The Wisdom of the Hebrew Alphabet* by Rabbi Michael L. Munk, p. 207

[717] Ibid.

[718] *Hebrew Word Pictures* by Dr. Frank T. Seekins, p. 94

[719] *The Wisdom of the Hebrew Alphabet* by Rabbi Michael l. Munk, p. 211

[720] Ibid.

[721] *Understanding the Alef-Beis: Insights into the Hebrew Letters and Methods for Interpreting Them* by Dovid Leitner. The Letter ש. p. 312

[722] *The discovery of the elements* (6th ed.) by Mary Elvira Weeks (1956). Easton, PA: Journal of Chemical Education.

[723] "Scandium". http://www.chemistryexplained.com/elements/P-T/Scandium.html

[724] Ibid.

[725] "Scandium". https://www.ducksters.com/science/chemistry/scandium.php

[726] *Dekker encyclopédia of nanoscience and nanotechnology* by James A. Schwarz, Cristian I. Contescu, Karol Putyera (2004). Vol. 3. CRC Press. p. 2274.

[727] "A Novel Low-Density, High-Hardness, High-entropy Alloy with Close-packed Single-phase Nanocrystalline Structures" by Khaled M. Youssef, Alexander J. Zaddach, Changning Niu, Douglas L. Irving, Carl C. Koch (2015). *Materials Research Letters*. 3 (2): 95–99.

[728] "The properties and application of scandium-reinforced aluminum" by Zaki Ahmad (2003). *JOM*. 55 (2): 35.

[729] "A batty business: Anodized metal bats have revolutionized baseball. But are finishers losing the sweet spot?" by Steve Bjerklie (2006). *Metal Finishing*. 104 (4): 61.

[730] "Easton Technology Report: Materials / Scandium" (PDF). EastonBike.com.

[731] *The Gun Digest Book of Smith & Wesson* by Patrick Sweeney (13 December 2004). Gun Digest Books. p. 34

[732] "History of Laser Dentistry" by Keyvan Nouri (2011-11-09). *Lasers in Dermatology and Medicine*. p. 464–465.

[733] "Visual colour rendering based on colour difference evaluations" by Norbert Sándor and János Schanda, (September 1, 2006), *Lighting Research and Technology*, 38 (3): 225–239

[734] *Lighting Control: Technology and Applications* by Robert S. Simpson (2003). Focal Press. p. 108.

[735] *Biochemistry of Scandium and Yttrium* by Chaim T. Horovitz and Scott D. Birmingham, (1999).

[736] http://www.chemistryexplained.com/elements/P-T/Scandium.html

[737] www.jewishencyclopedia.com

[738] Ibid.

[739] Ibid.

[740] 1 Enoch 6 – James H. Charlesworth editor, *The Old Testament Pseudepigrapha*

[741] 1 Enoch 7 – James H. Charlesworth editor, *The Old Testament Pseudepigrapha*

[742] *Genesis 6 Giants* by Stephen Quayle, p. 148-149

[743] Ibid., p. 150

[744] 1 Enoch 8 – James H. Charlesworth editor, *The Old Testament Pseudepigrapha*

[745] *Genesis 6 Giants* by Stephen Quayle, p. 150

[746] 3 Enoch 4 – Translated by Dr. A. Nyland

[747] *Abaddon Arising* by Tom Horn

[748] Ibid.

[749] Ibid.

[750] *The Wisdom of the Hebrew Alphabet* by Rabbi Michael l. Munk, p. 213

[751] Ibid.

[752] Ibid.

[753] *Mystic Mentoring* Group Ascension - Artists Ascension #66 on 08-15-2022

[754] Thymus. https://my.clevelandclinic.org/health/body/23016-thymus

[755] https://www.verywellhealth.com/

[756] Thymus. https://my.clevelandclinic.org/health/body/23016-thymus
[757] https://www.verywellhealth.com/
[758] *Thymus Gland* by BD Editors. Last Updated October 4, 2019. https://biologydictionary.net/thymus-gland/
[759] Ibid.
[760] Thymus. https://my.clevelandclinic.org/health/body/23016-thymus
[761] https://www.medicalnewstoday.com/articles/thymus
[762] *The Wisdom of the Hebrew Alphabet* by Michael L. Munk, p. 215
[763] Ibid.
[764] *Understanding the Alef-Beis: Insights into the Hebrew Letters and Methods for Interpreting Them* by Dovid Leitner. The Letter ת. p. 317
[765] Ibid., p. 316
[766] "A Brief History of Titanium". https://titek.co.uk/brief-history-titanium/
[767] "I Am Titanium". https://chemistrytalk.org/titanium-element/
[768] "Titanium". *Encyclopædia Britannica*. 2006.
[769] "I Am Titanium". https://chemistrytalk.org/titanium-element/
[770] "Titanium". https://www.ducksters.com/science/chemistry/titanium.php
[771] "Titanium". *Columbia Encyclopedia* (6th ed.). Columbia University Press. 2000-2006.
[772] https://www.ducksters.com/science/chemistry/titanium.php
[773] *CRC Handbook of Chemistry and Physics* (86th ed.) by D.R. Lide, ed. (2005).
[774] *Titanium Alloys: Russian Aircraft and Aerospace Applications* by Valentin N. Moiseyev (2006). p. 196.
[775] *The History and Use of Our Earth's Chemical Elements: A Reference Guide* (2nd ed.) by Robert E. Krebs (2006).
[776] "Titanium". *Nature's Building Blocks: An A-Z guide to the elements* by John Emsley (2001).
[777] *CRC Handbook of Chemistry and Physics* (86th ed.) by D.R. Lide, ed. (2005).
[778] "Titanium Fills Vital Role for Boeing and Russia" by Andrew E. Kramer (5 July 2013). *The New York Times*.
[779] "Titanium". *Nature's Building Blocks: An A-Z guide to the elements* by John Emsley (2001).
[780] "The development of 990 Gold-Titanium: its Production, use and Properties" by G. Gafner (1989). *Gold Bulletin*. 22 (4): p. 112–122.
[781] *Titanium: A technical guide* by Matthew J. Donachie, Jr. (1988). p. 11.
[782] "Titanium". *Nature's Building Blocks: An A-Z guide to the elements* by John Emsley (2001). p. 451
[783] "Titanium, Sinusitis, and the Yellow Nail Syndrome" by Fredrik Berglund & Bjorn Carlmark (October 2011). *Biological Trace Element Research*. 143.
[784] *ASM Handbook: Surface Engineering* (10th ed.). Catherine Mary Cotell, J.A. Sprague, & F.A. Smidt (1994). p. 836.
[785] Please refer to the "DNA is Light" section of the *DNA—The Multidimensional Code of Life* chapter.
[786] "ALL WHO ARE THIRSTY COME! Hoshanah Rabbah (The Great Salvation)" by Robin Main. https://sapphirethroneministries.wordpress.com/2017/10/07/all-who-are-thirsty-come-hoshanah-rabbah-the-great-salvation/
[787] Refer to the "Redemption of ZAYIN (ז)" section of the ZAYIN (ז) chapter.

ABOUT THE AUTHOR

Robin Main is a prophetic artist, author, speaker, teacher and mentor who equips people to be the unique and beautiful creation that they have been created to be. She flows in love, revelation and wisdom with her SPECIALTY being kingdom enlightenment.

Her MISSION is to enlighten the nations by venturing to educate and restore the sons of the Living God.

Her CALL is a clarion one to mature sons, and the pure and spotless Bride of Christ who will indeed be without spot or wrinkle.

Her ULTIMATE DESIRE is that everyone be rooted and grounded in love, so they can truly know the height, width, breadth and depth of the Heavenly Father's love.

OTHER BOOKS BY ROBIN MAIN

The Book of Creation is written in the language of the Hebrew Living™ Letters. When ALEF-TAV (את) built the House of Creation, it began with the quantum Hebrew letters of the Word of God. You are invited to discover for yourself the deepest energetic system of Creation, which God used as the building blocks for everything in heaven and earth. It is recommended that you start either with the *Dwelling in the Presence of the Divine: A Commonplace Book of the Hebrew Living™ Letters* or *ALEF-TAV's Hebrew Living™ Letters: 24 Wisdoms Deeper Kingdom Bible Study*; and then, advance to *Quantum 22™: The Hebrew Living™ Letters*. The ALEF-TAV Bible Study book serves as an on-ramp for the *Quantum 22™* Highway of Holiness. The *Dwelling in the Presence of the Divine: A Commonplace Book of the Hebrew Living™ Letters* is meant to be a companion to the Deeper Kingdom Bible Study—*ALEF-TAV's Hebrew Living™ Letters* and *Quantum 22™*; however, its presence-filled pages can be contemplated and savored all by themselves.

Additional books by Robin Main can be found at sapphirethroneministries.com

www.ingramcontent.com/pod-product-compliance
Lightning Source LLC
Chambersburg PA
CBHW080411170426
43194CB00015B/2771